Television Production & Broadcast Journalism

Phillip L. Harris
Springfield, Virginia

Second Edition

Publisher
The Goodheart-Willcox Company, Inc.
Tinley Park, Illinois
www.g-w.com

2

Library of Congress Cataloging-in-Publication Data

Harris, Phillip L.
  Television production and broadcast journalism /
  by Phillip L. Harris -- 2nd ed.
    p. cm.
  Includes index.
  ISBN 978-1-60525-350-3
    1. Television -- Production and direction. 2. Broadcast journalism.
I. Title.

PN1992.75.H37 20012
791.4502'32 -- dc22   2010040430

Introduction

Television Production & Broadcast Journalism introduces the basic skills needed to enter the television production industry. The text includes an overview of the equipment, job responsibilities, and techniques involved in both traditional studio production and remote location work. The second edition has been extensively updated to address the digital technology used to create television programming. Additionally, new broadcast journalism sections and chapters have been incorporated to support schools that air student-produced newscasts.

This text strongly emphasizes the importance of vocabulary and the correct use of industry terms. When a term is presented, it is set in *bold italic* type, immediately defined, and is used consistently throughout the chapters that follow. *Talk the Talk* features explain the difference between common consumer terms and professional terms, and provide examples of the industry's use of terms. Knowing the proper use and meaning of professional terms is a significant requirement when beginning a career in the television production industry. Oftentimes, the respect you receive from industry professionals is based greatly on the way you speak. Correctly using professional terms is required at all times in our industry.

Principles involved in camera operation, picture composition, scriptwriting, lighting, remote shooting, directing, and many other areas are discussed with illustrated examples and explanations. *Production Note* features are found within each chapter and provide additional information or tips that expand on or reinforce a particular topic. *Visualize This* features also reinforce concepts by providing examples intended to help students create a mental picture of a concept or scenario. These features provide relatable examples, which helps ensure students gain the knowledge and skills necessary to be successful.

Even though the technology involved in television and television production is evolving, the basic knowledge and skills required to enter this field remain consistent. **Television Production & Broadcast Journalism** is designed to provide a foundation of information and skills on which a rewarding career can be built.

Television Production & Broadcast Journalism Chapter Components

Talk the Talk. These features explain the difference between consumer and professional terms and provide examples of appropriate uses of industry terms.

Assistant Activity. These features contain activities students can complete independently, outside the classroom. The activities enhance comprehension of chapter content and concepts through experience.

Sample page 276:

graphics: All of the "artwork" seen in a program, including the paintings that hang on the walls of a set, the opening and closing program titles, computer graphics, charts, graphs, and any other electronic representation that may be part of a visual presentation.

Talk the Talk
In some facilities, the terms "visuals" and "graphics" are used interchangeably.

Copyright

Any picture taken from a magazine or book, Web site, or a motion picture still frame is almost always copyrighted. This means that these images may not be used in a video program without the copyright owner's permission. The simplest way to obtain images for a production is to create your own—take original photographs or make unique pieces of art. If existing copyrighted works must be used, find the copyright holder and get permission. Refer to the copyright information in Chapter 12, *Legalities: Releases, Copyright, and Forums.*

Still Photos

If used sparingly, still photography can work well in a video program. Excessive use of still photography, however, makes a television program look like a slide show. To make interesting use of still photos, move the camera around on the picture to create a sense of motion.

Assistant Activity
Watch a few documentaries to see this "roaming the camera on a still photo" technique. Also notice that sound effects and music are added while the camera roams across a still image. The net result is quite effective.

Photos may be used in a video production, if certain precautions are taken. Take the picture holding the camera horizontally. A horizontally-oriented picture more closely matches the shape of the television screen than a vertically-oriented picture, **Figure 14-1**. A horizontal picture is a rectangle with the long side on the top and bottom, just like a television screen. If using a photo print, use a satin finish instead of glossy. Glossy photo paper reflects the glare of lights into the lens of the video camera.

Photographic slides may be used instead of printed photos, which eliminate the issue of lighting glare. A photographic slide must be oriented in the slide projector horizontally, rather than vertically.

With the prominence of digital video, many non-linear editors (discussed in Chapter 24, *Video Editing*) accept image files in various formats, such as .jpg, .tif, and .gif. There are also computer programs available that allow the user to crop and change a photo in many ways before sending it to a non-linear editor, **Figure 14-2**.

Sample page 35:

Chapter 2

Working in the Television Production Industry

Objectives
After completing this chapter, you will be able to:
- Explain how the responsibilities of each production staff position are dependent on the functions of other production staff positions.
- Identify the primary responsibilities of each production staff position.
- Recall the activities in each step of a production workflow.

Introduction
To understand an individual role in the broadcasting industry, you must be familiar with all aspects of the production process. Each production area is interconnected to many others, with the interrelationships resembling a spider web, **Figure 2-1**. To learn proper camerawork, you must understand proper lighting technique. Proper lighting technique is dictated by the colors used on the set and on the costumes. The colors of the set and costumes directly affect the kind of special effects used in the program. Special effects are created in the special effects generator, but must be edited. Knowing the tools and techniques of editing is also required. To learn television production, you must have a solid understanding of all the contributing roles.

Professional Terms

anchor	makeup
assignment editor	makeup artist
assistant director (AD)	news director
audio	photographer (photog)
audio engineer	photojournalist
camera operator	post-production
cast	pre-production
CG operator	producer
content specialist	production
crew	production assistant (PA)
cue	production manager
director	production switching
distribution	production team
editing	production values
editor	reporter
executive producer (EP)	robo operator
floor director	scenery
floor manager	scriptwriter
frame	special effects
framing	staff
gaffer	talent
graphic artist	video
grip	video engineer
lighting director	VTR operator
maintenance engineer	video operator

35

Objectives. A list of learning objectives is included at the beginning of each chapter. The objectives provide an overview of the chapter topics and explain what each student should know or be able to do upon completion of the chapter.

Professional Terms. Each chapter begins with a list of terms that are introduced and defined in that chapter. The terms in this list appear in *blue bold-italic type* when first presented in the chapter.

Production Note. The information in these text features may provide additional information that relates to a chapter topic or may provide professional production tips.

Visualize This. These features help students create a vivid picture in their mind to assist in fully understanding a topic or concept.

Safety Note. The information contained in these features provides important cautions related to equipment, environment, and the well being of all individuals involved in a production.

Running Glossary. Definitions for Professional Terms are provided in the margin on the page that the term is first introduced.

134

Television Production & Broadcast Journalism

PRODUCTION NOTE

In broadcast journalism, a reporter can record voiceover narration at the studio in a sound booth or conduct an interview in a quiet location (to ensure good sound quality) about a topic which includes action. In each of these situations, B-roll shots with nat sound become very important in the editing process to provide visuals and sounds of the action associated with a story. The B-roll video with nat sound audio of related action shots can be mixed with the voice track from the A-roll. As the clips with nat sound are inserted into the story during the editing process, the nat sound audio level can be controlled to ensure the volume doesn't overpower the reporter's voiceover. A talented and experienced reporter can plan/pace their voiceover to incorporate the nat sound during pauses in speech (at commas or periods in scripted narration). This way, the nat sound, which was recorded because it is an important part of the topic/event covered, becomes an important piece of raw material the reporter can use to build the story.

Nat sound is environmental sound that helps call attention to what a reporter is saying and entices the viewer to continue paying attention to the story. Nat sound is only the environmental sound that supports the story. Where things get tricky is the processing that must sometimes be done to make sure relevant nat sound is present in a recording, without the unwanted background sound.

VISUALIZE THIS

A busy roadway with holiday weekend traffic, horns honking, loud music playing, and motors revving are all part of the background sound to a traffic report. The reporter and photog must be very careful to make sure the reporter's audio is in the foreground and the traffic sounds are in the background. However, this same traffic scene can be shot without the reporter, taken back to the studio, and used as nat sound behind the reporter's voiceover. The traffic sounds can be coordinated with the reporter's voiceover, so that the honk of a car horn happens right when the reporter finishes a sentence and acts as an "exclamation point" for the story. When natural sound is used this way, it is placed with the visual that accompanied it in reality.

If nat sound is extracted from a video recording and manipulated, ethical issues arise about modifying reality. Using a shot and its accompanying sound to illustrate the reporter's narration is ethical. The only acceptable alteration of audio in news is to reduce the volume of nat sound to better hear the voices of speaking.

Room tone is the sound present in a room, or at a location, before occupation. Room tone is the "sound of silence" in the shooting environment. If shooting on location, it is important to clear the set for a few minutes after the equipment is set up. Once all the talent and crew are off the location set, turn on the recorder and record at least three minutes of the existing environmental sound. Having the environmental

---

302

Television Production & Broadcast Journalism

Safety Note

Do *not* use wooden dowels or cardboard to make a flag. When exposed to the heat of a lighting instrument, these materials become a fire hazard.

Other ways to remove a light hit include:
- Spray the item with dulling spray available from photo supply stores. The spray can be removed with a damp cloth after the shoot.
- Spray the item with inexpensive hair spray. Hair spray is also water soluble for easy removal.

Fluorescent Lamps

The types of instruments discussed to this point in the chapter use incandescent lamps. *Incandescent lamps* contain a filament inside the lamp that glows brightly when electricity is applied. Incandescent lamps used in television production are usually tungsten, tungsten halogen, or quartz halogen.

A *fluorescent lamp* functions when electricity excites a gas in the lamp, which causes the material coating the inside of the lamp to glow (fluoresce) with a soft, even light. Older fluorescent lamps were unsuitable for use in television production environments due to the bluish or greenish color temperature of the lamps. Professional television lighting fluorescent lamps are available in various shapes, sizes, and color temperatures, **Figure 15-9**. The most important color temperature in the television industry is 3200° Kelvin. Color temperatures are discussed in detail later in this chapter.

incandescent lamp: Type of lamp that functions when electricity is applied and makes a filament inside the lamp glow brightly.

fluorescent lamp: Type of lamp that functions when electricity excites a gas in the lamp, which causes the material coating the inside of the lamp to glow (fluoresce) with a soft, even light.

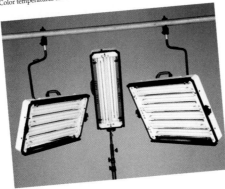

Figure 15-9. Fluorescent instruments can hold multiple lamps and can be hung from a grid or placed on lighting stands. *(Photo courtesy of Lowel-Light Mfg., Inc.)*

6

Chapter 8 Scriptwriting **185**

Wrapping Up

Organizing your ideas and developing a script, however brief, helps to focus your thoughts. Never shoot a program without a script of some kind. When this rule is broken, the crew inevitably ends up reshooting on location because the first shoot lacked a plan. Few people would attempt a cross-country auto trip without planning the trip on a map ahead of time. At the same time, people don't often strictly adhere to the original plan. Traffic backups, taking side trips on a whim, and road construction are just some of the things that may sidetrack a journey. The same is true for a script. Few scripts are shot exactly the way they are written. They do, however, provide the backbone structure to hold the director's creative vision together. Deviations from the script are common during the shooting process, but the basic structure of the program is constant because a script exists.

Review Questions

Please answer the following questions on a separate sheet of paper. Do not write in this book.

1. What are *nod shots*? How are they used?
2. What items are included in a program proposal?
3. What is a script outline?
4. How is a program treatment developed?
5. List the three types of scripts used in television production and the unique characteristics of each.
6. Why are television scripts written using informal language?
7. What is a *montage*?

Activities

1. For each of the program formats listed below, name a television show currently on the air that serves as a format example:
   - Lecture
   - Lecture/Demonstration
   - Panel Discussion
   - Interview
   - Newscast
   - Magazine
   - Drama
   - Music Video
   - PSA

   Be prepared to explain the characteristics of the selected television show that qualify it as an example of the corresponding program format.
2. Record an episode of your favorite sitcom and create an outline for the program. Remember that an outline for this type of program breaks each major event in the story into the fewest number of words possible and progresses chronologically.

Wrapping Up. Each chapter concludes with a summary of or brief discussion related to the topics presented.

Review Questions. Questions designed to reinforce the chapter material are presented at the end of each chapter.

Activities. End-of-chapter activities provide students with an opportunity for additional experience with chapter concepts.

STEM and Academic Activities

Chapter 16 Studio and Remote Shooting **345**

STEM
Integrated Curriculum

Science
1. Evaluate the environmental impact of a typical remote shoot. What environmental aspects should be considered when performing the location survey (pollution, waste disposal, etc.)?

Technology
2. Identify some advances in technology that have affected remote shoots in terms of efficiency, convenience, and cost-effectiveness.

Engineering
3. Design the floor plan for a television news studio. Indicate the placement of all production equipment and specialized areas.

Mathematics
4. Compare the costs involved in shooting a program in the studio to the costs of shooting the same program at a remote location.

Language Arts
5. Watch several evening news programs and list instances of ENG footage and EFP footage used in news stories. Explain why each piece of footage is ENG or EFP.

STEM and Academic Activities. Student activities that integrate chapter topics and concepts with STEM (science, technology, engineering, and math) areas, as well as social science and language arts.

Brief Contents

8

Expanded Contents

Chapter 7
Connectors

Chapter 8
Scriptwriting

Chapter 9
Broadcast Journalism

14

From the Author

Television Production & Broadcast Journalism grew out of a need to provide my students with an up-to-date high school textbook on the subject of television production. After trying to find such a textbook for several years, I finally decided to write one myself. This text is fundamentally a written version of my lecture notes, combined with information from trade magazines, over 34 years of my experiences as a videographer and instructor, and interaction with other broadcasting, communications, and television production instructors as a consultant, workshop leader, and convention speaker.

This book could not have been written without considerable help from many people. I must thank all of my students for their ideas, suggestions, and consideration, as they were the guinea pigs with my draft of the first edition of this text.

I must thank my good friends Dick Blocher and Brian Franco for their patient willingness to answer any technical question I put to them. Chapter 25, *Getting Technical* has been extensively revamped thanks to their technical knowledge. They have helped me explain digital video in clear and correct terms.

Carol Knopes, of the Radio Television Digital News Foundation, strongly urged me to include broadcast journalism in the second edition and has been encouraging and supportive throughout the entire project.

Janet Kerby is an educational consultant/trainer for high school broadcast journalism. Her extensive teaching experience includes high school and graduate-level courses for broadcast journalism teachers. She is my good friend and partner in www.video-educator-training.com. She has had a profound impact on this book and I can't thank her enough. When I first presented her with the idea of adding broadcast journalism to a television production textbook, she was a bit uncertain it could be done. I knew it was possible because she is the living, successful example of the two disciplines being combined, as they often are in high school classes. And, that is exactly how she teaches. She provided me with invaluable insight into and advice on the world of broadcast journalism. The chapters specifically devoted to broadcast journalism would not exist without her input and editing.

Eric Drucker, a lighting expert from Lowel-Light, helped me revamp Chapter 15, *Lighting* to include the new generation of fluorescent lighting instruments and the technique of lighting with these instruments.

Randy Jacobson (digital photography instructor at Fairfax Academy for Communications and the Arts) and his students provided an enormous amount of the photography work.

Adam Goldstein, Broadcast Attorney Advocate for the Student Press Law Center, graciously donated an entire morning to my queries on everything in Chapter 12, *Legalities: Releases, Copyrights, and Forums*. I compiled dozens of questions and scenarios concerning releases, copyrights, music, and forums posted by teachers on the Radio Television Digital News Foundation and Student Television Network LISTSERVS. Adam answered every question and agreed to review the material I wrote based on the interview. We both wanted to be sure that all the material in the *Legalities: Releases, Copyrights, and Forums* chapter is correct, as of the copyright date of this text.

Finally, I must thank all the equipment manufacturers who have, so graciously, allowed me to include pictures of their gear in this text.

Phillip L. Harris

About the Author

Phil Harris received a Bachelor of Science degree from East Carolina University and Master of Arts in Technology Education from George Mason University. He is also a graduate of Imero Fiorintino Lighting Seminars. Mr. Harris's professional production experience includes a wide range of freelance videography (from weddings to commercials), over 25 years as a freelance theatrical makeup designer and artist, and experience directing more than 25 plays and musicals for community theater since 1979.

Mr. Harris brings over 34 years of teaching television production to this book. As part of the program he taught, Mr. Harris created Digital Wave Productions, a school-based enterprise that allows students to gain professional work experience and raise funds, while producing video projects for clients.

Mr. Harris retired from teaching in 2006, but remains passionate about sharing his successful curriculum and facility design tips with fellow television production and broadcast journalism instructors. He is well-known in the career education field as a convention speaker and is a session presenter and contest judge at many conventions and conferences, including SIPA, JEA, STN, ITEA, ACTE, and ASPA. Through his consulting business, Mr. Harris assists school districts and individual schools design, develop, and implement television and video programs, and provides training for teachers of these courses. Mr. Harris can be contacted through the Video Educator Training website (www.video-educator-training.com).

Chapter 1

The Television Production Industry

Objectives

After completing this chapter, you will be able to:

- Identify the various areas within the television production industry and recall the unique characteristics of each.
- Explain the roles of networks and affiliates in the process of scheduling programming.
- Summarize how the cost of an ad is determined.

Introduction

There are many different types of television production companies and more are forming all the time. Because the future of television is strongly tied to developments in digital technology, no one can predict how much more the industry will explode. We know that the television production industry is growing incredibly fast and that jobs are plentiful. The topics presented in this chapter provide a brief idea of the various areas within the industry.

Professional Terms

ad
affiliate
broadcast
closed circuit television
  (CCTV)
commercial broadcast
  television
corporate television
educational television
home video
industrial television
large-scale video
  production company
local origination
network
small-scale video
  production companies
spot
subscriber television
surveillance television
syndication

The Growth of Television Technology

The idea of sending a picture over a wire or through the air is an old one. As early as 1862, a still picture was transmitted through a wire. Moving images were not successfully sent for another 65 years. On April 9, 1927, the first moving images were transmitted via television between Washington, DC and New York City. The next year, Charles Jenkins of Maryland was issued a license for the first television station, W3XK. In 1930, Jenkins broadcast the first television commercial.

By 1936, there were 200 television sets in the United States. At the 1939 World's Fair in New York City, the Radio Corporation of America (RCA) sponsored the first televised Presidential speech, delivered by Franklin Delano Roosevelt. This was the viewing public's introduction to RCA's line of television sets. Seven years later, the first practical color television system was demonstrated. Color broadcasts became increasingly common by the mid-1950s, **Figure 1-1**.

The number of television sets in use in the U.S. passed the one million mark in 1948. In the same year, Community Antenna Television (CATV) was introduced in mountainous rural areas of Pennsylvania where broadcast television signals could not normally be received. This system would become what we now refer to as cable TV.

For the first forty years of its existence, television was mostly "live." Programs were broadcast as they were being performed. Programs recorded onto film were very poor in quality. In 1948, however, the Ampex Corporation introduced the first broadcast-quality magnetic tape recording system, the Video Tape Recorder (VTR). A practical videotape recording system for home use was not available until 1976.

Satellite broadcasting was introduced in 1962. This development made it possible to send and receive television signals anywhere in the world. In 1969, satellite broadcasting allowed the world to watch live as television pictures were transmitted from the moon. By 1983, consumers could

Figure 1-1. In 1954, RCA introduced its first all-electronic color television. The CT-100 had a 12″ screen and sold for $1,000.00. *(RCA)*

subscribe to direct satellite systems for delivery of programming to their homes, instead of cable systems or conventional broadcast programming.

In 1995, the number of television sets in use worldwide passed the one billion mark. One year later, the Federal Communications Commission (FCC) approved the broadcast standards for high-definition television (HDTV). With the vast changes and improvements that digital technology offers, the FCC decided that standardization was necessary. In 2002, the FCC mandated that television manufacturers must equip all new televisions with tuners capable of receiving digital signals. All analog television broadcasts ceased on June 12, 2009. As of that date, all television broadcasts have been digital signals.

Evolution of the Industry

Television production became a thriving industry in the 1950s. The first generation of television production professionals learned the processes, techniques, and technology as they went. The learning process was natural and everyone was learning together.

The second generation of TV production personnel came into the field as it transitioned from black and white to color. This was a huge shift for the consumer, but many of the same production processes applied to both color and black and white television. Both the early black and white television and the first generation of color television used an analog television process.

The third generation entered the television production industry during the 70s and 80s. The professionals of this generation have many years experience and are earning sizeable salaries, but have been confronted with drastic changes in their field in recent years. Their experience lies mostly in analog technology, while the industry as a whole is implementing digital technology and processes in place of analog.

Because today's students have grown up with computers and technology, the digital technology that now prevails in the industry is easier for them to learn and use every day, **Figure 1-2**. Computer software production tools, Internet media productions (webcasts and podcasts), and digital recording

Figure 1-2. Today's students are very knowledgeable with computers and current technology.

and editing processes require production personnel to be informed and proficient with changing technologies. Employers are eager to hire knowledgeable and ambitious staff members who demonstrate competency with new equipment and resources.

Areas of Television Production

There are many different kinds of television production companies within the industry as a whole. Most consumers are familiar only with broadcast, satellite, and cable television. These forms of television comprise only a small portion of all the television produced. All the broadcast, satellite, and cable television produced in a year represents only about 5% of all the television made annually.

PRODUCTION NOTE

Which part of an iceberg is bigger, the part above the water or the portion below the water? The part of an iceberg that lies below the surface is vastly greater in size than the visible peaks above water. Following this example, understand that commercial broadcast/cablecast television is only the proverbial tip of the iceberg in the television production industry, **Figure 1-3**.

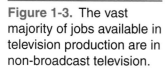

ad: A television advertisement for a product or service. Also commonly called a *spot.*

broadcast: The television signal travels through the air from one antenna to another antenna.

Commercial Broadcast Television

Consumers define a "commercial" as a television advertisement for a product. In the television production industry, an advertisement is an *ad* or *spot.* The industry definition of "commercial" merely refers to a business that is profit-generating in nature. *Broadcast* means that the signal travels through the air from one antenna to another antenna.

Figure 1-3. The vast majority of jobs available in television production are in non-broadcast television.

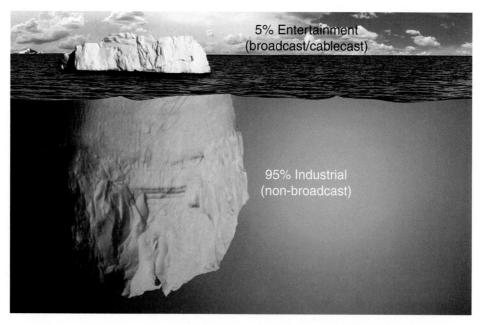

5% Entertainment (broadcast/cablecast)

95% Industrial (non-broadcast)

A *commercial broadcast television* facility is one that is "for-profit" and sends its signal via a transmission tower through the air, **Figure 1-4**. This signal is free and anyone with an antenna may pick it up. The signal is radiated out in a pattern that crosses city, county, state, and national boundaries.

commercial broadcast television: This type of television production facility is "for-profit." The television signal is sent via a transmitter tower through the air and is free for anyone with an antenna to receive it.

Talk the Talk

Do not get confused by the term "commercial." Industry professionals call television advertisements "ads" or "spots," not "commercials."

Subscriber Television

Subscriber television is fee-for-service programming, where customers pay scheduled fees based on the selected programming package. The signals for subscriber television are transported by satellite transmission, by underground cables, or a combination of both.

To receive a satellite television signal, special equipment must be installed inside the home and a small satellite dish is installed and positioned on the outside of the home. The channels and networks available vary among programming packages and satellite providers in each area.

subscriber television: Fee-for-service programming where customers pay scheduled fees based on the selected programming package. The television signals are transported by satellite transmission, by underground cables, or a combination of both.

Figure 1-4. A commercial broadcast station sends its signal from an antenna, through the air, to the viewer's antenna.

Most satellite packages also include local programming, such as the local morning and evening news and talk shows.

The signal for each cable television system is available only to a particular region. The cable programming that your home receives is likely different from the programming available to a neighboring town or county. Cable franchises are set up by local governments. Therefore, the available recipients of the programming package are predetermined.

PRODUCTION NOTE

Cable television used to be called CATV (Community Access Television), but the public just called it "cable TV" because it came into the home through a cable line. The name "cable TV" stuck. Many cable systems no longer use wires to carry the signal. These cable systems were upgraded to fiber optics, which enables more programming to be sent throughout the system.

How, then, can a broadcast station's signal be received through an antenna, a cable system, and a satellite system? The broadcast signal starts from a transmission tower, is sent through the air, and is grabbed by either the local cable company's receiving satellite dish or a transmission tower. Most often, the cable company sends the signal underground into its cable system. A satellite provider sends the signal to a satellite. Since the signal is first and foremost a free broadcast signal, it may also be received by anyone else with a receiver antenna.

Educational Television

educational television:
Television that aims to inform the public about various topics. This includes television programming that supports classroom studies and replays classroom sessions.

Educational television aims to inform the public about various topics and is usually considered nonprofit. Educational television is often broadcast (such as PBS), but a video recording about the Lewis and Clark Expedition shown in history class is also educational television. Most of the programming on educational or instructional television is funded by corporate or federal grants. Originally, this type of television programming was exclusively intended to support or replay existing classes. It has come to include programming designed to inform the public about any topic, in addition to nonprofit programming that supports and replays existing classes. Well-known preschool programs, like *Sesame Street*® and *Barney & Friends*™, are also considered educational television.

Industrial Television

industrial television:
Television that communicates relevant information to a specific audience, such as job training videos. Also commonly called *corporate television*.

Industrial television, sometimes called *corporate television*, communicates relevant information to a specific audience. For example, a company may use industrial television to train employees or to communicate within the company. Training examples may include videos that teach workers how to operate machinery, help travelers learn a language, or instruct soldiers on strategy. A manufacturer of photocopying machines may show a training video to its repairmen that instructs them how to repair a specific copier model. Auto dealers may show a video that informs mechanics of a specific repair issue. Rather than sending employees to an off-site training class, retail businesses can contract with a production facility to produce a training video. This video can be viewed on the store premises and be reviewed as often as necessary, **Figure 1-5**. A college may send informational or promotional videos to prospective students, showcasing the particular benefits and offerings of the college. The Internet is also an outlet for industrial television, with countless video programs posted on the web to be repeatedly streamed or downloaded by the public.

Closed Circuit Television

Closed circuit television (CCTV) is sent through wires and serves only an extremely small, private, predetermined area. For example, your neighbor cannot pick up a signal from your DVD player and watch the movie you are playing. This is because you have a closed circuit television system. Theoretically, you could string a connecting cable from your DVD player across the yard to their TV. Both televisions show the movie, but no one else in the neighborhood can receive the signal. Therefore, the person who creates the closed circuit also determines the size of the circuit.

Surveillance television is a form of CCTV that is usually, but not always, used for security purposes. Surveillance television is not really television <u>production</u>. It simply involves setting up a camera to watch an area, **Figure 1-6**. The surveillance cameras are always interconnected to a CCTV system. Surveillance television employs very few people, other than installers. After the system is installed, only a guard is necessary to monitor activity. Surveillance television systems help in protecting and securing banks, prisons, office buildings, apartment buildings, construction sites, and many other public and private locations. Surveillance television has also been used at traffic intersections to record images of traffic violators, and as dashboard cameras in police cruisers.

closed circuit television (CCTV): Television where the signal is sent through wires and serves only an extremely small, private predetermined area.

surveillance television: A form of CCTV that is usually, but not always, used for security purposes. The cameras used in the system are always interconnected to a closed circuit television system.

Figure 1-5. A recorded program provides a more economical and efficient option for training, compared to taking an employee off the job to attend or teach a class.

Figure 1-6. Surveillance cameras are primarily used for security purposes.

Assistant Activity

Police dramas on television frequently have detectives standing at a crime scene, looking at the buildings all around them, and asking someone to "pull the surveillance tapes from the security cameras there, there, and there," while pointing at cameras on the surrounding buildings. Most people are totally unaware of how often they are captured on surveillance cameras as they go about their daily activities.

Next time you are out walking in a public place, notice how many security cameras you can see when you're looking for them!

Home Video

home video:
Videotaped records of family events and activities taken by someone using a consumer camcorder.

Home video refers to someone using their consumer camcorder to record family events and activities, like a birthday party. While home video provides an archive for important family events, there is no realistic opportunity for financial gain. One in many thousands of home videos may be awarded a prize on a "silliest home videos" television program. Another possible source of financial gain for a home videographer is the unlikely event of recording something newsworthy while videotaping a family activity. One of the most famous examples of this is the Zapruder film of President Kennedy's assassination in Dallas, Texas. News agencies have been known to pay a great deal of money for newsworthy videos shot by enterprising consumers.

Video Production Companies

large-scale video production companies: Facilities with sufficient staff and equipment to produce multi-camera, large-budget programming shot on location or in studios for broadcast networks or cable networks.

Large-scale video production companies are facilities with sufficient staff and equipment to produce multi-camera, large-budget programming shot on location or in studios for broadcast networks or cable networks. Many of the programs you watch on CBS or other networks are not actually produced by network employees. Most networks produce only their own news, news magazines, and sports programming. The majority of what is seen on network television is actually produced by another company and sold to the networks for airing.

Small-scale video production companies are businesses with limited staff and equipment resources. They exist by the hundreds across the country. These companies thrive on producing videos of private events (**Figure 1-7**), commercials for local businesses, home inventories for insurance purposes, seminars, legal depositions, weddings, and real estate videos. A company of this type has a staff that rarely exceeds five people.

Television Program Origination

A *network* is a corporation that bundles a collection of programs (sports, news, and entertainment) and makes the program bundles available exclusively to its affiliates. The networks generally produce some of their own programming, but do not produce all of their own programs. Networks may produce sports and news oriented programming and some entertainment programming through a production division of the corporation. However, most of the dramatic programming (both dramas and comedies) is produced by large-scale production companies and sold to the networks.

small-scale video production companies: Businesses with limited staff and equipment resources. They thrive on producing videos of weddings, commercials for local businesses, home inventories for insurance purposes, seminars, legal depositions, and real estate videos.

network: A corporation that bundles a collection of programs (sports, news, and entertainment) and makes the program bundles available exclusively to its affiliates. Generally, networks produce some of their own programming, but do not produce all of their own programs.

Figure 1-7. Wedding videography is a growing market.

affiliate: A broadcast station that has aligned itself with a particular network. The network provides a certain number of hours of daily programming. The affiliate is responsible for providing the remainder of programming to fill the daily schedule.

An *affiliate* is a broadcast station that has aligned itself with a particular network. A typical contract between an affiliate station and the network stipulates that the network provides a certain number of hours of daily programming. The affiliate is responsible for providing the remainder of programming to fill the daily schedule. **Figure 1-8** is an example of a typical day at the fictitious Television Production Network (TPN).

During the 21½ hour broadcast day, the TPN network provides 4 hours of national news and 7½ hours of other entertainment programming. The local affiliate station must provide the remaining 10 hours of programming. It is not likely that a local station can produce that amount of programming on a daily basis. The schedule displayed in **Figure 1-8** indicates that the affiliate station produces local news for 5 of the 10 hours. The affiliate must either create its own programming or buy programming to fill the remaining 5 hours of scheduled broadcast time.

Figure 1-8. An example of TPN's daily program schedule. Notice that both the network and the affiliate provide programming.

TPN Weekday Schedule			
Time	**Program**	**Affiliate**	**Network**
5:30 a.m.–7:00 a.m.	Local News, Traffic, Weather, Sports	✔	
7:00 a.m.–8:00 a.m.	National Network News		✔
8:00 a.m.–10:00 a.m.	*Good Morning, USA*		✔
10:00 a.m.–11:00 a.m.	Syndicated Talk Show	✔	
11:00 a.m.–12:00 p.m.	Syndicated Reruns of *Friends* and *Seinfeld*	✔	
12:00 p.m.–12:30 p.m.	Local News at Noon	✔	
12:30 p.m.–1:00 p.m.	Syndicated Rerun of *Everybody Loves Raymond*	✔	
1:00 p.m.–4:00 p.m.	Soap Operas		✔
4:00 p.m.–5:00 p.m.	Syndicated Talk Show	✔	
5:00 p.m.–6:00 p.m.	Local Evening News	✔	
6:00 p.m.–7:00 p.m.	National Network News		✔
7:00 p.m.–8:00 p.m.	Syndicated Game Shows	✔	
8:00 p.m.–11:00 p.m.	Network Entertainment		✔
11:00 p.m.–11:30 p.m.	Local News	✔	
11:30 p.m.–1:00 a.m.	Late Night Network Show		✔
1:00 a.m.–3:00 a.m.	Late, Late Movie	✔	

Syndication

Episodes of former network programs that have been purchased and released for syndication are available to affiliate stations. These television programs are sold in blocks of a specified number of episodes or in blocks of time. For example, purchasing a particular program may provide one episode per week for 52 weeks.

If a network program ran for at least 3 years, there are enough episodes (26 episodes per year for a total of 78 episodes over three years) to make it available for syndication. *Syndication* is the process of making a specified number of program episodes available for lease to other networks or individual broadcast stations, after the current network contract for the program expires. Syndicated programs not only include those seen in primetime on major networks, but also some programs that were never picked up by a major broadcast network.

Usually, the production company that made the program leases the right to air that program to a network. It is commonly stipulated that the network may air that program a maximum of three times during the broadcast year (September through the following August). After that, the rights to the program revert back to the production company. The production company may then offer a lease of the program rights to any customer. Customers may include broadcast networks, subscriber networks, affiliates, or distribution companies that bundle the program with others to create a programming package. For example, a program bundle might include *I Love Lucy*, *Hogan's Heroes*, *McHale's Navy*, and *Gilligan's Island*. Another bundle might include *Friends*, *Fraiser*, *Seinfeld*, and *King of Queens*. This second bundle carries a higher leasing fee because the programs are newer than the first bundle and are typically in higher demand by leasing stations. Stations know that the second program bundle will draw a bigger audience. Given that each of the shows in both packages are 30 minute episodes, either package provides a two-hour daily package of four shows, with the rights to air the programs an unlimited number of times during the broadcast year.

The contract terms for syndicated programs vary greatly. Most contracts depend, to a certain extent, on the program itself and its marketability. A highly marketable show, such as *Everybody Loves Raymond*, may be placed in a package and made available only with the lease of the other three shows in the package. A bundle of this arrangement allows the distribution company to make more money than with the one show alone. Another highly marketable show, such as *Law and Order*, may be leased directly from the production company as a multi-episode contract of a single series.

Various types of programs are available for syndication, including:

- dramas
- comedies
- talk shows
- game shows
- cooking shows
- animated programs
- children's shows
- movies

syndication: The process of making a specified number of program episodes available for "lease" to other networks or individual broadcast stations, after the current network's contract for the program expires.

Shopping for Programming

Shopping for programming is usually done in person, over the phone, via fax, or on the Internet. When a local affiliate decides to purchase programming, a budget is set. Someone from the affiliate station must negotiate with vendors to get the highest quality program for the allotted money. A program's popularity and the population size of the broadcast area are among the factors to consider when shopping for programming. These factors directly relate to the purchase price of a program. For example, obtaining the game show *Jeopardy!* for Dead Gulch, Nevada with a population of 350 is not nearly as expensive as getting the same program for New York City.

Competition

There is some urgency in the decision-making process for programming. If another station in your broadcast area contracts with a vendor for a particular program before you do, they obtain exclusive rights to air the program in your broadcast area.

Many television stations in a single area compete for viewers. Each tries to choose programming that pulls viewers away from the competition, while the competition is doing exactly the same thing. To develop the best programming, you must examine the potential audience. Determine who is likely to be watching television at each particular time of day in the area. Knowing the demographics of your audience helps to develop programming that appeals to that audience. Statistics that are considered in demographics include age, gender, race, education, and economic level.

The reason networks run soap operas in the afternoon is that women caring for young children comprise a large portion of TV viewers during that time of day. The children are typically napping and the adults are taking a break from a busy morning keeping up with the kids. Many stations run children's programming early in the morning because children are likely to be watching at that time. Stations usually run programming that appeals to young, school-age children later in the afternoon when the children are home from school. Ultimately, stations must pay for the programs they buy and must consider the audience when deciding on these purchases.

Local Origination

local origination:
Programming made in a specific geographic area, to be shown to the public in that same geographic area.

Local origination is programming made in a specific geographic area, to be shown to the public in that same geographic area. For example, the evening news in New York City reports that traffic is backed up in the Lincoln Tunnel. Do the people watching the evening news in Mayberry, North Carolina hear about the Lincoln Tunnel traffic in New York? Of course not. Viewers in both areas are watching the local news. Local origination comes in many forms. The local evening news is an example of local origination programming. Local stations may also produce a program about a local sports team or televise "town hall" meetings and local telethons, which are other examples of local origination programming for a specific community.

Financing the Programming Decisions

The ads that run during programs pay for the purchase price of those programs. Any money earned by the ads above the cost of the program goes to station overhead (equipment, salaries, rent, etc.). Advertising on the radio or in print is always an option, but television ads are very effective.

The advertiser must first contract with a video production company to produce a television ad. Once the ad is made, the company approaches the television station or network and asks that the ad be aired. The station charges a fee each time the ad is aired. The fee is not a set amount. It changes based on the time of day and the day of the week that the ad airs. If a company wants their ad to air during an extremely popular program that is seen by the largest audience of the week, a substantially higher fee is charged than if the ad runs at 2:00 a.m. during the *Late, Late Movie*.

PRODUCTION NOTE

Companies clamor to purchase coveted ad time during the annually televised Super Bowl. The cost of a 30-second spot fluctuates depending on when the commercial airs during the event. Ads that air before half-time may be charged differently from those airing in the third and fourth quarters. In the span of a decade, the average price tag for a 30-second commercial to air during the Super Bowl has risen from $1.2 million to about $3 million.

Television stations or networks cannot rely on individual companies to approach them with product ads. Funding is required to buy programming. Assume that a station's research shows that the majority of potential viewers for programs airing from 11:00 a.m.–12:00 p.m. are either people with impaired health or who are retired. Programming of interest to this particular audience needs to be purchased. Once the programs are obtained, the sales staff is sent out to find organizations or companies to advertise their products or services during that program. If the station cannot find anyone to advertise during the program, they cannot afford to air the program.

Through a rating system called the Nielsen Ratings, a figure is determined that represents approximately how many people watch a program. If the numbers are too low, advertisers will insist that the advertising rates be lowered to reflect the smaller audience reached. If rates are lowered to the point that the program costs the network more than the ads bring in, the network must either continue to run the program at a financial loss or cancel the program. Sometimes a cancelled network program can find an extended life, with new episodes, in the cable industry. A competing network may even pick up a cancelled program and may be able to breathe new life into it.

PRODUCTION NOTE

The Nielsen Company has developed a media research system that estimates the size and demographics of the viewing audience for almost every program seen on television. To gather this information, the company distributes television diaries to selected Nielsen TV families and installs electronic television monitoring equipment in certain homes, **Figure 1-9**. These methods, along with random phone surveying and e-mail surveys, contribute to generating the ratings that are referenced for both advertising and programming decisions.

The Business of the Industry

Money is the driving force of the television industry. Remove any preconceived ideas that television is an art form. While art may sometimes occur, television is a business first and foremost. The success and failure of a business hinges on money. Every business decision made considers profits and losses.

Follow the trail of profit motivation in the following scenario:

- Widgets, Inc. has created the "Snapper," a device that opens the flue of a fireplace with the snap of a finger. Advertising on the radio or in print is an option, but the company has decided that the best way to reach their intended customer base is to advertise on television.
- Widgets, Inc. contracts with the video production company AdsRUs to produce a television ad. Once the ad is made, Widgets, Inc. writes a check to AdsRUs for their services.
- Widgets, Inc. now approaches a television station or network to ask if they would air the ad for their product.
- The station assigns a fee that Widgets, Inc. will pay each time the ad is aired. The fee depends on the time of day that ad is aired, the day of the week, and during which programs the ad is aired.

Figure 1-9. The People Meter is a box connected to an in-home television, which records both who in the household is watching television and what they are watching. Data collected by the People Meter is used to generate Nielsen ratings. *(The Nielsen Company)*

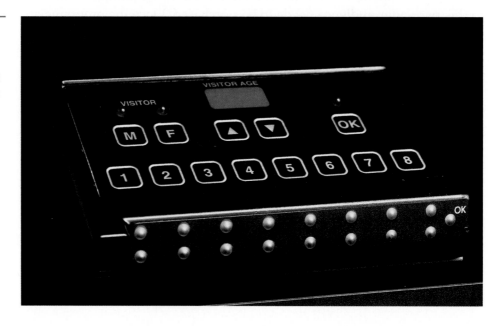

- The station uses the fee assessed for the airing of the ad to pay for the purchase of the scheduled programming.
- If no one advertises during a particular block of time or programming, the station cannot afford to continue airing the program(s).
- Perhaps this helps you understand the logic behind the cancellation of a television program.

Talk the Talk

We often hear that a program got good ratings or was highly rated. Most consumers think that "highly rated" means that critics think the program is quite good. In fact, "highly rated" has nothing to do with program quality. It only means that a large number of people watched the program. The influx of reality programming is watched by many, many people, but a television critic probably would not consider these programs to be of meritorious quality.

Wrapping Up

Very few people begin their careers in the entertainment, major market news broadcast, or cable television arenas. Most begin in non-broadcast television, perfect their knowledge and skills, and eventually move into the broadcast arena. Non-broadcast television has many more jobs available than broadcast or cable television, and the jobs are more secure than those in the entertainment industry. The television industry is extremely competitive. To be successful, you must work your way up from the bottom, through the ranks, and prove your worth the whole way.

Review Questions

Please answer the following questions on a separate sheet of paper. Do not write in this book.

1. How can a broadcast station's programming be received through both cable and satellite systems?
2. What are the differences between educational television and industrial television productions?
3. List six examples of closed circuit television systems.
4. Explain the relationship between a network and an affiliate station when scheduling daily programming.
5. What is local origination programming? Provide an example of local origination on a broadcast station and an example on a cable channel.
6. How do stations pay for original programming and syndicated programs that they purchase?

Activities

1. Select a 3-hour block of time from the television schedule listing in a local newspaper, online, or from another source. List the types of television programs aired during the selected block of time and determine the viewing audience for each program type. Make a list of the products that are best suited to advertise during each program. Record the programs and see how your list compares to the ads that actually run during that time period.
2. Create your own dictionary of chapter terms. Your instructor will provide an alphabetical list of all the terms defined in this book. Copy the entire list of terms into a separate notebook, leaving two or three lines after each for the definition. Update your dictionary with the corresponding definitions as you complete each chapter. This custom-made dictionary will serve as a reference tool for you throughout this course, future courses, and when working in the broadcasting industry.

STEM and Academic Activities

1. Create a presentation that explains the difference in how satellite television signals and cable television signals are transmitted and received.

2. Research the cost and options included in three similar subscription television packages, from different providers, available in your area. Determine which provider offers the lowest cost per channel.

3. Find a weekday programming schedule for a local television station. List the programs scheduled to air between 1:00 p.m. and 3:00 p.m. Identify the intended audience for this block of programming and explain why these programs appeal to that audience.

4. Explain both the positive and negative aspects related to the increase in available television channels over the last 20 years.

5. Make a map of the route you travel from your home to school. On the map, identify the places and areas where you may be under security surveillance.

www.careerpage.org A website sponsored by the National Alliance of State Broadcasters Associations that provides broadcasting industry professionals with job information and search tools.

Chapter 2

Working in the Television Production Industry

Professional Terms

anchor	makeup
assignment editor	makeup artist
assistant director (AD)	news director
audio	photographer (photog)
audio engineer	photojournalist
camera operator	post-production
cast	pre-production
CG operator	producer
content specialist	production
crew	production assistant (PA)
cue	production manager
director	production switching
distribution	production team
editing	production values
editor	reporter
executive producer (EP)	robo operator
floor director	scenery
floor manager	scriptwriter
frame	special effects
framing	staff
gaffer	talent
graphic artist	video
grip	video engineer
lighting director	VTR operator
maintenance engineer	video operator

Objectives

After completing this chapter, you will be able to:
- Explain how the responsibilities of each production staff position are dependent on the functions of other production staff positions.
- Identify the primary responsibilities of each production staff position.
- Recall the activities in each step of a production workflow.

Introduction

To understand an individual role in the broadcasting industry, you must be familiar with all aspects of the production process. Each production area is interconnected to many others, with the interrelationships resembling a spider web, **Figure 2-1**. To learn proper camerawork, you must understand proper lighting technique. Proper lighting technique is dictated by the colors used on the set and on the costumes. The colors of the set and costumes directly affect the kind of special effects used in the program. Special effects are created in the special effects generator, but must be edited. Knowing the tools and techniques of editing is also required. To learn television production, you must have a solid understanding of all the contributing roles.

Figure 2-1. In order to do your job properly, you must know the jobs of others on your production staff. This illustration depicts the spider web of interrelationships between jobs on the production staff.

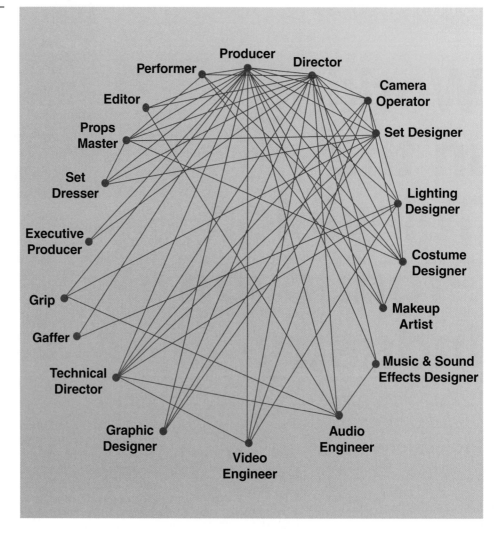

production team: Everyone involved in the production, both staff and talent.

staff: Production personnel that work behind the scenes and generally includes management and designers.

crew: Production personnel that are normally not seen by the camera, which generally includes equipment operators.

talent: Anyone seen by the camera, whether or not they have a speaking or any other significant role in the program, as well as individuals who provide only their vocal skills to the production.

Dividing Up the Work

All television production organizations, from the largest to smallest, divide the production workload among the company's employees. In large production companies, a different individual may be assigned to each job title. In smaller production companies, however, it is not unusual for a single person to fill multiple jobs on the same production. Some jobs are easier to combine than others. For example, it is difficult to imagine acting in front of a camera when you must also fill the role of camera operator. On the other hand, the person who painted the set could certainly be an actor because these two jobs do not take place at the same time.

The following sections address many of the main jobs included in the collective term *production team*, also called *production staff*. All production personnel can be divided into three categories:

- The *staff* works behind the scenes. These individuals work in the more creative levels of production management.
- The *crew* are generally equipment operators. They are not normally seen by the camera, but are integral to the production.
- Anyone seen by the camera, whether or not they have a speaking or any other significant role in the program, is *talent*.

PRODUCTION NOTE

There are exceptions to the "not seen by the camera" criteria for staff and crew. At the opening or closing of many newscasts, for example, there is often a long shot of the studio. The audience may see a shot that includes the studio's camera operators. Camera operators and technicians are regularly seen on sports programs, such as on the sidelines of a football field. These types of production personnel, who may be seen by the viewer, are not considered talent—they are considered part of the production environment.

The talent hired for a production also includes the individuals who provide only their vocal skills to the production. These positions include the on-screen actors, cast of extras, the narrator, voiceover talent, and announcers. *Cast* is the collective name given to all the talent participating in a production. It is important to remember that a program's talent includes more than actors and on-screen personalities.

cast: The collective name given to all the talent participating in a production.

Talk the Talk

When referring to multiple individuals hired as talent for a production, the correct plural form of the term is "talent." It is incorrect and unprofessional to say "talents."

Executive Producer

The *executive producer (EP)* provides the funding necessary to produce the program, but rarely steps foot on the set. There are times, however, when the EP is involved in every aspect of the production. The level of involvement varies from production to production. A single production may have several executive producers. The more expensive a program is to produce, the more likely it is to have multiple EPs. In some cases, an EP is merely an individual who invests a large sum of money in the program and, in return, is given a credit at the beginning of the program and portion of the profits generated by the sale of the program. The executive producer essentially puts the money for the production in the bank, hires a producer, and hands the bank account over to the producer.

executive producer (EP): The person, or people, who provides the funding necessary to produce the program.

Producer

The *producer* in a non-news environment purchases materials and services needed to create a finished program. The producer hires a director, designers, camera operators, a lighting director, sound engineer, and the talent. Materials purchased for the production include, but are not limited to, set construction items, costumes, and props. The producer also arranges travel plans, if necessary, for the staff and talent, including transportation, lodging, and meal catering. Because of the many facets of a producer's job, being successful requires extreme attention to detail and strong organizational skills. The producer is ultimately responsible for the program's successful completion.

The amount of input a producer has on creative decisions varies. Ideally, the producer hires a director with whom he works well. Together

producer: In a non-news environment, the producer purchases materials and services in the creation of a finished program. In a broadcast news facility, the producer coordinates the content and flow of a newscast.

pre-production:
Any activity on a
program that occurs
prior to the time that
the cameras begin
rolling. This includes
production meetings,
set construction,
costume design,
music composition,
scriptwriting, and
location surveys.

production: The actual
shooting of the program.

post-production:
Any of the activities
performed after a
program has been shot.
This includes music
beds, editing, audio
overdubs, titles, and
duplication.

distribution: The final
phase of production,
which includes DVD
authoring, DVD/
videotape duplication,
and distribution to the
end user.

the producer and director make the hiring decisions regarding the rest of the production team. As decisions are made, the producer and director must constantly be aware of the budget. Compromise is necessary to balance and successfully complete all aspects of the program production.

The producer interacts with a majority of the production staff on a day-to-day basis during all four phases of production—pre-production, production, post-production, and distribution. *Pre-production* refers to any activity on the program that occurs prior to the time that the cameras begin recording. This includes production meetings, set construction, costume design, music composition (**Figure 2-2**), scriptwriting, and location surveys. *Production* refers to the actual shooting of the program. *Post-production* involves any activities done after the program has been shot until the finished program is completed, including music beds, editing, audio overdubs, and titles. *Distribution* is the final phase of production and includes DVD authoring, DVD/videotape duplication, and distribution to the end user.

In a broadcast news facility, the producer coordinates the content and flow of a newscast and is very involved in the decision-making process during the daily morning production meetings. In a news environment, this is an extremely high-pressure, important position. Typically, a producer has earned this position by working for years as a reporter, and has a keenly developed "reporter's sense."

The producer is involved in deciding which stories will be aired, the order in which the stories will appear on the newscast, and in developing promotions and "ahead at 11" teasers for the upcoming newscast. The producer is the person who decides whether to interrupt a newscast in progress to report breaking news, and feeds the breaking news directly to the anchors through their earpieces.

A television newsroom may have several producers; reporters work for and report to their assigned producer. In the hierarchy of a television newsroom, the producers work for and answer to the news director, who makes the final determination on the content of the newscast.

Figure 2-2. Using a Musical Instrument Digital Interface (MIDI) computer program, a composer sends notes directly from the keyboard to the computer. The notes are placed on sheet music and can be printed, so the music can be performed, recorded, and placed into the soundtrack of the production.

News Director

The *news director* is responsible for the structure of the newsroom, for personnel matters (performance evaluations and hiring and firing employees), managing the budget, and the overall effectiveness of the newsroom. The news director's involvement in the actual newscast varies from station to station. However, the news director is the final authority on which stories will air during the newscast.

news director: The person responsible for the structure of the newsroom, for personnel matters (performance evaluations and hiring and firing employees), managing the budget, and the overall effectiveness of the newsroom. The news director is also the final authority on which stories will air during a news broadcast.

PRODUCTION NOTE

In a broadcast journalism class, the role of news director is commonly held by the instructor. This is because the instructor is responsible for teaching journalistic principles and ethics, news judgment, and adhering to the objectives of the broadcast journalism course. The grade assigned to students' work is equivalent to a performance evaluation in the broadcast journalism industry.

Director

The *director* is in charge of the creative aspects of the program and interacts with the entire staff. While directors are responsible for casting the program's talent, sometimes they must concede to the EP or producer's decisions. For example, casting Leonardo DiCaprio in a leading role and insisting on elaborate sets may exhaust the entire production budget on just those two items. The director must be willing to compromise for the sake of a successful and complete production.

The director reviews the program's script and visualizes the entire production, **Figure 2-3**. Those ideas must then be communicated to the staff

director: The person who is in charge of the creative aspects of the program and interacts with the entire staff.

Figure 2-3 The director reads a script and envisions the way the program's scenes should appear.

and talent. The director guides their performance to create an acceptable representation of his vision. During production, the director must coordinate and manage the staff and cast to keep to the production schedule and to ensure that all of the program's elements are properly incorporated.

Very few people begin their career as a director. Becoming a good director requires extensive experience. Most directors have worked their way up from a production assistant and, therefore, know each job on the staff quite well. That knowledge is key to communication with the staff and crew.

Production Manager

production manager:
The person who handles the business portion of the production by negotiating the fees for goods, services, and other contracts and by determining the staffing requirements based on the needs of each production.

The *production manager* handles the business portion of the production by negotiating the fees for goods, services, and other contracts, and by determining the staffing requirements based on the needs of each production. An additional responsibility is to ensure that programs and scripts conform to established broadcast standards. The production manager contributes to the successful completion of a production by managing the budget and available resources.

Production Assistant

production assistant (PA): The person who provides general assistance around the studio or production facility. The PA is commonly hired to fill a variety of positions when key personnel are sick, out of town, working on another project, or otherwise unavailable. In many facilities, the production assistant position is synonymous with the *assistant director (AD)* position.

In many television production companies, the titles *production assistant (PA)* and *assistant director (AD)* are interchangeable. However, these position titles are *not* interchangeable in the film industry. The PA serves as a jack-of-all-trades, but is a master of none. In some facilities, the PA is merely a "gofer." In most facilities, however, the PA is hired to fill a variety of positions when key personnel are sick, out of town, working on another project, or otherwise unavailable. Most people begin their career as a PA. From the PA position, motivated individuals can rise through the ranks of a facility until achieving the position they want.

On a daily basis, the PA provides general assistance around the studio or production facility. On the occasion that a staff position needs to be temporarily filled, an ambitious PA should offer to perform the tasks of that job. Competent performance will be noted by the production team and remembered for subsequent opportunities.

PRODUCTION NOTE

If there are several PAs in the company, it is vital that you volunteer before someone else and take an active part in the advancement of your career. A passive individual does not last long in a PA position—you must be active and aggressive. When a qualified person is in front of an employer eagerly saying, "I am qualified. I want to do the work. Give me a chance to prove it," why would the employer offer the position to a wallflower who is too shy to speak up? Aggressive, energetic, and enthusiastic go-getters populate this industry.

Instead of waiting long periods of time for a promotion at a company, PAs who have acquired significant skills and experience typically choose to move from company to company. This is the most common way to move up the ranks within production teams in the industry. Most professionals in this industry change companies 7 or 8 times within their first 10 years of working. Choosing to stay with one company comes with the risk of waiting years for someone above you to retire, transfer to another company, or otherwise leave the position. Although promotions do occur within companies, it is more likely to happen when moving laterally from one company to another.

PRODUCTION NOTE

Because changing companies is so common in the television production industry, it is strongly recommended that you maintain your own investment/retirement accounts. Starting an Individual Retirement Account (IRA) right out of college is one of the smartest and least expensive things you can do for your own future.

In addition to proven knowledge and skills, an important key to employment in this industry is networking. Job openings in television production are rarely found in the want ads of the local newspaper. Consider the following scenario: Bill has been an assistant camera operator for a significant amount of time at the XYZ Production Company. He feels he has developed the skills to become a camera operator and wants to move up to that position. However, there are three other people that hold the three camera operator positions in the company, and none of them has indicated a desire to retire or leave the company. Bill's prospects of advancement at his current company are slim, so he begins networking by telling virtually everyone he knows in the industry that he is looking for a camera operator position. Eventually, someone who knows of an opening hears that a talented guy is looking for a job as a camera operator. That person contacts Bill through the network to let him know that he should apply for the position. Nearly everyone in the industry maneuvers from job to job using this networking technique, and most people are willing to help other professionals advance.

Floor Manager

The *floor manager*, or *floor director*, is the director's "eyes and ears" in the studio. The floor manager wears a headset and relays the director's commands to all studio personnel, except the camera operators. The camera operators are usually in direct communication with the director via their headset intercoms. The floor manager is the only person in the studio who may say, "Cut," other than the director. When the floor manager says, "Cut," it is usually because the director has instructed them to do so.

floor manager: The person who is the director's "eyes and ears" in the studio. The floor manager relays the director's commands to the studio personnel. Also commonly called *floor director*.

PRODUCTION NOTE

In larger studios, the headset communication system has multiple channels so that the director can speak just to the camera operators, just to the floor manager, or to everyone at once.

The floor manager is responsible for making sure the set is ready for production, for initiating the program countdown, and for giving various cues to the talent. A *cue* is a signal that implies something specific is to happen. One familiar cue is the "cut" signal. The floor manager makes a cutting motion with his hand across his neck. There are many other hand signals and cues that are standard within the industry. The signals can be any action that the production team agrees upon and understands, **Figure 2-4**.

Figure 2-4. The floor manager gives silent signals, or cues, to the talent.

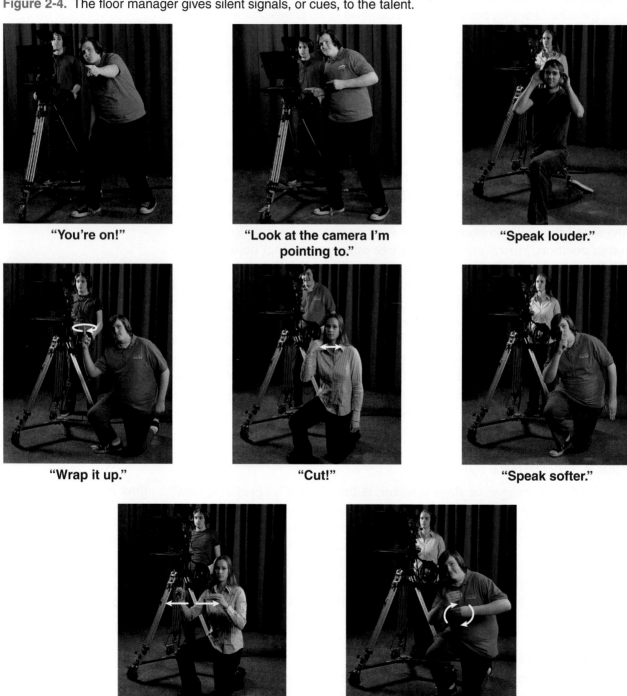

"You're on!"

"Look at the camera I'm pointing to."

"Speak louder."

"Wrap it up."

"Cut!"

"Speak softer."

"Stretch what you're saying— you're going too fast."

"Speed up your speaking— time is running out."

Camera Operator

The *camera operator* runs the piece of equipment that captures the video images of the program, **Figure 2-5**. Camera operators are responsible for framing shots that are visually pleasing to the viewers. The director may often call for a particular shot—the camera operator must not only provide the shot requested, but must frame the shot so that annoying or inappropriate background information does not detract from the image.

camera operator: The person who runs the piece of equipment that captures the video images of the program.

Photographer

The *photographer*, often called *photog* or "shooter," is the cameraperson who goes into the field on location with a reporter in a news operation. The photog's responsibilities include all things technical—transporting the camera, tripod, mic, all the cabling, and any batteries necessary. While shooting, the photog monitors the audio of both the reporter and the interviewee through headphones. Setting up and tearing down the equipment is also the responsibility of the photog (reporters often help, but these tasks officially fall to the photog).

photographer: The cameraperson in the field, on location with a reporter in a news operation. Also commonly called *photog* or *shooter*.

Photojournalist

The *photojournalist* is a photographer who regularly performs duties of both the photographer, as well as the reporter. A photojournalist is a one-man band. It is a good idea for anyone who wants to be a reporter to also obtain

photojournalist: A photographer who regularly performs duties of both the photographer, as well as the reporter.

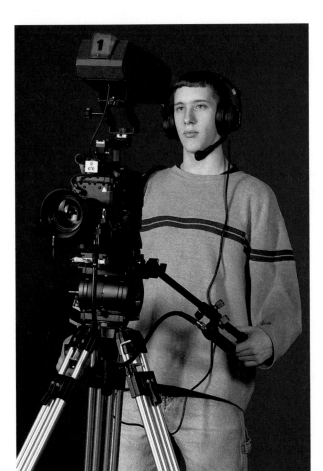

Figure 2-5. The camera operator captures the shots that the director requires. Camera operators never sit down when operating a camera mounted on a tripod and never let go of the pan handles without making certain the camera is locked and stable.

the skills to be a photog. When applying for a job in the news industry, the ability to successfully perform both reporter and photog roles provides an advantage over those who are unable to wear more than one hat.

Reporter

reporter: The individual responsible for gathering information from many sources, including research and interviews, for writing news stories, and often editing their own stories.

Reporters are responsible for gathering information from many sources, including research and interviews, for writing news stories, and often editing their own stories. The role of a reporter can vary from station to station, or even story to story. Sometimes reporters are on-screen throughout their story, but it is also common to see the reporter only during the introduction and closing of stories. Some stories include the reporter's voiceover throughout the story, in addition to other audio, with the story footage, but the reporter may never be seen on screen. A reporter may write a story and be present for shooting and editing the story, but an anchor may end up reading the script of a story written by the reporter. Reporters often find that the job entails erratic work hours—some days may require far more than the typical eight hours. The job of a reporter is physically and emotionally demanding.

Assignment Editor

assignment editor: The person who schedules necessary equipment and personnel to cover the stories for the day's newscast.

During the morning meeting, decisions are made regarding which stories reporters will undertake for the day's newscast. When the stories are chosen, the *assignment editor* schedules the equipment and personnel to cover the stories. The assignment editor pairs reporters and photogs, and schedules photojournalists if there are more stories to cover than available reporters and photogs. Typically, the assignment editor assigns each reporter two stories per day. Exceptions are made, however, when major news events occur.

Anchor

anchor: The person who delivers the news from the news desk set in a studio.

The *anchor* delivers the news from the news desk set in the studio. Delivering the news involves reading the news content displayed on a teleprompter, providing the intro and closing of taped stories that are inserted into a live telecast, and conducting conversations with reporters in the field reporting live. The greatest expectation of an anchor is to accurately read and relay the news and related information.

Video Engineer

video engineer: The person who manages the video equipment and is ultimately responsible for the technical quality of the video signal.

The *video engineer* is ultimately responsible for the technical quality of the video signal, **Figure 2-6**. A video engineer has extensive schooling in the electronics of video production and must keep current with the technology and changes in the video industry. This member of the production team is greatly valued and highly compensated in any production facility. In a studio environment, one of the video engineer's responsibilities is to ensure that the images captured by each of the studio cameras match exactly. This consistency is important when, for example, the director cuts from one camera to another. The video engineer's skills ensure that an actor's skin color does not change from normal, to pinkish, to greenish when cutting between cameras.

Figure 2-6. The video engineer uses specialized test equipment to examine and maintain consistent picture quality.

Audio Engineer

The *audio engineer* is responsible for the audio/sound quality on the production. The audio engineer often operates the microphone mixer, as well as the music and sound effects recorders/players, **Figure 2-7**. The audio engineer mics the talent and is responsible for maintaining the overall audio levels on the studio's master recorder.

audio engineer: The person responsible for the audio/sound quality on the production and related equipment.

Figure 2-7. The audio engineer maintains the quality and volume of the sound, so that viewers never have to adjust the volume on their television sets.

lighting director: The person who decides the placement of lighting instruments, the appropriate color of light to use, and which lamps should be used in the instruments.

gaffer: The lighting director's assistant who often does the actual hauling of heavy instruments up and down ladders.

scriptwriter: The person responsible for placing the entire production on paper.

Lighting Director

The *lighting director* decides the placement of lighting instruments, the appropriate color of light to use, and which lamps should be used in the instruments. In a television studio, as on the stage in a high school auditorium, there are an amazing number of lights hanging overhead from pipes on the ceiling. The lights are purposefully aimed in various directions with varying degrees of brightness and color. Determining the placement of the lighting instruments is the lighting director's job. The lighting director's assistant, a *gaffer*, often does the actual hauling of heavy instruments up and down ladders, **Figure 2-8.**

Scriptwriter

The *scriptwriter* is responsible for placing the entire production on paper. The script must meet the objectives of the producer and the message to the viewer must be clear. However, the scriptwriter is not often an acknowledged expert in the program's subject matter. Because of this, a content specialist is

Figure 2-8. The lighting designer tells the gaffer (on ladder) where to aim the lighting instruments.

usually hired to work with the scriptwriter. The *content specialist* is a person considered to be an expert on the program's subject.

Content specialists from the military are commonly hired to help scriptwriters and directors create authentic action scenes in movies with military action or plot lines. A content specialist would also assist a scriptwriter in writing the script for an instructional program detailing techniques of a new and innovative heart transplant technique. To ensure accurate information, the content specialist in this scenario would likely be the doctor who invented the technique. The content specialist reviews the entire script before production begins and is, ideally, present for the shooting and post-production to keep everything accurate.

Graphic Artist

The *graphic artist* is responsible for all the artwork required for the production. This includes computer graphics, traditional works of art, charts, and graphs. The graphic artist is usually very well versed in computer graphics applications, from the amazing animations seen in modern films to the charts and graphs included in an economics program.

VTR Operator

The *VTR operator* is in charge of recording the program onto videotape by correctly operating the VTR equipment, **Figure 2-9.** Producing a recording of the program with quality video and audio is an immense responsibility. The VTR operator must take every precaution to ensure that each piece of equipment is functioning properly to produce a quality recording of every scene.

Many newer television facilities have eliminated the use of videotape entirely, and record onto DVDs or directly to a hard drive. The job title "VTR operator" no longer applies in these facilities. In a tapeless environment, this job function is performed by the *video operator*.

content specialist: A person who works with the scriptwriter and is considered to be an expert in the program's subject matter.

graphic artist: The person responsible for all the artwork required for the production. This includes computer graphics, traditional works of art, charts, and graphs.

VTR operator: The person in charge of recording the program onto videotape by correctly operating the VTR equipment.

video operator: The individual responsible for recording the master video file in a tapeless television production environment.

Figure 2-9. The VTR operator places the entire program on the master videotape.

robo operator: The person who remotely operates all of the cameras and the robotic camera mounts from a single location in the studio or control room.

editor: The person responsible for putting the various pieces of the entire program together. The editor removes all the mistakes and bad takes, leaving only the best version of each scene, and arranges the individual scenes into the proper order.

Robo Operator

Some television production environments have eliminated the job of an in-studio camera operator. Cameras are placed on remote-controlled robotic camera mounts. All of the cameras are controlled by the *robo operator* from one location in the studio or control room, **Figure 2-10**.

Editor

The *editor* puts the various pieces of the entire program together. Individual scenes are arranged into the proper order, with all the mistakes and bad takes removed, leaving only the best version of each scene. The editor must be aware of the psychological effects involved with the theory of movement and passage of time, as different angles of the same person or scene are cut together. For example, a skilled editor makes two sides of a conversation, which were shot during two different recording sessions at different locations, flow together into a natural sounding conversation on one recording.

Figure 2-10. A robotic camera in use on a production studio set. A—A traditional video camera and teleprompter are attached to the robotic mount. B—Away from the studio floor, the robo operator remotely controls the robotic camera equipment from a control panel with joystick controls. *(Photos courtesy of Brian Franco)*

A

B

Talk the Talk

"Editor" is a television production term with dual definitions. This chapter addresses the person, called an editor, who arranges the individual pieces of a program to produce a complete product. To do this, the editor uses a machine that is also called an "editor," which is discussed in Chapter 24, *Video Editing*.

Makeup Artist

The *makeup artist* applies cosmetics to the face and body of talent, giving them the intended appearance in front of the camera, **Figure 2-11**. The cosmetics used may enhance facial features or change the talent's appearance entirely, as necessary to convincingly portray a particular character.

CG Operator

The *CG operator* creates the program titles using a character generator, **Figure 2-12**. Titles include the credits that appear before and after a movie

makeup artist: The person responsible for applying cosmetics to the talent's face and body, giving them the intended appearance in front of the camera.

CG operator: The person who creates the titles for the program using a character generator.

Figure 2-11. The makeup artist ensures that performers look natural under the bright studio lights.

Figure 2-12. The CG operator types the titles for the program.

or television show, as well as a news flash that crawls across the bottom of the television screen. The titles may be a single page, scrolling, or have animated letters. The CG operator must create titles that are accurate and appropriate for the program, and ensure that they are legible to viewers.

Grip

grip: A person who moves the equipment, scenery, and props on a studio set.

The *grip* is a person who moves the equipment, scenery, and props on a studio set. In theater productions, a grip is called a stagehand. Perhaps the job title comes from the fact that one must have a good grip in order to move anything large!

maintenance engineer: The person who keeps all the production equipment functioning at its optimum performance level.

Maintenance Engineer

The *maintenance engineer* keeps all the production equipment working according to "factory specifications." The maintenance engineer is not a repairperson, but may assist in troubleshooting if problems arise. The primary responsibility of this position is to ensure that each piece of production equipment functions at its optimum performance level, **Figure 2-13**.

Figure 2-13. The maintenance engineer keeps equipment running in top performance.

Program Production Workflow

The following is a general overview of the steps involved in producing a program. The timeframe to complete each step depends on the type of production. Completing most of the steps to produce a public service announcement, for example, would likely take considerably less time than to produce a one-hour, prime-time drama. One of the best ways to create a successful program is to have excellent production values in the program. *Production values* are the general aesthetics of a show. Most of this book aims to show you how to attain these high production values.

production values: The general aesthetics of the show.

Various terms are presented in the sections that follow, as well as in the proceeding chapters. Many of the terms have commonly known "consumer" definitions. Some words used in this business, just as in the English language, have multiple definitions. Memorize the professional definitions of terms and learn the difference between those with multiple meanings. Use the terms appropriately during class to help you get in the habit of using them correctly when working in the television industry.

Program Proposals

The first step in producing a program is to develop a program proposal (discussed in Chapter 8, *Scriptwriting*). This is a plan that includes the basic idea of the program, the program's format, intended audience, budget considerations, location information, and a rough shooting schedule. The program proposal is reviewed by investors and production companies for financing considerations and overall project approval.

Scriptwriting

Before a script is written for the program, a script outline is created. This outline contains comments noting the direction of the program and varies depending on the program format (drama, panel discussion, interview, or music video). Television scripts are usually written in a two-column format. The left column contains video/technical information and the right column contains audio and stage direction.

Producing

The day-to-day activities involved in producing a program ensure that the production process runs as smoothly as possible. Important decisions that affect the program's ultimate success are made throughout the production process, including coordinating schedules, acquiring the necessary resources, monitoring the activity and progress of various production teams, and weighing budgetary considerations.

Directing

Directing involves shaping the creative aspects of a program and interacting with the entire staff and cast to realize the director's vision of the production. In addition to verbally providing direction during production, many important pre- and post-production directing activities contribute to a program's success.

Lighting

When planning the lighting for a production, there should be sufficient light to meet the technical requirements of the camera and to produce an acceptable picture on the screen. Various lighting techniques are also used to meet the aesthetic requirements of the director. Accurate lighting in a program is necessary to create the desired mood, appearance, and setting. Most importantly, proper placement of lighting instruments contributes to creating three-dimensionality on a flat television screen.

Scenery, Set Dressing, and Props

Careful planning and consideration when choosing scenery, set dressings, and props helps create a believable environment for the program.

Figure 2-14. Scenery is anything, other than people, that appears behind the main object of the picture.

scenery: Anything placed on a set that stops the distant view of the camera. Outside the studio, scenery may be a building or the horizon.

The placement of these items on a set also contributes to creating three-dimensionality on a flat television screen. *Scenery* is something that stops the distant view of the camera (further discussed in Chapter 18, *Props, Set Dressing, and Scenery*). In a studio, the scenery may be fake walls, set furniture, or a curtain. Outside the studio, it may be a tree, building, or the horizon. The scenery is nearly everything behind the main object of the shot, **Figure 2-14**. Set dressing includes all the visual and design elements of a set, such as rugs, lamps, wall coverings, curtains, and room accent accessories. Props are any of the items handled by the performers, excluding furniture, **Figure 2-15**. Furniture may be a prop if used for something other than its apparent and intended use.

Costumes and Makeup

Costumes and makeup enable actors to look like the characters they portray. Even news anchors and other on-screen personalities who are not

Figure 2-15. Props are items, other than furniture, that are handled by performers.

"acting" wear makeup and have their wardrobe selected to ensure the best possible appearance on the television screen.

Costume selection is dependent on many existing factors, including plot, setting, set dressing, program format, and lighting arrangement. *Makeup* is any of the cosmetics applied to a performer's skin to change or enhance their appearance. The makeup may create a drastic change, such as aging, alien appearance, or injuries, or it may simply enhance the talent's natural features while in front of a camera.

makeup: Any of the cosmetics applied to a performer's skin to change or enhance their appearance.

Graphics

Graphics are all of the artwork seen in a program, including computer graphics, traditional works of art, charts, and graphs (discussed in Chapter 14, *Image Display*). When choosing or creating graphics for television, pay particular attention to the amount of detail in a graphic. Losing the fine detail in images is natural in the process of creating an analog television picture (digital technology is continually evolving and changing this limitation). For example, a beautifully detailed title font of medieval style writing may look wonderful on a computer screen, but will likely dissolve into mush on a television screen. The television screen requires bolder images than a computer screen. If the audience is unable to read what is written on the screen or cannot clearly see the information presented in a chart, then you are not effectively communicating.

Camera Operation

The portion of the program that you can see is called *video*. The camera operator is responsible for capturing the program images with a video camera.

video: The portion of the program that you can see.

Talk the Talk

The term "video" has different consumer and professional definitions. Consumers often use "video" to refer to the tape or the DVD you rent or purchase for viewing at home. Television production professionals use "video" to refer to the visual portion of a program; the part that is seen by the audience.

A *frame* is the actual edge of the video picture; the edge of the picture on all four sides, **Figure 2-16**. *Framing* a shot is the camera operator's responsibility and involves placing items in the picture by operating the camera and tripod. Shooting a vase of flowers sitting on a table seems simple until you realize there are an infinite number of ways to shoot it (long shot, close-up, from a side angle, from below, from above, zoom in, or zoom out). A good camera operator has the ability to frame shots effectively for the audience.

frame: The actual edge of the video picture; the edge of the picture on all four sides.

framing: Involves placing items in the camera's frame by operating the camera and tripod.

We have all seen home movies taken of someone else's family. Home movies are usually not tolerable to watch for long periods of time. One reason is camerawork—it is generally shaky and out of focus, **Figure 2-17**. An important production value is quality camerawork. Put the camera on

Figure 2-16. The edge of a picture on all four sides is the frame. An operator frames a shot by determining exactly what to include in the frame.

a tripod for stability and frame the shots correctly. Make sure that people in the video do not have oddly cut off body parts, **Figure 2-18**.

Audio Recording

Audio is the portion of a program that you can hear. Audio includes narration, spoken lines of dialog, sound effects, background music, and all other aspects of a program that are heard by the audience.

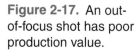

audio: The portion of a program that you can hear. Audio includes narration, spoken lines, sound effects, and background music.

Figure 2-17. An out-of-focus shot has poor production value.

Figure 2-18. A shot of a person "missing" important parts of the body is considered a poor production value.

Production Switching

When shooting in a studio, there may be three cameras shooting a scene from three different angles. Each of the cameras captures a different picture:

- Camera 1 is on one news anchor in a close-up.
- Camera 2 has a two shot of both anchors.
- Camera 3 is on the other news anchor in a close-up.

All three of these cameras connect to a production switcher, **Figure 2-19**. A cable coming from the switcher connects to a video recorder. By pressing buttons marked "1," "2," and "3," the picture from different cameras can be sent to the recorder. The process of cutting between cameras (from camera 1, to camera 3, to camera 2, and back to camera 1) is called *production switching*.

production switching: The process of cutting between cameras.

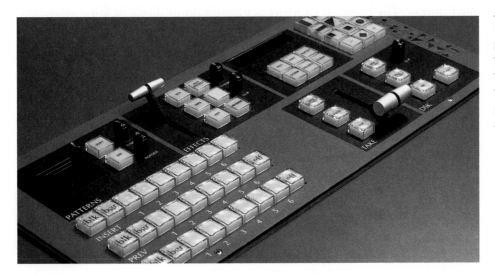

Figure 2-19. The production switcher allows the operator to send pictures from different cameras to the video recorder, simply by pushing a button or pulling a lever.

Special Effects

special effects:
Anything the audience sees in a video picture that did not really happen in the way it appears on the screen.

An entire book could be written on special effects alone. In simplest terms, *special effects* are anything the audience sees in a video picture that did not really happen the way it appears on the screen. Special effects alter the reality perceived by the viewer.

PRODUCTION NOTE

One of the cardinal rules of television production and filmmaking: "It does not have to be, it only has to appear to be!"

Editing

editing: The process of placing individual recorded scenes in logical order.

Placing the individual scenes in logical order on another tape is called *editing* (discussed in Chapter 24, *Video Editing*). When writing a research paper, you make notes on note cards. One of the next tasks in the process is to arrange the note cards in an order that makes the paper flow logically. The process of arranging note cards corresponds to editing a video program.

The scenes of a program are not usually shot in the order seen in the finished product. All the scenes that take place in one location are shot at the same time, even if they appear at different times in the finished program. Imagine that scenes 25, 41, and 97 of a movie take place in Egypt at the Sphinx, and scenes 24, 40, and 98 take place at the Eiffel Tower in Paris, France. To shoot the scenes in chronological order, all the people and equipment would need to be transported back and forth three times between these distant locations. The increased production cost in providing such travel arrangements is unreasonable and, most often, not possible. Setting up once at each location to shoot all the necessary scenes requires that the scenes be edited together in the proper order in post-production.

Duplication and Distribution

A master program is copied to multiple media formats, such as tape or DVD, for distribution and viewing. The programs may be individually sold by a retailer, used as informational material for a specific workforce or company, cablecast, or be broadcast and viewed on televisions in millions of homes. Programs may also be streamed and downloaded using the Internet. The finished product is viewed on millions of televisions and computer screens around the world almost instantly.

Assistant Activity

- Watch 20 minutes of the local or national news with the sound turned off. Can you still follow what the newscasters are communicating? Why?

- Watch 10 minutes of a sitcom with the sound turned off. Can you still follow the storyline?

- Try watching a commercial that you have never seen before with the sound turned off. What is the commercial trying to tell you about the product? Can you figure it out without the audio?

  Be prepared to discuss your experience with each of these scenarios.

Wrapping Up

The academic aspects of TV production must be learned to understand how all the elements of production fit together. Everyone needs to understand everyone else's job in order to fit into the matrix of production. There are hundreds of factors to consider when producing a television program, and this chapter presents only a few of the main jobs involved. An enormous number of people are typically involved, from the beginning to the end of a production. Remember that this text is an introduction to television production and broadcast journalism. It is not intended to be the end of your learning, rather just the beginning.

The approach this text uses to teach this complex subject is a vocabulary-based, progressive method. Concentrate on learning the industry terminology, and understanding the principles behind those terms should come naturally. The content touches every topic briefly at first and, as the chapters progress, continues to address the topics in increasing depth and detail. Progressive learning means that each day of class builds on a foundation created by all previous days and lessons—you cannot forget what has come before. If your brain is a hard drive, for example, you have one file called "Television Production" and you keep adding more information to that one file. Nothing can be deleted!

Review Questions

Please answer the following questions on a separate sheet of paper. Do not write in this book.

1. Explain the difference between the *talent* and the *staff* of a production.
2. What are the four phases of program production?
3. Describe how the director interacts with the program's producer.
4. What are the typical responsibilities of a photog?
5. What information is included in a program proposal?
6. What is the *frame* of a video picture?
7. Explain the process of production switching.
8. Why is it usually impractical to shoot all the scenes of a program in sequential order?

Activities

1. Record the final credits of your favorite television show. Play the credits back slowly and notice all the job titles listed. List any of the titles that are unfamiliar to you and research the responsibilities of each job. Be prepared to present this information in class.
2. Research basic hand signals used by floor managers on a production set by searching the Internet, checking reference material at the library or, best of all, by visiting a local TV station and interviewing a floor manager. Make an illustrated poster of the new signals you learn.

STEM and Academic Activities

Technology

1. Identify the positions on a production crew that have changed dramatically with technological advancements. Identify production crew positions that have changed little, or not at all, despite technological advancements in television production.

Engineering

2. Create a flowchart that depicts the television production process from beginning to end. Include the titles of staff, crew, and talent involved in each part of the process.

Mathematics

3. Calculate how much money you will accrue after 30 years if you invest $4500 every year into a retirement fund with 8% interest compounded annually.

Language Arts

4. Develop an idea for a new prime-time reality television show. Write a program proposal to pitch your program idea to the executive producers.

Social Science

5. In television, the four phases of production are pre-production, production, post-production, and distribution. Relate these phases to your life. Assign each of the activities you perform to complete a specific task or goal to one of these four phases.

Chapter 3

The Video Camera and Support Equipment

Objectives

After completing this chapter, you will be able to:

- Explain the differences between the various video cameras available.
- Identify each part of a video camera and note the corresponding function.
- Differentiate between the focal length and the focal point related to a zoom lens.
- Explain the interrelationship between f-stops, the iris, and aperture in controlling light.
- Identify the challenges and benefits involved in using hand-held camera shooting.
- Recognize the types of tripod heads available and cite the unique characteristics of each.
- Implement the proper procedures for cleaning and storing video equipment.

Introduction

The camera is one of the first pieces of equipment that new students gravitate toward because it appears to be the most central item in a television studio. Good camera operators must first learn the capabilities of their equipment. This chapter presents parts of the video camera, related support equipment, and basic operation procedures.

Professional Terms

aperture
auto-focus
auto-iris circuit
camcorder
camera control unit (CCU)
camera head
charge coupled device (CCD)
convertible camera
diopter adjustment
dolly
drag
fast lens
fluid head
focal length
focal point
focus
friction head
f-stop
gain
hot
iris
jib

lens
optical center
pan handle
pedestal column
pedestal control
remote control unit (RCU)
shutter
slow lens
studio camera
studio pedestal
subjective camera
target
tighten
tripod
tripod head
variable focal length lens
viewfinder
widen
zebra stripes
zoom in (ZI)
zoom lens
zoom lenses
zoom out (ZO)

PRODUCTION NOTE

In the classroom environment, it is not necessary to have "professional broadcast quality" cameras in order to effectively learn video camera operation. In making this decision for my own classroom, I discussed with a vendor whether I should spend a sizeable amount of money for one "broadcast quality" camera or the same amount of money for several "non-broadcast quality" cameras. My vendor's comment on the situation made very good sense, "You are teaching students to take pictures and, when you get right down to the bottom of things, all cameras point." As a result, I bought several good quality cameras rather than one high quality camera, which would not teach students anything more than the cameras I bought. The additional cameras also allow more students to get experience operating a camera without waiting in line for one to become available.

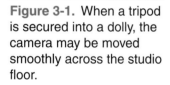

Types of Video Cameras

studio camera: A television camera placed on a tripod or studio pedestal for exclusive use within the studio.

tripod: A three-legged stand that supports a camera.

dolly: A three-wheeled cart onto which the feet of a tripod are mounted. A dolly allows smooth camera movements to be performed.

studio pedestal: A large, single column on wheels that supports the camera and is pneumatically or hydraulically controlled.

Several types of video cameras are available for professional use. Each camera type offers unique benefits and restrictions.

Studio Cameras

The *studio camera* is usually very large and too heavy to be used as a remote camera in the field. Because of its size, studio cameras may be placed on a three-legged stand, called a *tripod*, for support. To allow smooth camera movement, the feet of the tripod are placed into a type of three-wheeled cart called a *dolly*, **Figure 3-1**. A *studio pedestal* is another common type of camera support, **Figure 3-2**. The camera is attached to a large, single column on wheels that is pneumatically or hydraulically controlled. Due to the size, weight, and mount of studio cameras, they should not be taken out of the studio.

Figure 3-1. When a tripod is secured into a dolly, the camera may be moved smoothly across the studio floor.

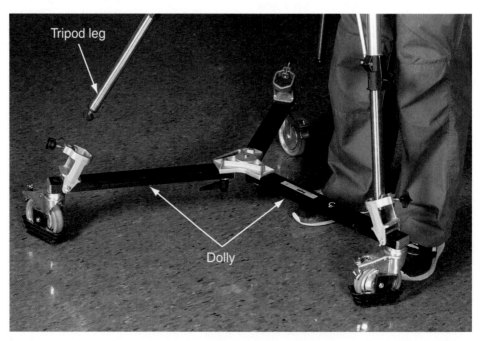

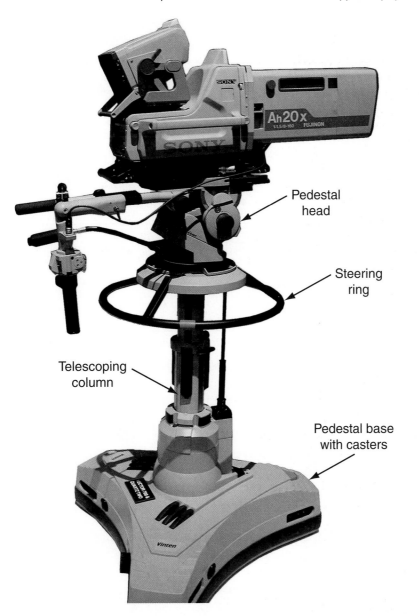

Pedestal head

Steering ring

Telescoping column

Pedestal base with casters

Figure 3-2. Placing the camera on a pedestal provides a steady and smooth shot while in the studio. *(Vinten Broadcast Ltd.)*

Talk the Talk

When referring to multiple camera dollys, the correct spelling of the term is "dollys." This rule applies only when making reference to this particular piece of equipment.

Each studio camera comes with a *camera control unit (CCU)*, sometimes referred to as a *remote control unit (RCU)*, **Figure 3-3**. The CCU is a piece of equipment that controls the video signal sent from the camera and is usually placed in the control room or the master control room. The CCU controls many signals from the camera, including the color, tint, contrast, and brightness. The video engineer manipulates the CCU controls to match the signal from each camera involved in the shoot, **Figure 3-4**.

camera control unit (CCU): A piece of equipment that controls various attributes of the video signal sent from the camera to the video recorder, and is usually placed in the control room or the master control room. Also commonly called a *remote control unit (RCU).*

Figure 3-3. The video engineer uses camera control units (CCU) to adjust the attributes of studio cameras from the control room. This provides a central location for one person to control all the cameras, rather than adjusting the settings on each individual camera on the studio floor.

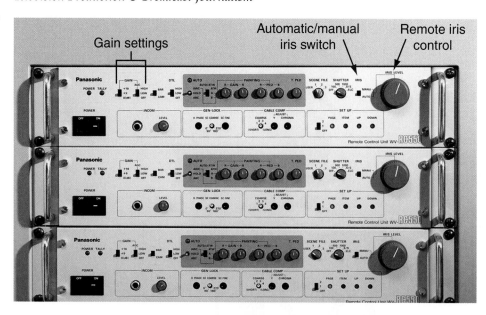

Gain settings   Automatic/manual iris switch   Remote iris control

Figure 3-4. A CCU matches the video signals when shooting with multiple video cameras. *(Jack Klasey)*

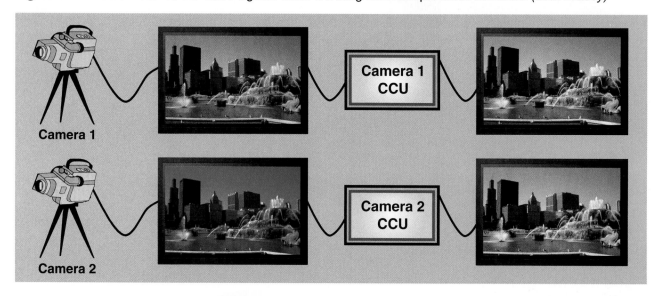

Camera 1

Camera 1 CCU

Camera 2

Camera 2 CCU

VISUALIZE THIS

The video engineer adjusts the settings on CCUs to match the signal from each camera to the others in the studio. The following scenario is likely to occur when cameras are not matched:

The on-screen talent is wearing a red dress and three cameras are shooting her. Every time the switcher cuts from one camera to another, the color of the dress changes from a shade of purple to orange to pink. This creates problems in the editing room during post-production.

Camcorders

camcorder: A portable camera/recorder combination.

Professional *camcorders* are lightweight, portable cameras, **Figure 3-5**, but are not quite as small as consumer camcorders. Professional models have many more internal components. The professional camcorder is a television camera and recorder in one unit and is relatively simple to take into the field. While in use, it is placed on the operator's right shoulder or on a field tripod. Newer camcorders record directly onto a DVD, mini DVD, memory stick, PCI card, internal hard drive, or external portable hard drive (**Figure 3-6**). Some cameras have USB or firewire ports to transfer captured video directly to a computer.

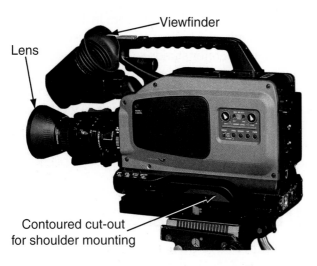

Viewfinder

Lens

Contoured cut-out for shoulder mounting

Figure 3-5. A professional camcorder can produce high quality pictures outside of the studio.

Figure 3-6. Portable hard drives are available in several configurations. A—One type of portable hard drive attaches to the back of the camera on the battery mount. The battery is attached to another mount on the opposite side. B—A hard drive unit is held in a holster on the camera operator's belt and is connected by cable to the camera.

A

B

Convertible Cameras

A *convertible camera* may be purchased with a variety of accessory packages that make it operational in a studio, as a portable field camera, or both. Many small-scale studios purchase convertible cameras because they are adaptable to a variety of situations and are often less expensive than larger studio cameras.

The studio package configuration of a convertible camera includes a CCU and a viewfinder (a small television monitor). Studio viewfinders usually measure at least 5″ diagonally. The camera operator stands several feet behind the camera, so the image must be large enough to be seen at that distance.

A remote camera package configuration usually includes a small view screen that folds out from the body of the camera and/or a viewfinder built into the camera body. The operator is likely to have the camera on his shoulder with his right eye pressed against the eyecup of the viewfinder, so a larger viewfinder is not necessary.

The Parts of the Camera

The camera, **Figure 3-7**, is comprised of three major parts:
- Camera Head
- Viewfinder
- Camera Lens

Camera Head

The *camera head* is the actual camera portion of the equipment, **Figure 3-7**. It contains all the electronics needed to convert the reflection of

Figure 3-7. Even a convertible camera, which may be configured either as a studio camera or a remote camera, has the same basic three components as other types of video cameras.

light from the subject into an electronic signal. The incoming light is split, usually by a prism, into individual red, green, and blue beams (**Figure 3-8**). Each beam hits the photosensitive surface, or the *target*, of the corresponding *charge coupled device (CCD)*, or "chip." There are hundreds of thousands of photosensitive elements on one side of the dime-sized CCD that convert light into an electronic, or video, signal. The electronic signal created by each photosensitive element varies, depending on the intensity and color of light that hits individual elements. The video signal passes through to the opposite side of the CCD and enters the rest of the camera. Professional cameras contain three CCDs—one for each colored light beam (red, green, and blue). Low cost cameras have only one CCD, which produces video that is considerably lower in quality than a professional camera.

Gain Control

Gain is the strength of the video signal. Some cameras have a "gain select" or "gain switch," while others may have this feature available through a menu option. On a studio camera, the gain control may be located on the CCU. If this function is available, you should be aware of its effect on the recorded image. Improper use of the gain switch can result in unusable footage. Adjusting this control allows the strength of the signal going from the camera to the recorder to be increased or decreased. The white level, black level, color, and tint are all equally affected when the gain setting is changed.

target: Photosensitive surface of a charge coupled device (CCD).

charge coupled device (CCD): A-component of the camera head into which light enters and is converted into an electronic, or video, signal. The video signal exits on the opposite side of the CCD and enters the rest of the camera.

gain: The strength of a video or audio signal.

PRODUCTION NOTE

The average consumer would say that the gain control adjusts the picture's brightness. In reality, gain is to brightness as a cubic zirconia is to a diamond. They look similar to an untrained eye, but there are vast differences between them. Adjusting the gain control changes the strength of the actual video signal. On the other hand, when the brightness is adjusted, only the amount of "white" in a picture is increased or decreased. In the realm of audio, "gain" is synonymous with "volume."

Figure 3-8. The prism block splits incoming light into three distinct beams (blue, red, and green) and sends each beam to a separate, dedicated CCD.

When shooting something that is dimly lit, the picture will be dark. In this respect, the camera is no different from your eye. It is difficult, sometimes impossible, for the human eye to see in the dark. A soldier on night maneuvers, for example, absolutely must be able to see in the dark. In this situation, night vision goggles are used. In recent years, news programs have commonly shown images of night vision from war zones. However, these images are not very clear. When a camera is shooting in the dark, increasing the gain may artificially brighten the picture.

As the gain is increased, the resulting image becomes increasingly grainy. This kind of picture is unusable in most professional productions. It is recommended that the gain switch never be moved from the "0" (zero) position. If an image is too dark, a light source should be added.

Viewfinder

viewfinder: A small video monitor attached to the camera that allows the camera operator to view the images in the shot.

A *viewfinder* is a small video monitor that allows the camera operator to view the images in the shot, **Figure 3-7**. Oftentimes, a camcorder has two viewfinders—one small fold-out screen on the side of the camera head and the other is usually much smaller (perhaps only an inch, if measured diagonally) and is a fixed part of the camera body.

The fold-out viewfinder is very convenient if the camera is on a tripod, and the larger screen is easier to view. On the other hand, the image displayed on a fold-out viewfinder may be difficult to see when the camera is in a well-lit environment. It may also be difficult to focus the image using the fold-out viewfinder, because it may not display a high-quality image.

diopter adjustment: A knob or lever that adjusts the magnifier on the viewfinder to compensate for differences in vision.

The smaller viewfinder has a magnifier in front of it, which makes the image easier to see. Because the viewfinder and magnifier are completely enclosed, the operator's eye must rest against a soft rubber eye piece to properly view the image. When using the small viewfinder, there is not a problem clearly seeing the image in bright light. However, operators who wear eyeglasses may find that their glasses get in the way of properly using the viewfinder. To allow for operation without eyeglasses and compensate for differences in vision, the magnifier on the smaller viewfinder is adjustable. This adjustment is called the *diopter adjustment*, and is usually a little knob that turns or a lever that moves, **Figure 3-9**. Move the diopter

Figure 3-9. The diopter adjustment allows camera operators to clearly see images in the viewfinder without their prescription eyeglasses.

Diopter adjustment level

adjustment until the writing on the viewfinder screen is clear. Then, adjust the focus of the camera normally.

A special feature on some viewfinders is *zebra stripes* (black and white diagonal stripes). When zebra stripes are displayed, the operator is made aware that an object in the shot may be too brightly lit. This is an extremely useful feature that should be engaged at all times, if available.

Camera Lens

In the early days of television, the imaging device on cameras was not a CCD, but a vacuum tube. Early cameras had several different lenses attached to a wheel called a "lens turret," **Figure 3-10**. Zoom lenses were not available on these early pieces of equipment. Technology has brought great changes and improvements to the imaging processes of television production.

The *lens* is an assembly of several glass discs placed in a tube on the front of a camera. Its primary purpose is to concentrate, or *focus*, the incoming light rays on the surface of the imaging device, or the target. A picture is considered to be "in focus" when the adjoining lines of contrast are as sharp as possible.

zebra stripes: A special function of some viewfinders that displays black and white diagonal stripes on any object in a shot that is too brightly lit.

lens: An assembly of several glass discs placed in a tube attached to the front of a camera.

focus: The act of rotating the focus ring on a camera lens until the lines of contrast in the image are as sharp as possible.

Talk the Talk

Both the individual glass discs and the tube-shaped piece of equipment that houses the glass discs are called "lenses." To differentiate these terms within this chapter, note that "lens" refers to the individual pieces of glass and "lens assembly" refers to the piece of equipment that houses the entire assembly of lenses. When working in the industry, both are referred to as a "lens" and are differentiated only by the context of the sentence.

Figure 3-10. This camera was used in the 1950s for both studio and field production. The attached lens turret rotates to allow the lenses to be changed from one size to another. *(Chuck Pharis Video)*

auto-focus: A common feature on consumer cameras that keeps only the center of the picture in focus.

Auto-focus is a common feature on consumer cameras that keeps only the center of the picture in focus. Because the average consumer usually places the most important portion of a picture in the center, this feature allows them to get an image that is in focus without adjusting any of the camera settings.

Auto-focus is not used on many professional cameras, because focus is a creative tool and professionals prefer to have creative control over the images. As the next chapter explains, the most important items in a shot should never be placed in the center of a frame. Therefore, the auto-focus feature keeps the wrong items in focus. Professionals should always turn the auto-focus option off.

Talk the Talk

Many people misuse the word "focus;" they incorrectly use it when they mean "zoom." For example, "focus in on the apple on the kitchen counter." In this context, "focus" actually communicates that the camera should zoom in on the apple on the counter. Use "focus" only when dealing with a picture that is blurry and in need of focus adjustment. Never say "focus" when you mean "zoom." To a professional, this misuse is the mark of an amateur.

Zoom Lenses

zoom lenses: Camera lens assembly that is capable of magnifying an image merely by twisting one of the rings on the outside of the lens housing. Also called a *variable focal length lens.*

Zoom lenses are a type of camera lens that can smoothly move from a close-up shot to a wide-angle shot, stopping anywhere between, all the while capturing usable footage. For example, a camera that is 15 feet away from a person can capture a very tight shot of the person's eyes. Most television camera lenses are zoom lenses, in that they are capable of magnifying an image merely by twisting one of the rings on the lens. A zoom lens may be operated at any speed, from extremely fast to so slowly the audience barely perceives that an object is getting larger or smaller. When the zoom ring on a lens assembly is rotated, an individual glass lens inside physically moves forward and back. The movement of this lens can be seen with the naked eye if you look into the front of the lens assembly while rotating the zoom ring. The movement of this single lens changes the type of shot captured by the camera. Rotating the zoom lens so that the center of the picture appears to be moving toward the camera is called *zoom in (ZI)* or *tighten*. Rotating the zoom lens so that the center of the picture appears to be moving away from the camera is called *zoom out (ZO)* or *widen*.

zoom in (ZI): The act of rotating a ring on the zoom lens so that the center of the picture appears to be moving toward the camera. Also called *tighten*.

zoom out (ZO): The act of rotating a ring on the zoom lens so that the center of the picture appears to be moving away from the camera. Also called *widen*.

It is very important to understand that a zoom shot does not produce the same effect for the audience as a shot where the camera physically moves toward the subject, or a dolly shot. A dolly shot, discussed further in the next chapter, takes the audience into the set in the same way a person moves through his environment. A dolly actually changes the perspective. The natural picture from a dolly shot, without a zoom, is three-dimensional and more realistic. When zooming in, the center of the picture gets larger because it is magnified. It does not appear as though the camera moves closer to the object, only that the center of the picture is larger. The zoom makes it possible to get a close-up of an object without physically moving the camera. With a zoom shot, however, the image takes on a flat appearance. A dolly shot provides more realism for the viewer than a zoom.

VISUALIZE THIS

Imagine that you are standing in the front of the classroom, facing the other students. From this vantage point, some of the students in the third row of desks, positioned horizontally to you, are not completely visible. Parts of their bodies, such as arms or hands, are blocked by students in the first and second rows. From your perspective, Rachel's left arm and hand are hidden. If you take a few steps down the aisle to stand even with the second row, you can see Rachel's arm on her desktop without a problem. As the camera (your eyes) moves into the set, the viewing perspective changes. Your body's movement is a dolly move. Another result of the dolly move is that Rachel gets larger in the picture because you are closer to her. This movement will "feel" realistic to the viewer.

Move back to the front of the class to examine this situation with a camera zoom. Rachel's left arm is blocked from view again because Bill, in the second row, is obstructing your view. Do not take a single step toward Rachel. Instead, pick up a pair of binoculars and view Rachel through them. She is larger in the picture, just like in the dolly, but you are still unable to see her arm. This is because Bill is larger now as well, and is still blocking your view. This movement is like a zoom shot with a video camera because it does not change the visual perspective. You will not see Rachel's left arm until either you move, Bill moves, or Rachel moves. This shot "feels" flat and unrealistic to the viewer.

When an image passes through a zoom lens, it is turned upside down, or is inverted. The physical location within the lens assembly where the inversion occurs is called the *optical center*. Another name for the optical center of the lens is the *focal point*. The optical center, or focal point, may not be in the center of the lens assembly as measured in inches, **Figure 3-11**. For example, the center is 3″ on a lens that measures 6″ long from front to

optical center: The physical location within the lens assembly where an image is inverted. Also called the *focal point*.

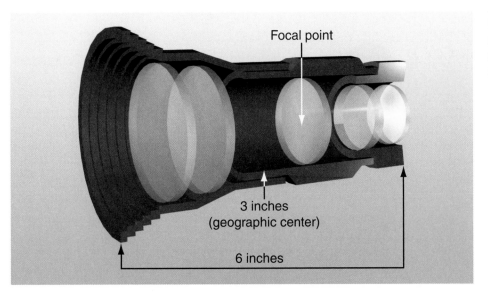

Figure 3-11. The optical center of a lens is not always in the physical center of the lens.

Focal point

3 inches
(geographic center)

6 inches

zoom lens: The particular piece of glass within the lens assembly that moves forward and back, magnifying or shrinking the image accordingly. This individual lens is the focal point, or optical center, of the zoom lens assembly.

focal length: The distance (measured in millimeters) from the optical center, or focal point, of the lens assembly to the back of the lens assembly.

back. The optical center is the point where the image is inverted, regardless of the physical location inside the lens assembly or the distance from the front or back of the lens assembly.

As the outside ring of a zoom lens assembly is rotated, one or more of the individual lenses inside the lens assembly moves backward or forward. You can see this movement by looking into a zoom lens as it is manipulated. As this piece of glass moves forward and back, the image is magnified or shrinks accordingly. This particular moving piece of glass within the lens assembly is called the *zoom lens*. This individual zoom lens is the focal point, or optical center, of the zoom lens assembly. The image is inverted wherever the zoom lens is positioned, within the range of the lens assembly, **Figure 3-12**.

Focal length is the distance (measured in millimeters) from the optical center (focal point) of the lens assembly to the back of the lens assembly, **Figure 3-13**. The "back" of the lens is the end of the lens assembly that attaches to the camera. The "front" of the lens assembly is the part closest to the subject being photographed or filmed. Camera lenses are classified by the focal length measurement. Since the optical center of a zoom lens

Figure 3-12. The individual zoom lens slides forward and backward within the zoom lens assembly. The focal point is located wherever the zoom lens is positioned.

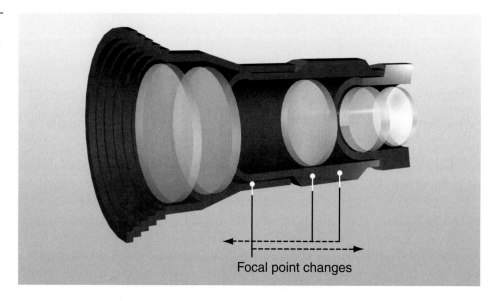

Focal point changes

Figure 3-13. The focal length is the distance (in millimeters) between the back of the lens assembly and the focal point.

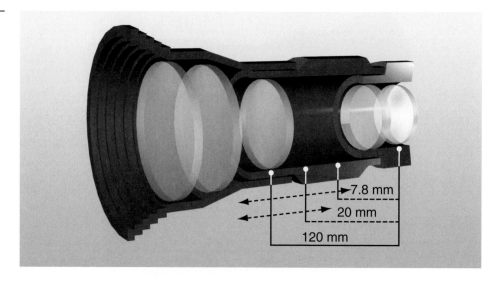

7.8 mm
20 mm
120 mm

can vary its position within the lens assembly, the focal length measurement varies as well. Therefore, a zoom lens is a *variable focal length lens*.

Controlling Light

There are at least three moveable rings on a professional camera lens assembly, **Figure 3-14**:

- The focusing ring is furthest away from the camera body. This ring adjusts the focus of the image in the frame of the picture.
- The zoom ring is in the middle of the lens assembly and moves the zoom lens forward and backward.
- The f-stop ring is the ring nearest to the camera. This ring is an external indicator of the amount of light passing through the lens and reaching the CCD. The f-stop ring regulates the amount of light that passes through the lens by controlling the iris and, therefore, the size of the aperture.

Three specific components of a lens assembly work together in regulating the light: aperture, f-stops, and iris.

variable focal length lens: A camera lens in which the optical center can vary its position within the lens assembly, varying the focal length measurement as well. Also called a *zoom lens*.

VISUALIZE THIS

When you enter a dark movie theater, your eyes dilate. The part of your eye that determines eye color, the iris, contracts. When the iris contracts, the pupil gets larger. The pupil is the black part in the center of the eye that is essentially a hole that lets light into the eye. With the pupil enlarged, more light can enter the eye to reach the rods and cones of the retina. This allows you to see in a darkened room. When you exit the theater and step into the bright daylight, you squint and the iris expands. This makes the pupil smaller and reduces the amount of light hitting the rods and cones. If the iris does not expand enough to sufficiently reduce the amount of light hitting the retina, you continue to squint until you get a headache or find sunglasses to further reduce the light hitting the retina. The television camera lens is expected to operate the same way as the human eye when reproducing colors and tones and reacting to lighting changes in the environment. Even though it valiantly tries, a television camera lens does not succeed in functioning as well as the human eye. The camera lens needs a human to help it operate.

Focus ring

Zoom ring

F-stop ring

Figure 3-14. A professional lens has at least three moveable rings: the focus ring, the zoom ring, and the f-stop ring.

iris: A component of a lens that is comprised of blades that physically expand and contract, adjusting the aperture size.

auto-iris circuit: A feature on many consumer and professional cameras that automatically examines the light levels coming into the camera and adjusts the iris according to generic standards of a "good" picture.

aperture: The opening, adjusted by the iris, through which light passes into the lens.

The *iris* is comprised of blades that physically expand and contract. The movement of these blades adjusts the size of the opening that allows light to pass through the lens, **Figure 3-15**. A camera's iris operates much like the iris of the human eye. As the size of the iris increases, light is blocked from passing through to the CCD. When the iris contracts, more light is allowed to pass through.

Many consumer and professional cameras have an auto-iris circuit, as well as a manual iris control. The *auto-iris circuit* examines the light levels coming into the camera and opens or closes the iris according to the generic definition of a "good" picture. The auto-iris is a useful feature for most circumstances in television production.

The *aperture* is the opening, adjusted by the iris, through which light passes. Aperture is nothing that can be touched; it is a hole.

Many cameras offer a manual iris control in addition to the automatic circuit. Adjusting the iris manually is accomplished by moving the f-stop ring. The *f-stop* setting determines the amount of light that passes through the lens by controlling the size of the iris. If the camera lens has a manual f-stop ring (some consumer cameras do not), numeric values are written on the corresponding moveable ring, **Figure 3-16**. When the f-stop ring is manually turned, the operator hears or feels a series of clicks or bumps

Figure 3-15. The size of the iris determines the size of the aperture. A large iris creates a small aperture; reducing the size of the iris produces a larger aperture.

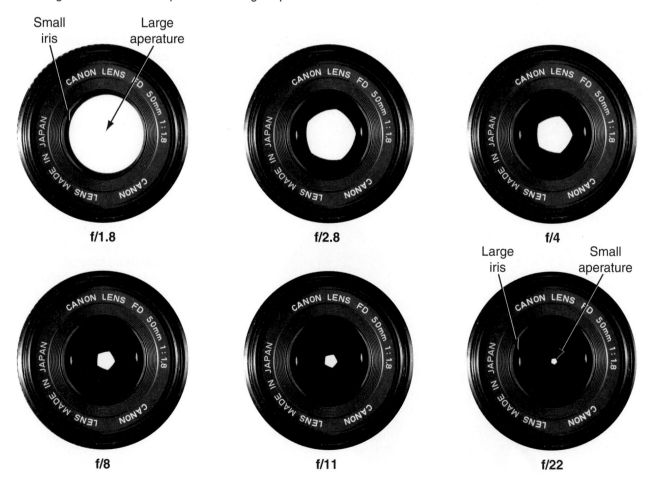

F-stop numbers

Figure 3-16. The f-stop ring is labeled with a series of numbers.

that indicate movement from one f-stop to another. Lower f-stop settings (numbers) allow a greater amount of light to pass through the lens. Higher f-stop numbers indicate that smaller amounts of light can pass through. The appropriate f-stop setting varies per situation, based on the lighting in the environment and the brightness of the object(s) in the shot. A lens that can produce a large aperture and let a great deal of light into the camera is considered a *fast lens*. A lens that is capable of small aperture settings lets little light into the camera and is considered a *slow lens*.

When shooting in high contrast situations, the auto-iris essentially becomes confused. It first adjusts to produce a good picture of the darker items, but the light items then begin to glow. When automatically adjusting for the light objects in the frame, the dark items lose all detail. The auto-iris should be disengaged in this type of situation. If this feature can be disengaged, manually adjust the f-stop ring to produce the best quality picture.

f-stop: A camera setting that determines the amount of light passing through the lens by controlling the size of the iris.

fast lens: A camera lens that can produce a large aperture and let a great deal of light into the camera.

slow lens: A lens that is capable of small aperture settings and lets little light into the camera.

PRODUCTION NOTE

It is important to remember how the aperture, f-stops, and the iris relate to each other. The f-stop indicates the size of the iris, which creates the size of the aperture.

Shutter

The *shutter* on video cameras is a circuit that regulates how long the CCD is exposed to light coming through the lens. As light hits the CCD, the photosensitive elements build up an electrical charge of varying strength depending on the intensity and color of light hitting them. This charge is sent to the camera processing circuits sixty times per second. When the charge is sent out, the photosensitive elements are discharged and begin collecting light again.

shutter: A circuit on a video camera that regulates how long the CCD is exposed to light coming through the lens.

Some higher-end cameras offer manual shutter speed settings. For example, the shutter speed on one camera can be increased exponentially from 1/100 to 1/8000. The higher the shutter speed, the clearer the footage is when played back in slow motion—often used in sports programming. However, higher shutter speeds "eat up" light, or require more light. To avoid a very dark picture, the amount of light must be increased dramatically if shutter speed is increased.

Mounting the Camera

There are two basic ways to support a camera while in use:
- Hand-held shooting
- Tripod shooting

Hand-Held Shooting

Many consumer cameras can just about fit in the palm of your hand and are easily held in the operator's hands while shooting. The size and weight of most professional cameras make it difficult to be held in the operator's hands for an extended amount of time. Professional cameras usually rest on the right shoulder of the operator, with both hands holding the camera lens steady. The right hand is positioned inside a strap holding it to the zoom lens control. The left hand holds the focus ring of the lens, **Figure 3-17**.

At first glance, the hand-held camera technique appears easy and the operator does not need to carry and set up a heavy tripod. However, hand-held camera operation quickly loses its appeal when gravity takes its toll. The camera operator's arms tire quickly—the heavier the camera is, the faster this happens. The result is very poor camerawork. An unsteady camera shakes, wiggles, tilts sideways, and eventually begins to point at the ground. Even if the camera is hand-held for a short time, the shot moves with every rise and fall of the operator's chest while breathing.

PRODUCTION NOTE

Professionals *do not* operate a camera with only one hand! The right shoulder bears the brunt of the weight of the camera. The right hand is positioned inside a strap holding it to the zoom lens control. The left hand holds the focus ring of the lens. Both hands should be on the camera when operating with the hand-held technique. A stable picture is virtually impossible if only one hand is used. Additionally, a $15,000 to $60,000 camera is very unlikely to fall off your shoulder when held with both hands.

If the lens is zoomed in on a person or object, the slightest shake or wobble is amplified. The resulting image is annoying to the audience. Any shaking is less noticeable with the lens zoomed out further. Therefore, always operate in the "zoomed out" position when hand-holding a camera. To get a close-up, move closer to the object; do not use the zoom.

If it is absolutely necessary to hand-hold a camera, brace yourself against a wall or tree, lean against a car door, or lie on the ground, **Figure 3-17**. Hold your breath to get the shot while steady, but realize that nothing will be usable beyond 5–10 seconds. You may see footage from a major news event with reporters from other news organizations in the background and other camera operators hand-holding their camera. There are many hand-held shots in the news, but the edited footage frequently cuts from one shot to another when reporting stories from the "field." The editor cuts out all the shaking shots.

The Glidecam™ and Steadicam® are examples of camera stabilization devices that attach to a harness worn by the camera operator, **Figure 3-18**. This harness is similar to that worn by a bass drummer in a marching band. A spring-loaded and shock absorbing arm is attached to the harness. The camera attaches to the arm using the same kind of mounting plate found on tripods. The weight of the camera is taken by the harness and, therefore, by the operator's entire torso. Because the arm is spring-loaded, the camera shot is kept steady even while the operator climbs steps, runs, or walks.

Figure 3-17. The camera operator can make use of items in the field to help steady a hand-held camera. A—One technique is to lean against a wall. This essentially makes the operator two legs of a tripod, with the wall as the third. B—An open car door can provide tripod-like support to the camera operator. The open door is one leg, the roof is another leg, and the operator's legs are the third leg of the makeshift tripod.

B

A

Figure 3-18. The Glidecam is a body mount that facilitates very smooth camerawork without using a tripod. *(Glidecam Industries, Inc.)*

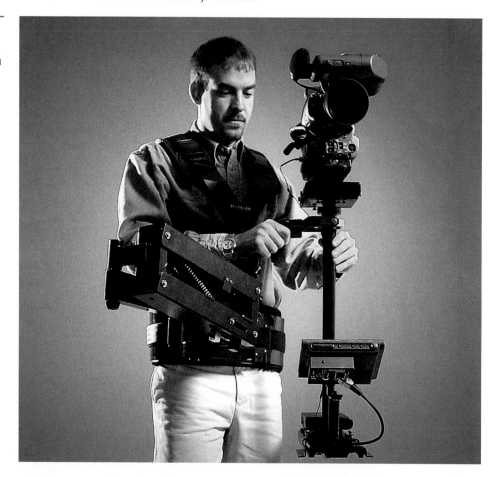

Assistant Activity

To help you understand how harness-type camera stabilization devices work:

1. Fill a 16 oz. drinking glass with water to 1/2″ from the top.
2. Hold the glass in your hand with your arm curved as it would be if you were holding the pole of a carousel horse.
3. Keep your arm in this position and walk, run, go up and down stairs, or dance without spilling a drop of the water.

How is it possible that the water does not spill? The muscles in your wrist, arm, elbow, and shoulder act as spring-loaded shock absorbers. If you hold the glass to your chest, with the knuckle of your thumb actually touching your chest, the water will spill almost immediately upon moving. The shock absorption has been removed and the glass is directly attached to the motion of your body.

The camera stabilization arm absorbs the shock of motion in very much the same way as your arm does in this activity. Visit either the Glidecam or Steadicam Web site to see the equipment in action!

subjective camera: A hand-held camera technique, in which the camera itself becomes the eye of one cast member. The viewers see the world through the eyes of that character.

Subjective Camera

Subjective camera is a special hand-held camera technique, **Figure 3-19**. The camera itself becomes the eye of one cast member. The viewer sees the world through the eyes of that character. Examples of this technique include:

- A camera is mounted in a stunt driver's car. As the car is driven up and down hills at high speeds, the audience's stomachs lurch as if they were actually riding in that vehicle.
- In a suspense film, the camera is positioned outside a house in the middle of the dark woods. "We" are looking through the branches of a bush into a window of the home. "Our hand" reaches up into the field of view of the camera and pushes away leaves on the bush to clear our view into the house.

Tripod Shooting

A tripod is the three-legged stand to which the camera is attached. The telescoping legs on most tripods allow the operator to position the camera at varying heights. The legs on all tripods spread out from the center. On most tripods, each leg operates independently. This is useful if the camera needs to be set up on sloped terrain, such as the side of a hill. Each leg can be extended to different lengths and spread out at different angles, which allows the camera head to be mounted level on an uneven surface. Most tripods and tripod heads are equipped with a leveling bubble that assists the operator in ensuring the camera head is level when mounted.

Tripods often have a column in the center, called a *pedestal column*, to raise or lower the camera. On the side of the pedestal column is the *pedestal control*, which is a crank that twists a gear to raise and lower the column. Turning the pedestal control to raise the column is to "pedestal up" and lowering the column is to "pedestal down" (discussed in Chapter 4, *Video Camera Operations*). This action does, however, cause considerable shaking of the camera. If the camera is *hot*, the audience sees every wiggle and shake. A camera is hot when the image captured by the camera is being recorded. Do not pedestal up and down on a tripod when the camera is

pedestal column: A column in the center of a tripod used to raise or lower the camera.

pedestal control: A crank on the side of the pedestal column that twists a gear to raise and lower the pedestal column.

hot: The state of a video camera when the image captured by the camera is being recorded.

Figure 3-19. In a subjective camera shot, the camera becomes the eyes of one character in the program.

hot. On the other hand, a camera mounted on a studio pedestal may pedestal up and down with great smoothness.

Mounting Heads

The *tripod head* is the assembly at the top of the pedestal column to which the camera attaches, **Figure 3-20**. The tripod head has several handles and knobs. These handles and knobs allow the operator to pan and tilt the camera while it is attached to the tripod head. The tripod head moves on the tripod in much the same way as your head moves on your neck—it can be tilted to point at the ceiling or the floor, or from side to side. One or two *pan handles* may be attached to the back of the tripod head, **Figure 3-21**. The pan handles allow the camera operator to move the tripod head while standing behind the tripod. There are two types of tripod heads available: friction head and fluid head.

A *friction head* is found on less expensive tripods and on almost all consumer tripods. The camera is stabilized by the pressure created when two pieces of metal are squeezed together by a screw. Releasing the pressure (loosening the screw) eliminates resistance between the pieces of metal and the parts slide easily against each other. The camera can then be tilted up and down using the handle. This type of tripod head is not usually found in a professional television setup because the resistance is either on or off, locked or completely loose. With the tripod head locked, the camera is frozen in place and produces images that may be boring to the audience. If the tripod head is unlocked, the camera is so loose that camerawork becomes obviously shaky. Either of these extremes can result in poor, unprofessional video images.

The *fluid head* is similar to the friction head, in that pressure between two pieces of metal restricts movement of the head. However, the fluid head has a thick fluid, such as oil or grease, between the two pieces of metal. This provides additional resistance to movement, called *drag*. The tripod head can be loosened, but is never completely free to move without resistance.

tripod head: The assembly at the top of the pedestal column to which the camera attaches.

pan handle: A device attached to the back of the tripod head that allows the camera operator to move the tripod head while standing behind the tripod.

friction head: A mounting assembly on some tripods that stabilizes the camera using the pressure created when two pieces of metal are squeezed together by a screw.

fluid head: A mounting assembly on some tripods that stabilizes the camera using the pressure between two pieces of metal and a thick fluid that provides additional resistance to movement.

drag: Resistance to movement created by tripod head mount.

Figure 3-20. The tripod head is located on top of the legs and pedestal column of the tripod, and includes the mounting plate or wedge.

Tripod head

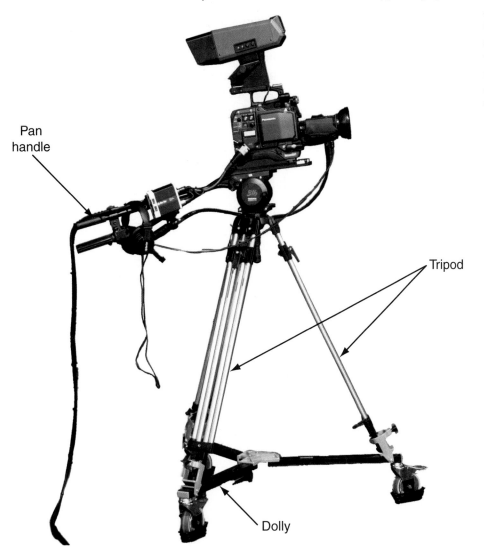

Pan handle

Tripod

Dolly

Figure 3-21. Pan handles allow the operator to perform camera movements while the camera is mounted to the tripod head.

VISUALIZE THIS

To help you understand how a fluid head provides greater resistance:

1. Swing one of your arms while standing in a room. Notice the free movement of your arm.
2. Imagine that you are standing in water up to your neck moving your arm the same way. The fluid (water) provides some resistance to movement and you have to work a bit harder to create the same motion.
3. Imagine standing in a pool of oil swinging your arm. It would be even more difficult to move.

Increasingly thicker fluids provide greater resistance to movement and therefore, cause smoother movement.

Experienced camera operators prefer more resistance to create smooth and stable camera movements. If the tripod head were completely resistance free, the camera would move with the slightest twitch or breath of the operator. A fluid head allows the camera operator to place fluctuating levels of pressure on the head, without moving the head until enough force is intentionally exerted. This prevents camera movement caused by slight touches or unintentional movement of the pan handles.

Many tripod heads have quick release plates. The plate can be removed and attached to a camera with a normal tripod mounting screw. The plate then slips into the tripod head and locks. The advantage of this feature is that a lever or button on the tripod head releases the mounting plate, and the camera can be separated from the tripod in a couple of seconds instead of having to unscrew the camera mount. Always carry the camera and tripod separately. Do not pick up a camera attached to a tripod by the camera handle. It is not strong enough to carry the weight of both items.

PRODUCTION NOTE

When operating a camera on a tripod, never let go of the pan handle without first locking the pan and tilt controls. Never sit while operating a camera on a tripod. You will be unable to react fast enough if the talent does something unexpected. It is unprofessional and may result in immediate dismissal.

Jib

jib: A type of camera mount that allows the camera to be raised high over the set and swung in any direction.

A *jib* is a type of camera mount that allows the camera to be raised high over the set and swung in any direction, **Figure 3-22**. It is a long pole on a lever with a camera mount on one end of the pole and a handle and camera controls on the other end of the pole. It operates much like a seesaw and captures crane-type shots (up to reasonable heights) with the operator safely on the ground.

The alternative to using a jib is a mechanical camera crane, which comes from the early days of filmmaking. On one end of the lever arm is a camera mount and a seat (with seat belt) for the camera operator. The other end of the lever arm is equipped with heavy counterbalancing weights. The device is mechanically operated by a second individual who raises and lowers the camera and operator on the end of the lever arm to capture the shots required. A third person is required to manage any necessary manual movement of the crane and monitor safety conditions. This device is very dangerous and has been known to nearly catapult camera operators if raised too quickly—the addition of a seat belt was a precaution against this hazard.

Camera Care and Maintenance

To help ensure the highest quality images, proper care and maintenance of video equipment is necessary. The recommended handling includes both appropriate cleaning and storage of equipment.

Cleaning a Dirty Lens

As with other pieces of video production equipment, the camera lens is delicate and requires special care when cleaning. Commonly used glass-cleaning solutions and materials are not appropriate for use on a camera lens.

Seeing little spots in a camera's viewfinder is not necessarily an indication that the lens is dirty. Perhaps the dirt is on the front of the viewfinder. Clean the viewfinder with a soft cloth. If this does not remove the spots, the lens should be cleaned. The following are some firm rules about cleaning lenses:

*   Never touch a lens with your bare fingers.
*   Never use a cloth or tissue moistened with saliva to wipe a lens clean. Saliva ruins the lens.

Figure 3-22. A jib arm allows the camera to be raised high above the set and swung in any direction. *(EZFX Inc.)*

- Wipe dirt away using photographic lens paper only.
- Use compressed air from a can to blow the dirt off a lens. Never try to blow the dirt off with your breath.

Post-Production Camera Care

While not in use, both studio cameras and camcorders should be stored in a protected and temperature-controlled location. All the related cables should be coiled and stored with the camera or camcorder.

Guidelines for care of a studio camera:

- Lock the pedestal and camera mounting head to prevent movement while not in use.
- Close the iris and attach the lens cap.
- Move the camera to a safe location within the studio.

Guidelines for care of a camcorder:

- Remove the tape from the camcorder, if present.
- Close the iris and attach the lens cap.
- Power-off all the camera functions (light, microphone, recorder).
- Detach the camera from the tripod when transporting the equipment.
- Place the camera in its case for storage and transport, **Figure 3-23**.

Figure 3-23. Place the video camera in its case for safe transport and storage.

Wrapping Up

The television production industry is labor and skill intensive. Careers in this industry require long hours of work. On the other hand, it is difficult to find anyone working in the television production industry who does not genuinely like his or her job.

Think about football players. They practice long hours in both wilting heat and humidity and freezing cold temperatures. Why? If you ask, you'll often hear, "I like to play football." Similarly, professional athletes don't say that they are going to "work." They are *players* and their job is to *play* a *game!* Most professional athletes, musicians, dancers, and actors feel the same way about their jobs. This also true of people working in the television industry. It is very demanding work with long hours. Yet, when you talk to industry professionals, you hear things like, "I have a shoot today," "I'm going to the studio," or "I'm starting to edit now." What you don't hear is, "I'm going to *work* today."

Review Questions

Please answer the following questions on a separate sheet of paper. Do not write in this book.

1. List the parts of a studio camera and note the function of each part.
2. How does the appearance of an image change when the gain is adjusted?
3. What is the *optical center* of a zoom lens?
4. Explain the significance of the numbers printed on the f-stop ring of a camera lens.
5. What are the challenges of hand-held shooting with a professional camera?
6. List the benefits of using a tripod when shooting outside of the studio.
7. What is the difference between a *friction head* and a *fluid head*?
8. What are the appropriate materials to use when cleaning a camera lens?

Activities

1. To illustrate the proper result of focusing a camera lens, perform the following:
    1. Lay a black sheet of paper on a flat surface and place a sheet of white paper on the right side of the black sheet.
    2. Point a camera at both pieces of paper.
    3. Move the lens so that the camera is out of focus.
    4. Notice that the left edge of the picture is clearly black and the right edge is clearly white. It is difficult to determine where the image turns from black to white, as the center of the picture is gray.
    5. Twist the focus ring of the lens, slowly bringing the picture into focus.
    6. The center of the picture becomes less and less gray and the image becomes sharper. When the picture is completely "in focus," the separation between black and white is as sharp as possible.

2. Create an analogy (a written paragraph or an illustration) that effectively explains the relationship between f-stops, the iris, and aperture.

3. To help you remember what kind of image various lens lengths can give you, try the following activity:

    1. Take a regular piece of 8 1/2″ × 11″ paper and roll it into a tight tube that is 11″ long and about 1″ in diameter.

    2. Close one eye and hold the paper tube to your other eye. Look through the tube to the other side of the room. Notice which of the items on the far wall you can see through the tube.

    3. Unroll the paper and make a new tube that is 8 1/2″ long by 1″ in diameter.

    4. Look through the shorter tube and note how many more items you can see on the far wall.

    5. Now place your thumb and index finger in the traditional "ok" sign and hold it to your eye. Notice how much more of the far wall you can now see.

The length of the three tubes affected how much of the wall you could see. The same relationship is true of short, medium, and long lenses, except that magnification also occurs. With a long lens, you see less real estate, but the details of what you do see fill the screen because the image is magnified. With a short lens, you see more real estate, but fewer details because less magnification is applied.

STEM and Academic Activities

Science

1. When an image passes through a zoom lens, it is turned upside down, or is inverted. The human eye perceives images the same way—images formed on the retina are upside down. Research human vision and explain why we do not see everything upside down.

2. The fluid used in a fluid head tripod creates resistance to movement. This resistance property of fluid is called *viscosity*. Research the viscosity of various fluids. Which fluids are most viscous? Which fluids are least viscous?

Technology

3. Research the evolution of the video camera. How has the video camera changed in size, quality, and cost over time?

Chapter 4

Video Camera Operations

Objectives

After completing this chapter, you will be able to:

- Understand how white balancing a camera affects the picture.
- Summarize how depth of field contributes to composing a good picture.
- Identify the composition of each type of camera shot.
- Illustrate a variety of camera movements.
- Explain how a videographer can psychologically and physically affect the audience.

Introduction

While learning to operate a camera is not complex, becoming a talented camera operator requires dedication and skill. Great camera operators:

- Know the basic rules of composition.
- Know the capabilities of their equipment.
- Know the basic process of production methodology.

Professional Terms

arc
arc left (AL)
arc right (AR)
bust shot
close-up (CU)
depth of field (DOF)
dolly
dolly in (DI)
dolly out (DO)
establishing shot
extreme close-up
　(ECU/XCU)
extreme long shot
　(ELS/XLS)
four shot
great depth of field
group shot
head room
high angle shot
knee shot
lead room
long shot (LS)
low angle shot
macro
medium close-up (MCU)
medium long shot (MLS)
medium shot (MS)
mid shot
minimum object distance
　(MOD)

narrow angle shot
nose room
over-the-shoulder shot
　(OSS)
pan
pan left (PL)
pan right (PR)
pedestal
pedestal down (PedD)
pedestal up (PedU)
pre-focus
profile shot
pull focus
rack focus
reaction shot
rule of thirds
selective depth of field
shallow depth of field
shot
shot sheet
three shot
tilt
tilt down (TD)
tilt up (TU)
truck
truck left (TL)
truck right (TR)
two shot
white balance
wide angle shot (WA)

Composing Good Pictures

Composing good pictures begins with learning some basic principles. These basic principles are the foundation on which experience is built, and only experience can perfect camera composition skills.

One of the principles of composition is maintaining constant control over the camera. The camera operator should never let go of the pan handles and should always have the pan and tilt unlocked during a shoot, but with sufficient drag engaged to handle any movement necessary.

Another major principle of composition is that anything not shown in the frame of the camera does not exist for the viewer. The frame of the picture defines what the viewer experiences. On a news program, for example, the audience sees a well-dressed news anchor sitting at a desk in the studio delivering important news to viewers. The anchor is dressed in a suit jacket, shirt, and tie, which helps establish his credibility with the audience. Outside the frame of the picture, the audience cannot see that the anchor is wearing Bermuda shorts instead of suit pants. This principle also applies to the set of a program. Can a wide sandy beach in Florida be used to shoot a scene that is set in the Sahara Desert? Yes! To make the shot realistic, the camera operator must be careful to avoid the ocean and condos on the shoreline in the frame of the picture. If it is not seen in the frame of the camera, then it does not exist for the viewer!

White Balance

white balance: A function on cameras that forces the camera to see an object as white, without regard to the type of light hitting it or the actual color of the object.

Each time a camera is powered up, it needs to be "told" what white is—this is called *white balancing* the camera. Every color is defined by its relationship to every other color. So, when the white balance is properly set, the camera "sees" all other colors correctly. Some cameras automatically perform a white balance, others require the white balance to be manually performed, and some cameras give the operator a choice of automatic or manual white balancing. When given the choice, always manually white balance the camera because it is usually more accurate. If white balancing is not performed, the recorded image of indoor scenes usually has a yellowish tint and outdoor scenes in daylight have a bluish tint.

The white balance settings are not stored by the camera when it is powered down. The camera must be "re-taught" next time it is powered up. To perform a white balance:

1. Zoom in on a white object on the set that is lit for the shooting.
2. Activate the white balance circuit on the camera.
3. Zoom back out and shoot normally.

Assistant Activity

1. Attach a color monitor to the video output of a camera.
2. Point the camera at a white object.
3. Notice that it does not appear white on the monitor. It may appear greenish, grayish, or even pinkish.
4. Press the white balance button on the camera.
5. Watch the monitor carefully to see the object transformed to a true white color.

Pre-Focusing Zoom Lenses

A zoom lens cannot be focused while it is in the "zoomed out" position. Focusing a zoom lens is a three-step process called *pre-focus*. To pre-focus a zoom lens:

1. Zoom in on the furthest object on the set that must be in focus in the shot. The furthest object that must be in focus might not be the background. For example, picture a cowboy on a horse in a prairie with the Rocky Mountains in the background. The furthest object in this shot that must be in focus is most likely the cowboy, not the mountains.
2. Focus the camera on that object.
3. Zoom the lens back out.

After a pre-focus is performed, everything from about 6′ in front of the camera to the furthest object focused on (in step 2) will be in focus. Everything remains in focus until the camera is moved toward or away from the object of the pre-focus, or until the lighting on the set is changed.

Many cameras offer a macro setting for the lens. The *macro* feature allows the operator to focus on an object that is very close to the camera, almost touching the lens. The relationship between a fully zoomed-in lens and a macro lens is similar to the relationship between a telescope and a microscope.

Depth of Field

The closest an object can be to the camera and still be in focus is the *minimum object distance (MOD)*. Minimum object distance contributes to depth of field. *Depth of field (DOF)* is the distance between the closest point to the camera that is in focus and the furthest point from the camera that is also in focus, **Figure 4-1**.

pre-focus: A three-step process to focus a zoom lens. 1) Zoom in on the furthest object on the set that must be in focus in the shot. 2) Focus the camera on that object. 3) Zoom the lens back out.

macro: A lens setting that allows the operator to focus on an object that is very close to the camera, almost touching the lens.

minimum object distance (MOD): The closest an object can be to the camera and still be in focus.

depth of field (DOF): The distance between the closest point to the camera that is in focus and the furthest point from the camera that is also in focus.

Figure 4-1. The depth of field is the area in front of the camera, regardless of the distance, in which objects are in focus.

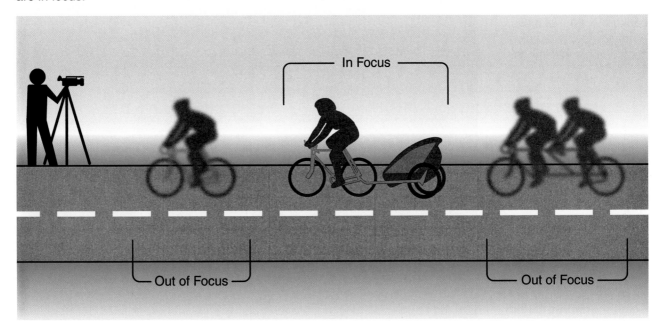

Assistant Activity

To help clarify this concept, find the MOD of your eye.
1. Close or cover one eye.
2. Hold up your index finger about 12″ away from your face.
3. With one eye open, look at your fingerprint; you should be able to clearly see it.
4. Slowly move your finger toward your face.

As you move your finger closer, there comes a point when your eye can no longer focus on your finger and you are unable to clearly see the fingerprint. This point is the minimum object distance (MOD) of your eye.

great depth of field: When a camera's depth of field is as large as possible.

shallow depth of field: A depth of field technique that moves the audience's attention to the one portion of the picture that is in focus.

Most of the time, a camera's depth of field should be as large as possible. This is called *great depth of field*. When using a great depth of field, zooming and some camera movements (such as a truck or arc) do not cause the image to go in and out of focus. However, when every element in the picture is in focus, no one particular item stands out for emphasis.

Using *shallow depth of field* moves the audience's attention to the one portion of the picture that is in focus. A shallow depth of field allows the program's director to control exactly what the viewer looks at within the frame of the picture. For example, scenes on television and in movies where the foreground is in focus and the background is out of focus direct the viewer's attention to the item or action in the foreground, **Figure 4-2**. The reverse is commonly used as well—the background is in focus and the foreground is out of focus. When using a shallow depth of field, the camera operator must refocus the camera if the talent moves toward or away from the camera (even slightly), or if any camera movements are performed.

Figure 4-2. The pain this young woman feels is more powerful with the background out of focus. A shallow depth of field compliments this image.

Selective depth of field is the technique of *choosing* to have a shallow depth of field in a shot or scene. One dramatic effect that results from this technique is changing the camera's focus from the foreground to the background (or the reverse) while the camera is hot. The attention of the audience may be intently concentrated on a foreground image, but the camera gradually brings something unexpected from the background into focus. The process of changing focus on a camera while that camera is hot is called *rack focus*, or *pull focus*. Keep in mind that selective DOF loses its impact when overused in a program.

selective depth of field: A technique of *choosing* to have a shallow depth of field in a shot or scene.

rack focus: The process of changing focus on a camera while that camera is hot. Also called *pull focus*.

VISUALIZE THIS

The following is a powerful example of the use of selective depth of field. The scene described is an anti-war spot that was used during a Presidential campaign in the 1960s.

A little girl wearing a yellow dress chases a butterfly around a beautiful field of flowers. She giggles and is obviously having a grand, happy time. The background is an out of focus greenish color. The viewer simply assumes that the background contains vegetation of some kind. The camera moves toward a shot of the girl's smiling face, with her two small hands reaching toward the butterfly a bit closer in the foreground. Right before the viewer's eyes, the camera's focus shifts and brings the background of the shot into focus. The background vegetation becomes a line of fifty or more soldiers with rifles ready to fire, stealthily moving out of the trees and toward the camera. The little girl is standing between the advancing soldiers and the camera/viewer and, therefore, in apparent danger.

The use of selective depth of field makes the line of soldiers surprising background material and increases the impact of the scene.

Factors Affecting Depth of Field

- Aperture—The size of the opening in the lens that allows light into the camera.
- Subject to camera distance—The distance between the camera and the subject of the shot.
- Focal length—The amount the lens is zoomed in or out.

PRODUCTION NOTE

Remember: The f-stop indicates the size of the iris, which creates the size of the aperture.

More movement in each of these three areas creates a more pronounced effect for either shallow or great DOF, **Figure 4-3**. For example, the effect on DOF produced by zooming in and increasing aperture size is not as great as when the camera moves closer to the subject *in addition to* zooming in and increasing aperture size.

Camera lenses are operated by camera operators. Even though depth of field involves manipulating light, the lighting designer does not have a part in this process. The camera operator creates depth of field by manipulating the

Depth of Field			
Director's Goal	**Zoom Technique**	**Dolly Technique**	**F-Stop Setting**
Obtain a shallow depth of field.	Zoom In	Dolly In	Use a lower f-stop value: • Reduces the iris • Increases the aperture; more light passes through the lens
Obtain a great depth of field.	Zoom Out	Dolly Out	Use a higher f-stop value: • Enlarges the iris • Decreases the aperture; less light passes through the lens

lens. Set lighting does not affect depth of field. Depth of field is affected by subject to camera distance, focal length, and aperture (not light). Understanding how to effectively use depth of field is a valuable tool that can greatly affect the impact and power of a scene for the viewer.

Since the majority of scenes in typical programs are shot using a great depth of field, a smaller aperture is more commonly used. A smaller aperture requires higher light levels to capture a good quality picture. This is why there are so many bright lights on the ceiling of a production studio. Studio sets are saturated with light, which allows the aperture of cameras to be reduced when necessary without affecting the picture quality.

Lines of Interest

rule of thirds: A composition rule that divides the screen into thirds horizontally and vertically, like a tic-tac-toe grid placed over the picture on a television set. Almost all of the important information included in every shot is located at one of the four intersections of the horizontal and vertical lines.

The *rule of thirds* for television production divides the screen into thirds horizontally and vertically; like a tic-tac-toe grid placed over the picture on a television set, **Figure 4-4**. Almost all of the important information included in every shot is located at one of the four line intersections. Studies have shown that the human eye is drawn first to those four intersection points on any picture and not to the center of the screen, as is commonly assumed.

The most common shots on television are the close-up and medium close-up. To follow the rule of thirds when shooting, the talent should be positioned so that their eyes are 1/3 of the way down from the top of the screen, or on the upper horizontal line of the tic-tac-toe grid. In broadcast television, the important people and objects in a shot are slightly to left or right of the center. The center of the tic-tac-toe box rarely contains the main subject or important object of the shot.

Action

Nearly every shot in broadcast television includes some kind of action. Either the main subject matter (objects or talent) in the picture provide

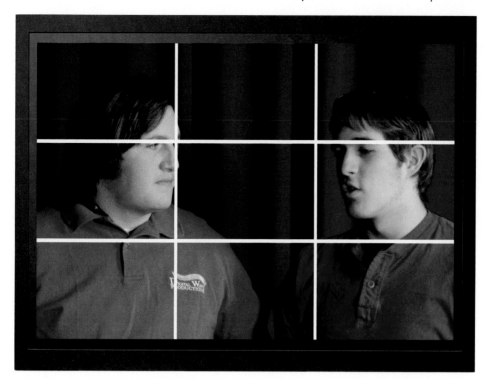

Figure 4-4. The rule of thirds states that the most interesting aspects of a picture should be positioned near the four intersecting points on a tic-tac-toe grid.

action or movement, the camera moves to provide a moving shot, or both the camera and subject matter move. There is rarely a shot without some type of action. Audiences today have become accustomed to seeing one action shot after another and typically lose interest quickly if this is not the case.

ASSISTANT ACTIVITY

Rent a movie you recently saw and thought was boring. Watch it again and notice the percentage of shots having little or no action. Does the lack of action contribute to your overall feeling of boredom with the movie?

Head Room

The space from the top of a person's head to the top of the screen is called *head room*, **Figure 4-5**. This space should be kept to a minimum, unless something important is going to happen above the head of the talent.

Nose Room

Nose room, or *lead room*, is the space from the tip of a person's nose to the side edge of the frame. Novice videographers often make errors in framing with respect to nose room. The natural tendency is to place the talent in the center of the screen with equal space on either side. That is acceptable only if the talent is directly facing the camera. The more the talent looks to the right or the left, the more room should be placed between their nose and that same edge of the screen, **Figure 4-6**.

head room: The space from the top of a person's head to the top of the television screen.

nose room: The space from the tip of a person's nose to the side edge of the frame. Also called a *lead room*.

Figure 4-5. Head room is an important consideration when framing a shot. A—Excessive head room usually indicates that something is about to happen above the subject's head. B—Correct head room spacing ensures that the audience's attention is not diverted from the main subject of the shot.

A

B

Correct use of nose room corresponds with the rule of thirds, in that the most important portion of the image (the faces) are positioned at the intersection of the gridlines instead of the center of the picture. Framing a shot in this way creates a more interesting shot for the viewer. If the talent is walking parallel to the camera, for example, sufficient space should

A

B

Figure 4-6. Nose room is another consideration when shooting. A—With too little nose room in a shot, the audience expects something to happen behind the subject. B—Correct nose room framing leaves sufficient space between the talent and the edge of the shot.

be placed in front of the person as they walk. If the scene is shot without enough nose room in this scenario, the talent appears to be pushing the frame of the picture with their nose as they walk. Additionally, if the person is moving laterally, the shot should be enlarged to at least a mid shot. Otherwise, the talent is likely to walk right out of the camera's view.

VISUALIZE THIS

Imagine a scene in a horror movie. One of the characters is about to be attacked by a vampire approaching from behind. It would be appropriate to leave space behind the talent in the frame so the vampire can enter the picture. When space is left behind someone in a horror movie, however, the audience expects a monster to jump in behind the talent. To mislead the audience, place the space behind the talent in the frame and either:

- make sure nothing happens from *that* direction, or

- have action occur on the unexpected opposite side of the screen.

Doing this once or twice is effective and increases the reaction of the audience. Using this technique too often reduces its effect and will be laughed at by the audience.

Shot Sheets

shot: An individual picture taken by a camera during the process of shooting program footage.

A *shot* is an individual picture taken by a camera during the process of shooting the program footage. In a typical studio shoot with three cameras, the output from each camera runs into a switcher in the control room. The director must decide which image to place on the master program recording and which camera to pull the image from. To do this, buttons on the switcher are selected to cut from one camera to another. For example, close-up shots of individual characters may be needed at various times in the program. To capture the shots necessary, the director must know what is going to be said and what actions are going to happen before they occur on the set. This way, the camera operators can be directed into position to capture the necessary shot when it happens. This kind of planning requires that the director be mentally 5–10 seconds ahead of the performers at all times.

Far less stress would be placed on the director if he did not have to think far enough ahead in the program to tell the camera operators to move. It would also be more efficient if the camera operators knew in advance what their next shot is supposed to be. This would allow them to execute the camera move before the moment arrives for their camera to be hot. Using a shot sheet relieves some of this stress and makes directing a three-camera shoot easier. A *shot sheet* lists each shot in a program numerically. The list given to each camera contains only the shots that particular camera needs to capture during the program. Each camera used in a production receives a completely different shot sheet.

shot sheet: A numerical listing of each shot to be captured by each camera in a multi-camera shoot. Shot sheets are developed specifically for each camera.

To use a shot sheet, the director reviews the script before the shoot, plans each camera shot, and assigns a sequential number to each shot. The numbered shots, with corresponding brief descriptions, are divided per camera and written on separate sheets of paper, **Figure 4-7**. Again, only the shots each particular camera is responsible for are on that camera's shot sheet. On the day of the shoot, the shot sheets are taped to the side of the corresponding camera. During the shoot, the director can simply say, "Take shot 4" instead of, "Camera 2, I want you to have a close-up of Mary next so get your shot ready while I'm still on camera 3."

As soon as the director cuts from camera 3 (shot 3) to the shot of Mary on camera 2 (shot 4), the camera 3 operator looks at his shot sheet and sees

Figure 4-7. Shot sheets are developed for each camera involved in a shoot.

Camera 1	**Camera 2**	**Camera 3**
2–Mid shot John/Mary 5–Close-up John 7–Close-up of phone	4–Close-up Mary 6–Three shot John, Mary, Bill 9–Close-up on fireplace	1–Close-up John 3–Wide shot entire set 8–Mid shot Mary

that his next shot is shot 8. The operator reviews the brief description and readies the shot without being told to do so. Camera 1 was assigned shot 5 and had the shot set-up and ready to go. Camera 2 has shot 6 and shot 7 is back on camera 1 again. Using shot sheets makes a multi-camera shoot much more efficient and less stressful for everyone.

Calling the Shots

There are many different types and sizes of camera shots that can be taken of a person standing in a studio. It is imperative to learn the names of individual shots and what each shot incorporates. Unfortunately, all professional television facilities do not use exactly the same terms. While working in the industry, it is important to know how your facility defines its terms. The sections that follow present the most common definitions of various "person" shots, but the terms are not universal. Obviously, there are times when shots do not contain any people. The shot names still generally apply to the object(s) in the shot that is the main item.

Wide Shots

The *extreme long shot (ELS/XLS)* is also known as a *wide angle shot (WA)*. This shot includes a person's entire body from head to toe, and as much surrounding information as the camera can capture by dollying and zooming out. This is generally considered to be the biggest shot a camera can capture of the subject matter, **Figure 4-8**. Overusing the extreme long shot, however, can prove ineffective. An extreme long shot of a crowd that is viewed on a small television screen appears to be an image of a multi-colored wheat field waving in the breeze. A shot that is too "long" creates a picture without detail.

An *establishing shot* is a very specific type of extreme long shot. The establishing shot is used to tell the audience where and when the program takes place. For example, if the opening shot is of a dusty town with dirt roads, cowboys riding horses, and a stagecoach approaching, the audience can assume the program is set in the Old West, and not onboard the starship Enterprise. Directors periodically return to an establishing shot during a scene to reinforce the location and to prevent confusion.

A *long shot (LS)* captures a person from the top of the head to the bottom of the feet, **Figure 4-9**. Much less of the surrounding details are included, compared to the extreme long shot.

extreme long shot (ELS/XLS): The biggest shot a camera can capture of the subject matter. Also called a *wide angle (WA) shot.*

establishing shot: A specific type of extreme long shot used to tell the audience where and when the program takes place.

long shot (LS): A shot that captures a subject from the top of the head to the bottom of the feet and does not include many of the surrounding details.

Figure 4-8. An extreme long shot is the largest shot the camera can get. The ELS is usually a shot of a person from head to toe and includes as much detail of the subject's surroundings as possible.

Figure 4-9. A long shot includes the subject from head to toe only.

medium long shot (MLS): A shot that includes the top of a subject's head to a line just above or just below the knee. Also called a *knee shot*.

medium shot (MS): A shot that captures a subject from the top of the head to a line just above or below the belt or waistline. Also called a *mid shot*.

Individual Subject Shots

A *medium long shot (MLS)*, sometimes called a *knee shot*, includes the top of a person's head to a line just above or just below the knee, **Figure 4-10**.

The *medium shot (MS)* is also referred to as a *mid shot*, **Figure 4-11**. This shot captures a person from the top of the head to a line just above or below the belt or waistline.

Figure 4-10. The bottom edge of a medium long shot is just below or just above the subject's knee.

Figure 4-11. The medium shot, or mid shot, includes the subject's head to just above or below the waistline.

medium close-up (MCU): A shot that frames a subject from the top of the head to a line just below the chest. Also called a *bust shot*.

close-up (CU): A shot that captures a subject from the top of the head to just below the shoulders. Also called a *narrow angle shot*.

A *medium close-up (MCU)*, also called a *bust shot*, frames a person from the top of the head to a line just below the chest, **Figure 4-12**. This is the type of shot usually seen of newscasters on daily news programs.

A *close-up (CU)* shot is also known as a *narrow angle shot*. For a person, this shot captures the top of the head to just below the shoulders, **Figure 4-13**. When framing a close-up shot, it is important to include the

Figure 4-12. A medium close-up captures a person from head to just below the chest.

Figure 4-13. The close-up shot includes a subject's head and neck, and *must* include the top of their shoulders.

extreme close-up (ECU/XCU): A shot of an object that is so magnified that only a specific part of the object fills the screen.

top of the shoulders. If the shoulders are not included, the image is a disembodied head at the bottom of the screen, **Figure 4-14**.

An *extreme close-up (ECU/XCU)* is a shot of a specific body part, **Figure 4-15**. This may be used, for example, in a makeup ad showing how mascara enhances the appearance of the eyes.

Figure 4-14. A close-up shot that does not include the subject's shoulders leaves a "floating" head in the frame.

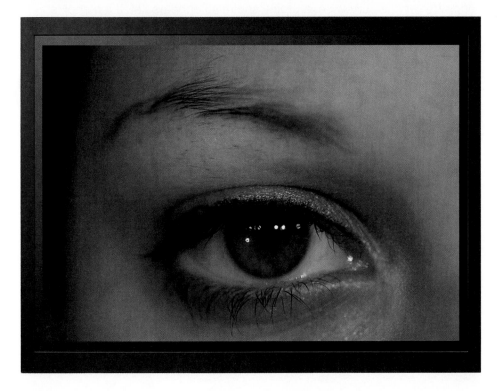

Figure 4-15. On a person, an extreme close-up is a shot of a specific body part or feature.

Multiple Subject Shots

- A *two shot* includes two items of primary importance. A shot of two news anchors sitting at the news desk is an example of a two shot.
- A *three shot* frames three items. For example, a sportscaster joins the two news anchors at the news desk.

two shot: A shot that includes two items of primary importance.

three shot: A shot that frames three items.

four shot: A shot that captures four items.

group shot: A shot that incorporates any number of items above four.

reaction shot: A shot that captures one person's face reacting to what another person is saying or doing.

profile shot: A shot in which the talent's face is displayed in profile.

- A *four shot* captures four items. Picture a meteorologist joining the news anchors and sportscaster at the news desk.
- A *group shot* incorporates any number of items above four. The shot of a basketball team after winning a game is an example of a group shot.

Specific View Shots

A *reaction shot* captures one person's face reacting to what another person is saying or doing. This is a very powerful type of shot. For example, in a scene where a policeman delivers sad news to a distraught parent, the shot should be of the parent hearing the news, not the policeman delivering the news.

A *profile shot* is generally considered to be a bad shot, **Figure 4-16**. The talent's face in profile appears completely flat on the screen and creates an unflattering picture.

PRODUCTION NOTE

The television screen is flat. A videographer must arrange shots in a way that creates the illusion of three dimensions and depth when displayed on a flat screen. When framing an individual shot of an object, whether it's as small as a person or as large as a building, try to shoot it at an angle. A straight-on shot of a person with their nose pointed at the camera lens appears very flat. Likewise, a profile shot also appears flat. If the shot is taken at an angle, somewhere between a profile and a straight-on shot, three dimensionality and depth are achieved. The most common shot is an angle that includes all of one side of the face and enough of the other side to see the cheekbone or eyebrow. When shooting a building, try to shoot it from a corner that includes two sides of the building instead of just one side.

Figure 4-16. The profile shot produces a very flat appearance.

The *over-the-shoulder shot (OSS)* is an extremely common shot on any program, **Figure 4-17**. The back of one person's head and top of their shoulder is in the foreground of the shot. A face shot of the other person in the conversation is in the background of the shot. One OSS is usually followed by another OSS from the other side of the conversation. It is a more interesting shot than just a close-up of each person speaking or listening.

over-the-shoulder shot (OSS): A shot in which the back of one person's head and shoulder are in the foreground of the shot, while a face shot of the other person in the conversation is in the background.

PRODUCTION NOTE

When framing shots of people, never allow the edge of a picture cut at the joint of the human body (ankles, knees, waist, wrists, elbows, or neck). The person pictured in the shot will appear to have amputated body parts, **Figure 4-18**. This is especially important to remember if your facility uses the terms "bust shot" and/or "knee shot." These shot names seem to "ask" for a poorly composed shot.

Camera Movement

It is important to understand *how* to move a camera when it is mounted on a tripod. Beginning from a still shot, slowly start the camera move, speeding up gradually until the move is nearly complete, and then gradually slow down until the move is completed. When performing camera moves, position your body where it needs to be at the end of the shot and twist to the position needed to begin the shot. As your body straightens to return to a normal standing position, the camera move is smoothly completed. This camera movement technique also applies to hand-held shooting.

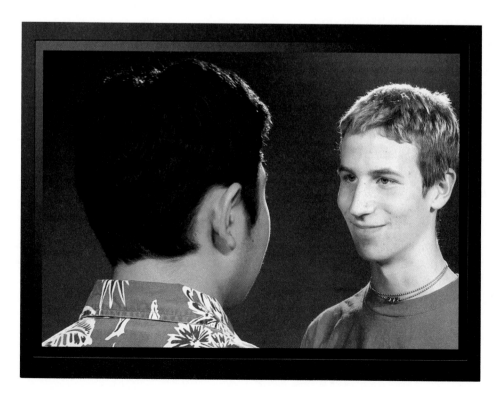

Figure 4-17. An over-the-shoulder shot adds three dimensions to an otherwise flat two-person conversation.

Figure 4-18. Never frame a shot that cuts off a person at a natural joint of the body.

dolly: Physically moving the camera, its tripod, and dolly perpendicularly toward or away from the set.

dolly in (DI): Smoothly pushing the camera directly forward toward the set.

dolly out (DO): Pulling the camera backward while facing the set.

truck: Moving the camera, its tripod, and dolly to the left or right in a motion that is parallel to the set.

truck right (TR): To move the camera, its tripod, and dolly sideways and to the camera operator's right while facing the set.

There is a specific term to indicate every type of camera movement possible. Being familiar with these terms is important to effectively communicate within the industry. Camera directions are always given in respect to the camera operator's point of view, not the talent's point of view. Unlike theatrical stage directions, camera movement commands in television production are intended for the camera operators. Illustrations of each camera movement defined are presented in **Figure 4-19**. The camera operator may use these camera movements in conjunction with zooming to create the director's intended effects.

- *Dolly*. Physically moving the camera, its tripod, and dolly perpendicularly toward or away from the set. Smoothly pushing the camera directly forward toward the set is *dollying in (DI)*. *Dollying out (DO)* involves pulling the camera backward while facing the set.

Talk the Talk

The word "dolly" has two meanings in the television production industry:

- Noun: It is the wheeled cart in which the tripod sits, enabling the tripod to be smoothly rolled around the studio.
- Verb: It is a camera movement in which the camera tripod, and dolly move perpendicularly toward or away from the set.

- *Truck*. Moving the camera, its tripod, and dolly to the left or right in a motion that is parallel to the set. To *truck right (TR)*, move sideways and to the camera operator's right while facing the set. This image is

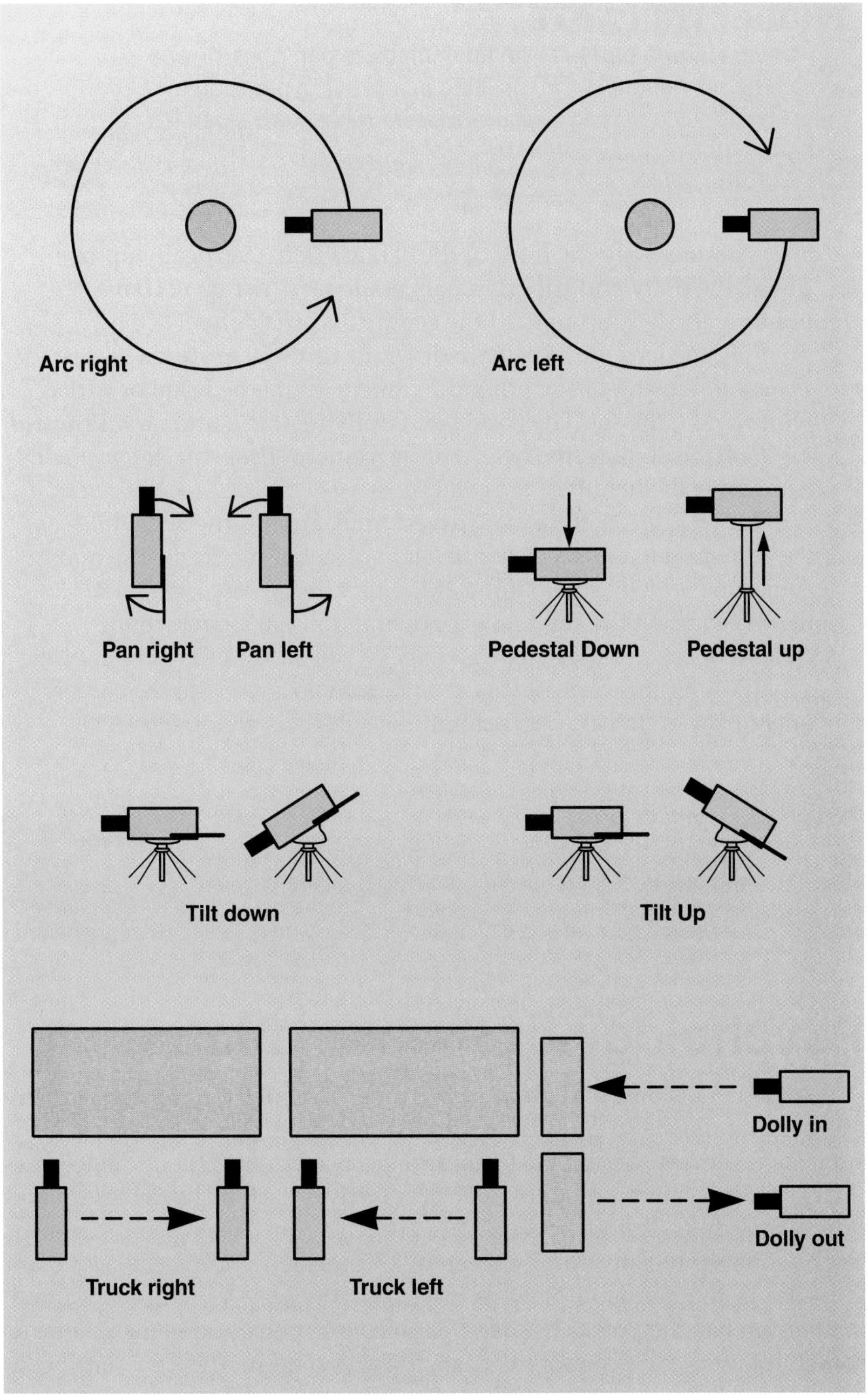

Arc right

Arc left

Pan right   **Pan left**

Pedestal Down   **Pedestal up**

Tilt down

Tilt Up

Dolly in

Dolly out

Truck right   **Truck left**

Figure 4-19. Illustrations of the camera movements defined.

truck left (TL): To move the camera, its tripod, and dolly sideways and to the camera operator's left while facing the set.

pan: Moving only the camera to scan the set horizontally, while the dolly and tripod remain stationary.

pan left (PL): Moving the camera to the camera operator's left to scan the set, while the dolly and tripod remain stationary.

pan right (PR): Moving the camera to the camera operator's right to scan the set, while the dolly and tripod remain stationary.

much like looking at people standing on a station platform while you are on a train pulling away from the station. To *truck left (TL)*, move sideways and to the camera operator's left, while facing the set.

- *Pan*. Moving only the camera to scan the set horizontally; the dolly and tripod remain stationary. *Pan left (PL)* is when the camera scans to the camera operator's left, and *pan right (PR)* is when the camera scans to the camera operator's right.

tilt: Pointing only the front of the camera (lens) vertically up or down while the dolly and tripod remain stationary.

tilt up (TU): Pointing the camera lens up toward the ceiling, while the dolly and tripod remain stationary.

tilt down (TD): Pointing the camera lens down toward the ground, while the dolly and tripod remain stationary.

pedestal: Raising or lowering the camera on the pedestal of a tripod, while facing the set. The tripod and dolly remain stationary.

pedestal up (PedU): Raising the camera on the pedestal of a tripod, while facing the set. The tripod and dolly remain stationary.

pedestal down (PedD): Lowering the camera on the pedestal of a tripod, while facing the set. The tripod and dolly remain stationary.

arc: Moving the camera in a curved truck around the main object in the shot—the main subject never leaves the frame of the picture.

arc right (AR): Rolling the camera, tripod, and dolly in a circle to the camera operator's right (counterclockwise) around the subject of a shot.

arc left (AL): Rolling the camera, tripod, and dolly in a circle to the camera operator's left (clockwise) around the subject of a shot.

PRODUCTION NOTE

Never follow a pan left with an immediate pan right, or vice versa. The movement in the resulting image is not pleasing to the viewer. You can cut to a different camera between a pan left and a pan right without ill effects.

- *Tilt.* Pointing only the front of the camera (lens) vertically up or down; the dolly and tripod remain stationary. *Tilt up (TU)* by pointing the lens up toward the ceiling and *tilt down (TD)* by pointing the lens of the camera down toward the ground.
- *Pedestal.* Raising or lowering the camera on the pedestal or tripod while facing the set. The tripod and dolly remain stationary. *Pedestal up (PedU)* is to raise the height of the camera. *Pedestal down (PedD)* is to lower the height of the camera.
- *Arc.* Moving the camera in a curved truck around the set, while the camera remains fixed on the main object in the shot—the main subject never leaves the frame of the picture. An *arc right (AR)* involves rolling the camera, tripod, and dolly in a circle to the camera operator's right (counterclockwise) around the subject of the shot. Rolling the camera, tripod, and dolly in a circle to the camera operator's left (clockwise) around the subject is an *arc left (AL)*.

VISUALIZE THIS

Think of an arc camera move like circling a car that you'd like to buy. While looking at the car, you walk all the way around it while facing the car as you walk.

Psychology of Presentation

Some television production techniques, if used properly, can actually cause the audience to physically "feel" something. An example of this is subjective camera, described in the previous chapter. The audience sees images from a camera mounted in a stunt driver's car as he drives up and down large hills at high speeds. The audience can "feel" their stomachs lurch as the car rockets down a steep hill. The videographer can also plant attitudes in the minds of viewers merely by the way a picture is framed. A program has the power to shape the viewers' perception of someone or something without expressly verbalizing an opinion. This is a significant power to have over a large number of people. An experienced and talented camera operator can influence an audience without the majority of individuals even realizing their opinion has been manipulated. This kind of talent comes with great responsibility as well.

A *low angle shot* is created by placing the camera anywhere from slightly to greatly below the eye level of the talent and pointing it upward toward the talent, **Figure 4-20**. The talent appears to be above the audience. Tilting the camera up while shooting a character makes the audience see the character as powerful, feel respect for the character, and possibly fear the character. On the other hand, tilting the camera down causes the audience to feel superior to the character. The character is perceived as weak and insignificant. Shooting talent with the camera higher in the air and pointed down at an angle is called a *high angle shot*, **Figure 4-21**.

low angle shot: A shot created by placing the camera anywhere from slightly to greatly below the eye level of the talent and pointing it up toward the talent.

high angle shot: Shooting talent with the camera positioned higher in the air and pointing down at an angle.

Figure 4-20. In a low angle shot, the camera is placed low to the ground and looks up at the subject.

Figure 4-21. The camera is high off the ground in a high angle shot and looks down on the object in the shot.

A greater degree of tilting up or down heightens or lessens the degree of emotion and perception felt by the audience. Making the audience feel inferior or superior only occurs with consistent use of low or high angle shots. Randomly using low or high angle shots on a character does not evoke the same emotions from the audience. If this technique is performed with extreme degrees of tilting, the entire effect is obvious to the point of being comedic. Always experiment before recording the final image.

When doing news programming, getting shots using perception-manipulating camera techniques is extremely unethical. Using shots like this is an intentional attempt at manipulating viewers to adopt the opinions of the news producers. This is sometimes called yellow journalism, propagandizing, or brainwashing.

To impart a neutral feeling, the camera should be placed at the talent's eye level. News sets in professional television studios are on raised platforms. The goal of news programming is to have newscasters relate to the audience and be believable as they report. Since newscasters sit in chairs, the cameras would have to look down on them as they report. It is not practical to have the cameras pedestal down to the eye level of the talent. Camera operators would have to bend over to see through the camera's viewfinder for the entire duration of the shoot. Building platforms for the newscasters is much less expensive than the medical care required for camera operators with back ailments from being bent over lowered cameras for extended periods of time.

Wrapping Up

Ultimately, the camera operator is responsible for framing each shot that is recorded for a program. In addition to camera focus and zooming, the camera operator must consider camera movements, specific shots and angles, and following the impromptu instructions of the director. An important section of this chapter addresses selective depth of field and the factors, controlled by the camera operator, that create selective depth of field. Only the camera operator can affect depth of field by manipulating the lens. Remember that the camera's aperture affects depth of field, not lighting. If the camera operator closes the iris down some, for example, the depth of field increases, but the picture is dark. In this case, the lighting designer may be asked to add additional lighting to the set. The set design and lighting contribute to the production goals, but the camera operator must capture all the program elements to realize the director's vision.

Review Questions

Please answer the following questions on a separate sheet of paper. Do not write in this book.

1. How does white balancing affect the images recorded by a camera?
2. List the steps in pre-focusing a zoom lens.
3. Explain why a camera's depth of field should most often be as large as possible.
4. Why is shallow depth of field used in a program?
5. How does the rule of thirds affect picture composition?
6. What is nose room?
7. How is a shot sheet created and used during production?
8. What is the purpose of an establishing shot in a program?
9. Describe a scene in which an over-the-shoulder shot would likely be used.
10. Explain the difference between a dolly camera movement and a truck camera movement.
11. How does the camera angle affect the audience's perception of a character?

Activities

1. Create a shot sheet for a three-camera production instructing viewers on how to make a peanut butter and jelly sandwich. The shots must vary. No single shot should last more than three seconds.
2. Choose one category of camera shots discussed in this chapter (wide shots, individual subject shots, multiple subject shots, or specific view shots). Create a display that illustrates each of the shots included in the selected category.

3. Use your own body to demonstrate the camera movements described in this chapter.

- Pan Left: Stand perfectly still and turn your head to your left.

- Pan Right: Stand perfectly still and turn your head to your right.

- Tilt Up: Stand perfectly still and point your nose to the ceiling of the room.

- Tilt Down: Stand perfectly still and point your nose to the ground between your feet.

- Pedestal Up: Rise up on your tiptoes while facing forward (toward the set).

- Pedestal Down: Squat down while facing forward (toward the set).

- Dolly In: Smoothly walk forward, directly toward the set.

- Dolly Out: Smoothly walk backward while facing the set.

- Truck Right: Walk sideways to the right while facing the set.

- Truck Left: Walk sideways to the left while facing the set.

- Arc: Walk in a circle around an object, keeping your eyes fixed on that object. Walking to your right (counterclockwise) is an arc right. Walking to your left (clockwise) is an arc left.

STEM and Academic Activities

1. Investigate how macro lenses work. Explain how the different millimeter designations of macro lenses affect how the lenses are best used.

2. Print screen shots of 10 individual scenes from various television shows, newscasts, or product spots. To follow the rule of thirds when shooting, the on-screen talent should be positioned so that their eyes are 1/3 of the way down from the top of the screen, or on the upper horizontal line of the tic-tac-toe grid. Using the rule of thirds model, draw a grid on each printout. Of the 10 scenes, how many followed the rule of thirds for talent placement? What percentage of scenes made proper use of the rule of thirds?

3. Research "yellow journalism" and choose one case of yellow journalism that interests you. Write a paper that explains the story and why it is considered an example of yellow journalism.

4. Review several still photos and note the portion of the photo that your eye is drawn to first. Draw a grid representing the rule of thirds on a sheet of vellum or transparency film. Lay the grid over each of the photos. Which quadrant of each photo do you look at first? What do your findings tell you about placement of the most important information contained in the television image?

Chapter 5

Videotape, Video Media, and Video Recorders

Professional Terms

1/2″ tape	helical scan
1″ tape	input
2″ tape	Mini-DV
artifacts	monitor
Beta SX	monitor/receiver
Betacam	output
Betacam SP	P2
control track	quadruplex (quad)
D-9	receiver
deck	RF
Digital Betacam (Digi-Beta)	RF converter
Digital S	slant track
digital video recorder (DVR)	Super VHS (S-VHS)
	tail
dropout	test record
dubbing	tracking control
DVCam	VHS
DVCPRO	VHS-C
DVCPRO50	video heads
DVCPRO100	video noise
DVD	VTR interchange
head	Y/C signal

Objectives

After completing this chapter, you will be able to:

- Illustrate the process of cleaning video heads.
- Identify professional quality videotape formats, as well as other video media types available.
- Summarize the function of the control track in regulating the playback speed of videotape.
- Recall the purpose of and process for performing a test record.
- Explain the role of an RF converter in television's use of audio and video signals.

Introduction

There are many formats of videotape and videotape recorders. "Format" refers to the size of the tape in terms of width, the materials used to make the tape, and the way the signal is placed on the tape by the video recorder. This chapter discusses many of the most common formats and related information that applies to professional video recorders.

As the video industry continues to keep pace with and incorporate evolving digital technology, new methods of recording video emerge. Using videotape as a primary recording medium will probably not be practical in the near future—facilities will soon be completely tapeless. Currently, the most common tapeless recording options are recording directly to portable hard drives, miniature DVDs, full size DVDs, and solid-state memory. The greatest advantages to tapeless options include,

- Instant access to any point in a recording during playback mode, rather than shuttling back and forth on a taped recording.
- No loss of video quality when copying video from tapeless media to another media, such as during editing and duplication. When copying a videotape, however, the copy is always lower quality than the original.
- Importing video into an editor takes mere seconds when using tapeless media. Importing from a videotape is a real-time process, which means two hours of recorded video takes two hours to import.

Even though changes in technology continually affect video media and equipment, most facilities do not update their equipment and technology until the equipment in use becomes inoperable beyond repair. To accommodate older equipment and media still in use, manufacturers gradually phase out the production and support of older equipment, supplies, and technology. For example, a company may discontinue manufacturing a particular analog model video recorder, but continue to offer customer support and replacement parts for several years.

Videotape Quality

The quality of videotape is directly related to how the videotape is made, **Figure 5-1**.

General steps in the manufacturing process include,

1. Placing an adhesive on a long strip of plastic.
2. Sprinkling an oxide material over the adhesive that is capable of holding a magnetic signal.
3. Pressing the oxide into the adhesive and rolling the tape onto a spool.

Both the plastic and the oxide material used on consumer VHS tapes are relatively standard. The most common area for variances in quality (and retail prices) of consumer VHS videotapes is the grade of adhesive used on the tape.

"Gold" and "Platinum" designations on videotape packaging are not indicators of the videotape's quality. Several trade magazines and Internet sites regularly research the various brands and types available and rate the products. An inexpensive tape often has low-cost, low-grade adhesive holding the magnetic recording material to the plastic base. Once the adhesive fails, the magnetic material comes loose from the plastic. Pieces of the magnetic material then stick to the video heads of the VCR or fall into the bottom of the VCR. The result is either tape heads that need to be cleaned or circuit boards that are littered with magnetic material.

Once the medium falls off the videotape, a tiny white dot is seen on the television screen when the video head passes over the "empty spot" on the tape. This dot is called *dropout*. Once a tape starts showing dropouts, it should be discarded. Dropouts are often confused with dirty heads. When the heads of a VCR are dirty, the whole screen is affected by hundreds of white spots. Dropouts, on the other hand, are individual white spots that appear randomly.

dropout: A tiny white dot seen on the television screen when the medium has fallen off an analog videotape and the video head passes over an "empty spot" on the tape.

Figure 5-1. All videotape is manufactured using relatively similar processes. The one variable is the quality of the adhesive.

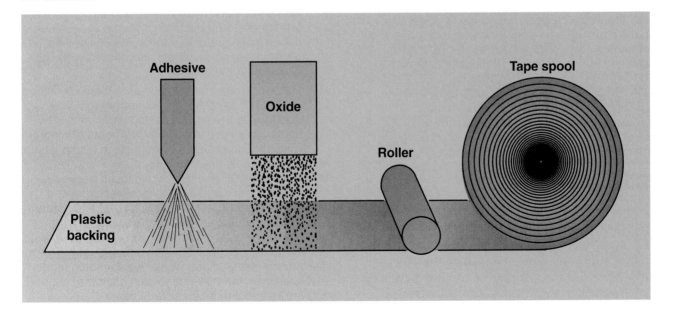

Video Heads and Helical Scan

Video heads are the mechanisms inside a VCR, or *deck* that lay down the video signal onto a tape when in record mode. When the VCR is in playback mode, video heads pick up the video signal from a tape. Consumer VCRs typically have 2, 3, or 4 video heads located around the head drum. Professional VCRs may have several more video heads depending on the format and quality of the machine. The video head drum assembly inside the VCR casing is tilted, while the tape transport system is horizontal. As the video heads spin, the video signal is placed onto the videotape in a slanted pattern, **Figure 5-2**. This slanted video track allows much more information to be placed on the tape than if the heads were horizontal. All videocassette tape formats utilize this *slant track* method of placing the signal on a tape. Another name for the slant track system is *helical scan*. The word "helical" comes from the Greek form of the word "helix," which means something in a spiral shape. The videotape is wrapped around the video head and, because the head is slanted, a spiral effect is created.

Dirty Video Heads

Video heads should be cleaned only when they are dirty, not necessarily according to the manufacturer's schedule. The very process of cleaning the heads has an effect on their life span. Do not clean a VCR's heads unnecessarily.

Usually, only two of the video heads around the head drum are used for normal recording and playback. The others are used for still-frame and/or slow-motion playback. Dirty video heads do not pick up the signal well. On VHS and S-VHS recorders, dirty heads produce many white spots on the television screen commonly referred to by consumers as "snow," **Figure 5-3**. The white dots seen on a screen when the videotape is blank or the heads are dirty is professionally referred to as *video noise*. The quantity of white dots is related to how dirty the heads are. If both heads are dirty, the entire screen is noisy but the audio can still be heard. If the audio is muffled or wavering, the audio head is dirty.

video heads: The components inside a VCR that lay down the video signal onto a tape when in record mode. When a VCR is in playback mode, the video heads pick up the video signal from a tape.

deck: The common term used for a video recorder/player.

helical scan: The pattern in which a video signal is placed onto a videotape. The videotape is wrapped around the video head and, because the head is slanted, the video signal is recorded diagonally on the tape. Also called *slant track*.

video noise: The white dots seen on a screen if the videotape is blank or if the heads are dirty.

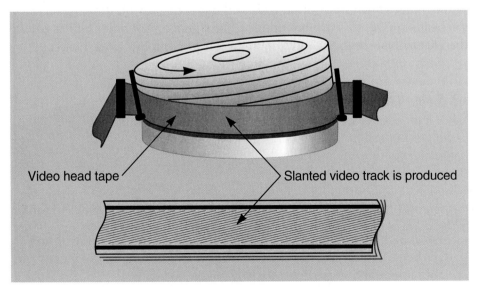

Video head tape
Slanted video track is produced

Figure 5-2. The video head drum is tilted, yet the tape moves across it horizontally. This causes the signal to be placed on the entire width of the videotape in a series of "slant tracks."

Figure 5-3. The picture created by dirty heads contains a great deal of video noise.

Talk the Talk

The correct use of industry terminology is considered an initiation test when working in the television production industry. For example, using the consumer term "snow" to describe the white dots displayed on a screen is not appropriate. This incorrect usage may cause peers and superiors to question your knowledge. The correct term is "video noise" or simply "noise."

Verify that the video noise is not isolated to a particular videotape by ejecting the tape and inserting another tape that is known to have clear video images and good sound recorded. If the display on the screen clears, the heads on the machine are fine and the signal on the other tape is bad. If the video noise persists, however, it is time to clean the video heads.

Talk the Talk

A tape with clear video images and good quality sound has a "clean signal." A tape with a clean signal does not produce any video or audio noise when played on a VCR.

Several factors can cause the heads to get dirty, but the four biggest culprits are:

- Smoking near the machine. Tobacco smoke is actually made of tiny particles that are very sticky and adhere to anything in the vicinity. In this case, the tiny particles stick to the video heads.

- Using poor quality videotape. When low-cost, low-grade adhesive fails, the magnetic medium comes loose from the plastic tape. Pieces may then stick to the video heads of the VCR.
- A family pet resting on top of the warm VCR. Pet hair and dander fall into the machine.
- Fingerprints on the surface of the videotape. Never touch the surface of videotape. Skin has oils on the surface that transfer to the tape. From the tape, the oils are transferred to the video heads.

Cleaning Video Heads in VHS and S-VHS Recorders

Even though fewer and fewer VHS and S-VHS recorders and players are being manufactured, hundreds of thousands of these decks are still in use. The video head cleaning information in this section applies only to VHS and S-VHS recorders and players.

Some head-cleaning kits are safe and effective. However, some head-cleaning cassettes damage the heads in the process of cleaning them. Video heads are very thin and fragile, and the best way to clean them is by hand. Professionals in a production facility are very unlikely to use head-cleaning cassettes; they manually clean the video heads. The sections that follow present this cleaning process, step by step.

PRODUCTION NOTE

Suppose a friend's eyeglasses were dirty and you offer to clean them. The friend gratefully hands them to you to clean and you pull out a piece of sandpaper to clean the eyeglass lenses. What do you think your friend would say? "Stop!" Of course! There is no doubt that dirt can be removed from the lenses using sandpaper, but the lenses would be scratched horribly in the process. This scenario illustrates the potential problem that exists when using some head-cleaning kits available in the marketplace. Some head-cleaning cassettes will certainly clean the heads, but may also damage them in the process.

Remove the Top Casing

All VCRs are designed to be opened quickly by technicians. Perform the following to remove the top casing from a VCR:

1. Unplug the unit from the wall.
2. Remove the screws in the top, sides, and back of the VCR that hold the top onto the machine, **Figure 5-4**. Some manufacturers may engrave the chassis with small arrows pointing to these particular screws. Do not unscrew every screw on the outside case. Loosening every screw will cause the internal parts to fall away from the chassis.
3. Remove the top casing.
4. Place the screws in a secure place.

If you encounter any resistance while removing the top, STOP. The top is designed to be easily removed. Check to make sure each of the screws

Figure 5-4. The arrows indicate the most common places where the top casing is secured with screws onto a VCR.

that secure the top have been unscrewed. Some tops do not lift directly off. Some may slide back a little and then lift up.

Locating the Video Heads

With the cover off, the bright silver video head drum should be visible. The drum looks like two drums placed on top of each other with a tiny crack between them. The bottom drum is stationary, but the top drum spins. Gently turn the upper drum and look for a little protrusion, about the size of the tip of a ballpoint pen, in the space between the drums, **Figure 5-5**. This protrusion is a video head. There is another video head 180° on the other side of the drum. If the machine has more than two heads, look for the others around the circumference of the drum.

Cleaning Supplies and Steps

Video head cleaning fluid can be purchased at most electronic specialty supply stores. "Wood" alcohol or "denatured" alcohol may also be used, but never use isopropyl (rubbing) alcohol. Do not use cotton swabs

Figure 5-5. The video heads are located on the head drum assembly. This assembly rotates at a very high rate of speed.

Video head

to clean video heads. The fibers stick to the corners of the heads and can cause problems worse than before cleaning began. Use foam swabs, with a foam rubber tip, instead of cotton swabs, **Figure 5-6**. Swabs with a small piece of chamois on the tip are also available for video head cleaning.

To clean the video heads:

1. Dip the foam swab into the cleaning fluid.

2. Rub the swab on each video head in a *sideways* motion, **Figure 5-7**. WARNING: Rubbing the swab up and down will cause the head to break.

After heads are cleaned, replace the outside cover of the VCR and secure each screw that was removed. Plug the machine into an electrical outlet and insert a tape with a clean signal recorded. If the image is clean when displayed, the video heads have been successfully cleaned. If the picture is still noisy, repeat each step. This process applies to both consumer and professional VHS and S-VHS machines.

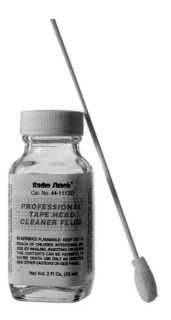

Figure 5-6. Use only video head cleaning fluid or "denatured" or "wood" alcohol and foam or chamois swabs to clean video heads.

Figure 5-7. Rub the swab over the video heads in a horizontal direction only.

Cleaning Video Heads in Digital Recorders

The head drum in digital recorders is considerably smaller and more fragile than the drums found in VHS and S-VHS recorders. In digital recorders, the drums are usually difficult to reach and are often covered by a delicate circuit board, which would need to be removed to manually clean the video heads. The potential for physically damaging parts of the deck in trying to reach the heads by hand is greater than the possible effects of commercially-made cleaning cassettes. The video heads in a digital recorder can be cleaned quite successfully if a high quality head cleaning cassette from a well-known manufacturer is used, and the instructions provided are carefully followed, **Figure 5-8**.

Before cleaning the video heads, verify that the display problems are not isolated to a particular videotape by ejecting the tape and inserting a previously viewed tape that is known to have good quality video and sound recorded. If the video displayed is clear, the heads on the machine are fine and the other tape is bad. However, if the video displays artifacts on screen, the video heads probably need cleaning. *Artifacts* are tiny, rectangular distortions that appear on the screen when a portion of the digital signal is corrupted in some way. The two most common causes of artifacts are dirty heads or a tape that has come to the end of its useful life and needs to be replaced.

To clean the video heads in a digital recorder:

1. First, read the instructions on the head cleaning cassette.
2. Insert the cleaning cassette and press "Play." The instructions typically recommend that the cleaning tape run for 7 to 15 seconds. Follow the instructions provided.
3. Press "Stop" and eject the head cleaning cassette.
4. Insert and play the previously viewed "good" tape to ensure the display is satisfactory. If artifacts are still present on the display, repeat the cleaning procedure <u>one</u> more time. If the tape does not play satisfactorily after a second cleaning, the machine must be serviced. Do not continue cleaning the heads yourself.

artifacts: Tiny, rectangular distortions that appear on the screen in digital video formats when a portion of the digital signal is corrupted.

Figure 5-8. A mini-DV digital head cleaning cassette.

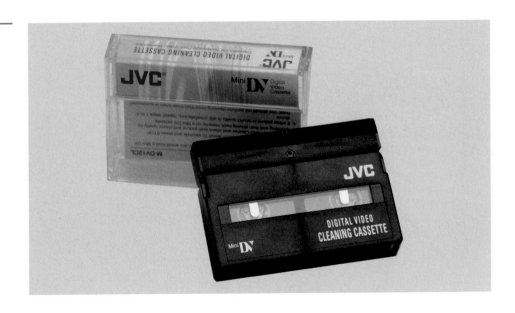

Videotape Widths and Formats

Several videotape formats are available; each is best suited for certain applications or effects. Videotape formats are further categorized by the actual width of the tape. There are definite differences between the consumer and professional varieties of videotape.

Videotape Reels

In the early days of video recording, videotape was not packaged in tidy cassettes, as it is today. Videotape was on reels, much like film reels. The tape on reels came in three widths:

- *2″ tape* is used on machines called *quadruplex*, or *quad*, recorders. These older machines are between the size of an oven and a refrigerator. Quad recorders are rapidly being phased out and replaced with newer, smaller machines of much higher quality.
- *1″ tape* comes in three formats: Type A, Type B, and Type C. Type C was the most common format.
- *1/2″ tape* is found only in low-end, industrial equipment and is relatively inexpensive. This type of videotape has been totally phased out.

2″ tape: A reel format videotape used on older machines called quadruplex recorders.

quadruplex (quad): A very large, older videotape recorder that uses 2″ tape.

1″ tape: A reel format videotape available in three formats: Type A, Type B, and Type C. Type C was the most common format.

1/2″ tape: A reel format videotape found only in low-end, industrial equipment.

Talk the Talk

When referring to reel formats, only the size designation is used: 2″ or 1″. "That program was recorded on 1″." The words "videotape" or "reel tape" are understood and, therefore, not actually spoken when industry professionals use these videotape format terms.

Videocassettes

Videocassettes replaced reel-to-reel formats to become the industry standard. There are, however, many types of videocassettes available. Several of the types discussed are pictured in **Figure 5-9**. Many of the videotapes discussed are "upwardly compatible." This means that a lower-end tape may be played in a higher-end machine, but a higher-end tape may not necessarily play in a lower-end machine.

- *VHS* (Video Home System) is a 1/2″ format that emerged in the 1970s as the preferred standard for consumer VCRs. VHS tape labeled with a "T" and a number, such as T-120, indicates the tape's run time on a VCR's highest speed setting. In the example "T-120," the tape lasts for 120 minutes at the highest speed setting. VHS tape can be purchased in a variety of lengths, including T-160, T-120, T-90, T-60, and T-30. Other formats of videotape also use this method to indicate tape length. The VHS format is being phased out and replaced with newer technology.
- *VHS-C* is the same tape format as VHS, but the tape is shorter and packaged in a smaller cassette. The smaller cassette fits into some consumer camcorders that are too small to handle a full-sized VHS cassette. The "C" stands for "Compact." These camcorders are no longer manufactured, and the VHS-C format is being phased out.

VHS (Video Home System)*:* A 1/2″ videotape format that emerged as the preferred standard for consumer VCRs.

VHS-C: A 1/2″ videotape format that is shorter than regular VHS and is, therefore, packaged in a compact cassette.

Figure 5-9. Pictured are some of the most common types of videotapes.

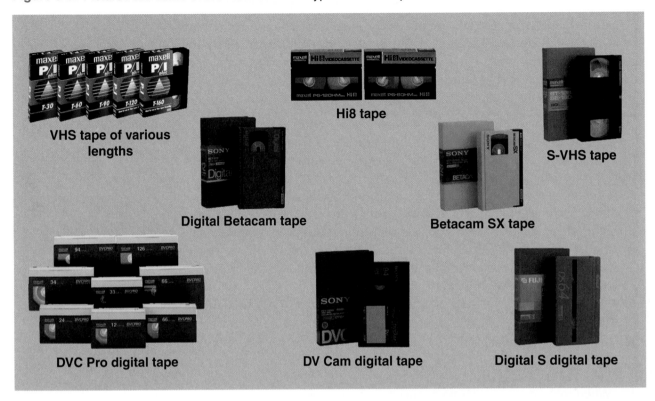

VHS tape of various lengths

Hi8 tape

S-VHS tape

Digital Betacam tape

Betacam SX tape

DVC Pro digital tape

DV Cam digital tape

Digital S digital tape

Super VHS (S-VHS): A low-end, industrial 1/2″ videotape format that is superior to VHS.

Y/C signal: The professional name for the signal placed onto S-VHS videotape.

Betacam: A 1/2″ format, broadcast-quality videotape.

Betacam SP: A 1/2″ videotape format that used to be the best format for professional television use, but digital video formats are challenging this format in professional markets.

- *Super VHS (S-VHS)* is a 1/2″ videotape format that is far superior to VHS. Many professionals consistently use this low-end, industrial format. *Y/C* is the professional name for the signal placed on S-VHS tape. An S-VHS videocassette does fit into a VHS recorder, but the Y/C signal recorded onto the videotape does not necessarily play on a consumer VCR. These decks are no longer manufactured.
- *Betacam* is a 1/2″, broadcast-quality tape. Betacam is upwardly compatible with Betacam SP, but is somewhat lower in quality and cost than Betacam SP.
- *Betacam SP* was considered to be the best format for professional television use. However, digital video formats currently challenge Betacam SP in professional markets. A Betacam SP deck is backward compatible with Betacam tapes, but these decks are no longer manufactured.

Digital Formats

There are two categories of digital recording formats—those that use digital videotape and those that use no tape at all. Recording formats that do not use videotape are called "tapeless" and record video directly onto a hard drive or solid-state memory device.

Digital Videotape Formats

Digital videotape formats can suffer degradation problems similar to dropout on analog tapes. Artifacts in digital media directly correspond to dropouts in analog media.

- *Digital Betacam (Digi-Beta)* is a 1/2″ tape with higher quality than Betacam SP. The cassette is the same size as Betacam SP, but allows quality recording of digital signals, instead of analog signals.
- *Beta SX* is a 1/2″ tape that uses digital MPEG compression. Beta SX equipment also plays Betacam SP tapes.
- *Mini-DV* is a 6mm digital video format used by many industrial video producers. The mini-DV tape is metal evaporated tape that is upwardly compatible with DVCPRO and DVCPRO50.
- *DVCPRO* is an excellent 6mm professional digital video format. The DVCPRO tape is metal particle tape that is upwardly compatible with DVCPRO50. A DVCPRO deck also plays mini-DV tapes.

PRODUCTION NOTE

Mini-DV tape is the size of a box of matches found at many restaurants and costs about as much as S-VHS tape. This tape format may be reused a maximum of nine times before artifacts begin to appear. A DVCPRO tape costs about four times as much as a mini-DV tape, but may be reused up to 100 times. DVCPRO can hold up to twice as much video as mini-DV and fits snugly into the front pocket of a men's dress shirt. DVCPRO is more economical in the long run than mini-DV. Most importantly, the DVCPRO's metal particle tape holds a much stronger signal than the metal evaporated tape used in mini-DV tape.

- *DVCPRO50* is a 6mm digital format with even higher quality than DVCPRO. A DVCPRO50 deck plays both DV and DVCPRO tapes.
- *DVCPRO100* is the high definition format of DVCPRO tape.
- *DVCam* is a 6mm digital format that is proprietary for Sony Corporation.
- *Digital S* is a 1/2″ digital format tape that is broadcast quality, also known as *D-9*. Decks for this format are no longer manufactured.

Tapeless Formats

The most familiar tapeless format is the DVD. However, other solid-state devices are also commonly used to record video footage, such as flash drives and flash memory cards (**Figure 5-10**). Granted, these small cards do not hold a tremendous amount of video but professional video memory cards will record a substantial amount of video. Tapeless formats offer all the advantages previously mentioned for digital television, with the added benefit of having no moving parts to wear out or get dirty.

- *DVD* (Digital Video Disc) is an optical disc that can store a very large amount of digital video data, as well as text and/or music. DVD is currently the standard distribution medium.
- Flash memory devices
- *P2* is a static memory card that is proprietary to Panasonic and used in certain high end cameras, **Figure 5-11**. Depending on the recording format, a 32 gigabyte P2 card can hold between 32 and 128 minutes of video.

Digital Betacam (Digi-Beta): A 1/2″ videotape with higher quality than Betacam SP and the capability of recording of digital signals instead of analog signals.

Beta SX: A 1/2″ videotape that uses digital MPEG compression.

Mini-DV: A metal evaporated tape, 6mm digital video format used by many industrial video producers.

DVCPRO: A 6mm, metal particle tape used as a professional digital video format.

DVCPRO50: A 6mm digital format with even higher quality than DVCPRO.

DVCPRO100: The high definition format of DVCPRO tape.

DVCam: A 6mm digital format that is proprietary to Sony Corporation.

Digital S: A 1/2″ digital videotape format that is broadcast quality. Also known as *D-9*.

DVD (Digital Video Disc)*:* An optical disc that can store a very large amount of digital video data, as well as text and/or music.

P2: A static memory card that is proprietary to Panasonic and used in certain high end cameras.

Figure 5-10. Both consumer and professional flash memory devices are solid-state and can hold varying amounts of video data, depending on the capacity of the device.

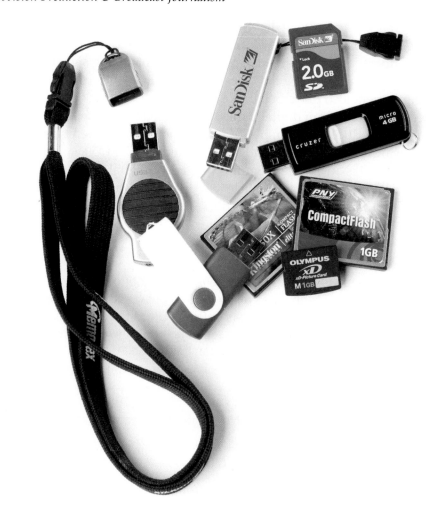

Figure 5-11. A P2 card is a lightweight, durable, and reusable solid-state recording medium. *(Panasonic Broadcast and TV Systems)*

Compatibility

VTR interchange refers to the ability of a tape recorded on one machine to be played back on another machine. As long as both machines are the same format, regardless of brand or model, tapes can be successfully played on both machines.

DVD is the current standard distribution medium, **Figure 5-12**. However, some duplication still occurs in VHS format. VHS format is the most popular videotape distribution format for industrial and commercial programming. While VHS tape does not provide a very robust signal, the quality is sufficient for viewing on any television screen. Because of this, nearly all videotape duplication begins with a high-quality master and is then *dubbed*, or copied, to VHS.

Control Track

The *control track* is a series of inaudible pulses recorded onto a tape that regulates the speed of the tape in playback. The function of the control track generally applies to VHS and S-VHS formats only. A circuit in the video recorder puts a little pulse signal, or blip, onto the tape 30 times per second while recording. These blips create the control track. If the machine is set for 2-hour speed (SP), the blips are spaced farther apart because the tape moves relatively quickly through the machine. If the machine is set for 6-hour speed (EP), the blips are much closer together because the tape moves more slowly through the machine. A blip is placed on the tape every 1/30th of a second, without regard to how fast the tape is moving, **Figure 5-13**.

VTR interchange: The ability of a tape that was recorded on one machine to be played back on another machine.

dubbing: The process of copying a video recording.

control track: A series of inaudible pulses recorded onto a tape that regulates the speed of the tape in playback.

Figure 5-12. This all-in-one disc publishing system can duplicate DVDs and print on the discs in full color. *(Primera Technology, Inc.)*

Figure 5-13. The control track pulses are placed further apart when recording in SP mode compared to EP mode.

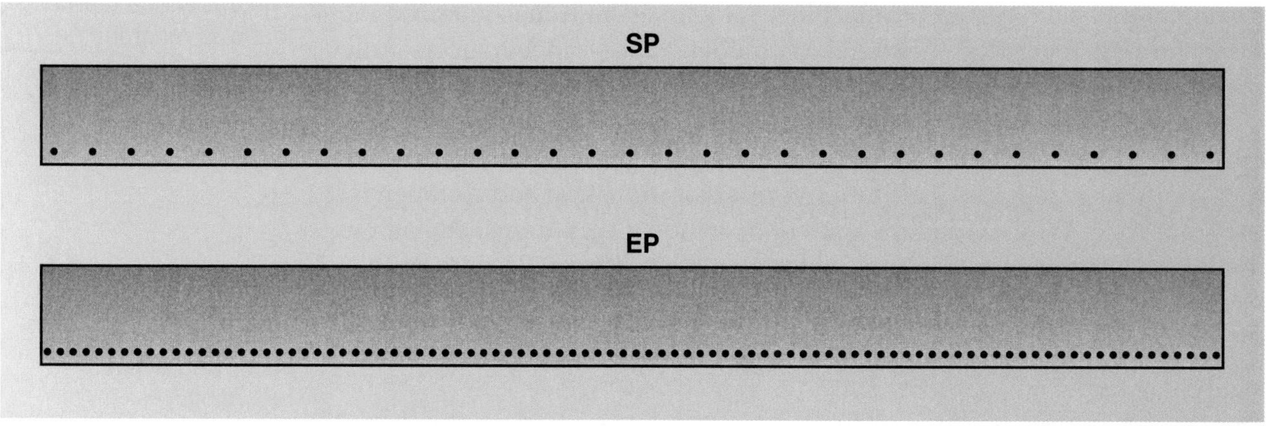

VISUALIZE THIS

Imagine you are driving a car on a flat, straight stretch of deserted highway. You are driving down the center of the road and the white line appears as a series of white dashes that disappear under the hood of the car. It would take some practice, but you could regulate the accelerator so that 1 dash disappears under the hood every 2 seconds. The white dashes may come in too fast at first, but you slow the car down until exactly 1 dash goes under the hood every 2 seconds. The control track circuit operates the same way. It speeds up and slows the tape down until the pulses occur at 1/30th of a second intervals.

Common recording speeds are:

- **SP** (Standard Play) records 2 hours of programming onto a T-120 tape. The "T-120" designation indicates the tape's run time on a VCR's highest speed setting. The vast majority of pre-recorded VHS movies are recorded at SP speed.
- **LP** (Long Play) is a 4-hour speed that is twice that of SP, but has been phased out.
- **EP** (Extended Play) records 6 hours of programming onto a T-120 tape, which is three times that of SP. The quality of an EP recording is noticeably less than an SP recording.

When a machine is in playback mode, the same circuit that placed the pulses on the tape now "listens" for the blips or pulses. The tape is sped up or slowed down accordingly to time the pulses at 1/30th of a second. For example, a VCR that is manually set to EP speed will play an SP tape perfectly because the circuit monitors the control track. The tape speed is adjusted to correctly play the tape. Depending on the VCR, the audio may cease and the screen may turn blue while this adjustment is made.

When watching a tape on a machine other than the one the tape was recorded on, a series of nearly horizontal white lines may sometimes appear on a portion of the screen, **Figure 5-14**. The white lines may even pulse, or rapidly appear and disappear. In this case, the control track circuit is not able to compensate enough to correct the display. The circuit only has a certain

Figure 5-14. If horizontal white lines persistently appear on the screen, manually adjust the tracking control.

range in which it can speed up or slow down the tape. If using a professional VCR the *tracking control* knob must be manually adjusted until the lines disappear. The tracking control should be adjusted back to its normal position when finished watching the tape, otherwise every subsequent tape will display the white lines. Many newer consumer VCRs have a built-in, automatic tracking control function that cannot be manually adjusted.

The control track circuit is extremely important. Some compare the importance of the control track pulses to the sprocket holes found on motion picture film—each is vital to viewing a program.

tracking control: A knob on a professional VCR that is used to manually adjust the tape tracking speed.

Digital Video Recorders

To this point in the chapter, information provided on digital (videotape) recorders involves units that require videotape as the recording media. A *digital video recorder (DVR)* records a digital signal directly onto either a hard drive or a solid-state memory module inside or connected to the DVR unit. DVRs do not use videotape, and are, therefore, "tapeless" recorders. These devices can record a large amount of video and audio in both standard and high definition video.

Some units record onto hard drives attached to or installed within the DVR, while others record using solid-state memory. Solid-state memory

digital video recorder (DVR): A device that records a digital signal directly onto either a hard drive or a solid-state memory module inside or connected to the DVR unit.

devices, such as flash drives, flash memory cards, and P2 cards, have no moving parts and are usually considered more durable than hard drives. A solid-state memory device is inserted into the designated slot on a camera to record video and is removed after shooting. The card or drive can be inserted directly into an editing system with corresponding slots to read the device or into a separate reader accessory that is connected by a cable to a port on the computer.

The DVR is rapidly becoming the recording system of choice in the professional broadcasting industry. Professional units have many more options and functions than consumer DVRs, which allow the DVR to process various video formats and "spruce up" the video and audio signals using extended digital controls.

Recording Audio and Video

The very first thing a video operator must do before any recording session begins, is label the videotape or recording media with appropriate identifying information: title of the program, director, scenes to be recorded, etc. Labeling the recording media is very important and should become second nature for everyone. Unlabeled recordings can be easily misplaced or important footage may be accidentally recorded over because the unlabeled tape was assumed to be blank.

If videotape is the recording media to be used, let the tape roll forward in record mode for at least 2 minutes. Never record anything important in the first 2 minutes at the beginning of a tape. If a tape is going to break, 9 times out of 10, it will break at the very beginning of the tape. By leaving the first 2 minutes of a tape blank, the tape may be reattached to its hub if a break occurs without losing any important footage.

Once the videotape has been labeled and the tape has rolled forward for 2 minutes, the video operator must perform a test record. Performing a test record before each recording session verifies that the audio and video signals are not only reaching the recorder, but are also being properly recorded. During a recording session, a monitor displays only what is going *into* the machine, not what is actually *being recorded*. This step avoids certain disasters that may be discovered in the editing room weeks after shooting.

Most recorders have a video AGC (automatic gain control), **Figure 5-15**. This control automatically adjusts the video signal coming into the machine to the best levels for recording. Inexperienced operators should leave the video AGC circuit active.

A signal comes into the video recorder through an *input* on the back of the deck, such as the "audio in." The signal leaves the deck and travels to another piece of equipment through an *output*, such as the "video out." This may seem obvious, but it is very easy to be careless and accidentally attach a CD player to the "Audio Out" connector on a recorder.

Test Recordings

A *test record* is the process of using the video recorder to record audio and video signals before the session recording begins. Many recording sessions are lost and must be reshot because the test recording step was omitted. For

input: A port or connection on a video device through which a signal enters the device, such as the "audio in" port.

output: A port or connection on a video device through which the signal leaves the deck and travels to another piece of equipment, such as the "video out" port.

test record: The process of using the video recorder to record audio and video signals before the session recording begins to ensure the equipment is functioning properly and to indicate any necessary adjustments.

Figure 5-15. The video AGC should remain active most of the time.

the technician, a test record indicates if the equipment is functioning properly and if any adjustments are necessary. For example, a test record lets the technician know if the video heads are dirty before program taping begins.

To make a test record once all the equipment is connected and powered on:

1. Activate the "Record" function on the machine.
2. Record any available signal for 1 minute.
3. Re-cue the recorder to the beginning of the clip and play it back.
4. Listen for appropriate audio and watch for appropriate video.
5. Make any necessary adjustments before beginning the recording session.

PRODUCTION NOTE

To avoid wasting tape or hard drive space, the video recorder is usually placed in the "record" and "stand-by" mode. The record circuit is open and the images are seen in the monitors, but no signal is actually being recorded—the hard drive is not recording and/or no tape is actually moving. The video recording operator must remember to take the recorder out of "stand-by" when the director calls for a recording session to begin. If the recorder is left in "stand-by" mode, the scene is not recorded. To complicate the situation further, some inexperienced directors skip reviewing the recorded scene before moving on. In the editing room weeks later, the director may realize there is no recorded footage of a particular scene!

It is important to be certain that the recording machine is properly set to record each scene. Footage should be reviewed immediately to ensure that the scene has recorded and is of acceptable quality.

Heads and Tails

The beginning of every take has a "lead-in" of at least 15 seconds, called a *head*. The head usually consists of the display of the slate and the

head: A 15 second "lead-in" recorded at the beginning of every take.

tail: A 10 second "lead-out" recorded at the end of each scene.

countdown (discussed in Chapter 20, *Directing*). Each scene should have a minimum of a 10-second "lead-out" at the end, called a *tail*. While recording the lead-out, performers simply continue their action without dialog for an additional 10–15 seconds until the director calls, "Cut." The talent does not ad lib lines during the tail, but continues the mood of the scene through silent body language. The recorded head and tail on each take is very important in the editing process, **Figure 5-16**.

PRODUCTION NOTE

If you are using videotape, do not simply fast-forward the tape at the end of a take instead of recording a tail. The portion of tape that was fast-forwarded does not have control track. This causes problems in the editing room and may prevent a successful edit of the program.

A tail also serves as a safety feature. When a tape is stopped, it retracts slightly into the cassette. If the tape is stopped immediately at the end of a scene, it is quite possible that the beginning slate of the next scene will record over the end of the scene just shot. Additionally, most camcorders and recorders have a feature called "automatic backspacing." This guarantees that the tape is about 3–5 seconds behind the initial stop point. As a result, when recording begins again, the end of the previous scene will be recorded over. A tail ensures that if any part of the previous scene is recorded over, it is the tail and not the actual scene.

Even in tapeless facilities, the heads and tails practice is still followed. The head and tail are quite necessary for clean editing on digital editing platforms.

Audio Levels

The video operator, or camcorder operator, is in control of the audio levels that are recorded onto the tape. In the studio, the audio engineer mixes the audio signals into a single signal and sends it to the video recorder. The video recorder displays the recorded audio levels on VU meters (see Chapter 6, *Audio Basics*). The audio levels for an analog recording should

Figure 5-16. Heads and tails must be shot for all scenes.

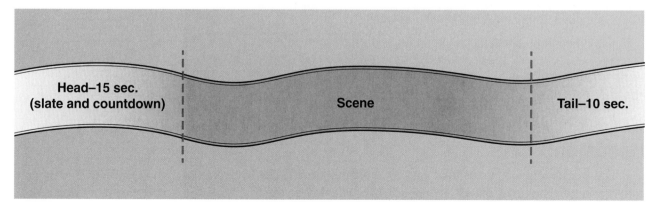

Head—15 sec.
(slate and countdown)

Scene

Tail—10 sec.

fluctuate between –3 and +3 dB on the meters. On a digital recorder, the levels should hover near –20. The recorded audio levels are a crucial portion of the program and cannot be fixed in the editing room if they are recorded improperly.

Radio Frequency

Video recorders create pure video and pure audio. Many consumer television sets, however, cannot receive pure video and pure audio on two separate cables. The video and audio must be combined into one cable and converted into an *RF* (radio frequency) signal by a small box inside the video recorder called an *RF converter*, **Figure 5-17**. Once combined as a radio frequency, a cable carries the signal from the "Antenna Out" on the video recorder to the "Antenna In" on the television set.

A television set that can receive only RF signals is a *receiver*. A television set that can receive only pure video and audio signals is a *monitor*. A *monitor/receiver* is a hybrid television that can receive both pure video and audio, as well as the RF signal. The types of connectors on the back of the television indicate the television type, **Figure 5-18**. A monitor television has RCA and/or BNC connectors. Connectors are discussed in Chapter 7, *Connectors*. An F-connector for coaxial antenna cable or two screws for a flat-lead antenna may be found on receiver televisions. Monitor/receiver televisions have both RCA and/or BNC connectors and an F-connector or the two screws.

Most new, higher-end consumer digital televisions are equipped with multiple inputs. A signal can be connected to the RF connector and to "Video In" and "Audio In" connectors. These multiple inputs may be used for video games, additional VCRs, and DVD players. Television sets with multiple inputs have a button or a menu option that enables the user to switch from one input to another.

RF: Radio frequency signal that is a combination of both audio and video.

RF converter: A small module inside the VCR that combines pure video and audio into one radio frequency.

receiver: A television set that can receive only RF signals.

monitor: A television set that can receive only pure video and audio signals.

monitor/receiver: A hybrid television that can receive pure video and audio signals, as well as RF signals.

Figure 5-17. The RF converter changes pure video and pure audio signals into one radio frequency.

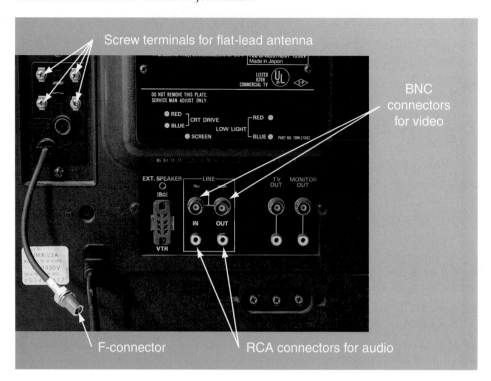

Figure 5-18. A monitor/receiver has connectors capable of receiving RF, as well as pure video and pure audio.

Wrapping Up

Due to the many different formats of videotape and videotape recorders, it is important to remember that tapes created in one format are usually not suited for playback on a machine of another format. The formats of digital media continue to evolve. However, it seems certain that videotape will soon become obsolete.

Review Questions

Please answer the following questions on a separate sheet of paper. Do not write in this book.

1. What are the possible causes of white spots appearing on the screen while viewing a videotape?
2. What are some common causes of dirty video heads?
3. List the appropriate materials to use for cleaning video heads.
4. What is an upwardly compatible videotape?
5. How is the control track related to the playback speed of a videotape?
6. What is the purpose of a test record?
7. What is the purpose of the head and the tail at the beginning of each scene?
8. How do monitors, receivers, and monitor/receivers differ from each other?
9. What is the function of an RF converter?

Activities

1. Research the evolution of video media formats and create a timeline that includes important dates, innovations, and events.
2. Inspect the television sets in your home and determine if each is a receiver, a monitor, or a monitor/receiver.

STEM and Academic Activities

1. What are some of the environmental impacts of discarding old videocassettes?

2. Research the evolution of recording media, from film to today's digital formats. Create a linear time line that notes the date, name, and price of each recording media development.

3. In a small group, brainstorm ideas for a new method or product that would improve the way video is recorded.

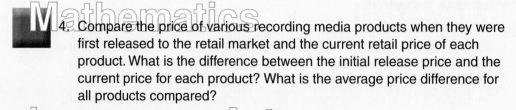

4. Compare the price of various recording media products when they were first released to the retail market and the current retail price of each product. What is the difference between the initial release price and the current price for each product? What is the average price difference for all products compared?

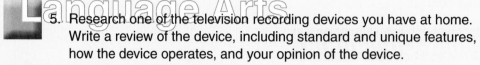

5. Research one of the television recording devices you have at home. Write a review of the device, including standard and unique features, how the device operates, and your opinion of the device.

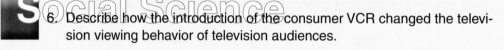

6. Describe how the introduction of the consumer VCR changed the television viewing behavior of television audiences.

Chapter 6

Audio Basics

Objectives

After completing this chapter, you will be able to:

- Explain the function of audio for television productions.
- Identify the most common use of each type of microphone presented.
- Understand the importance of the pick-up pattern classification when selecting a microphone.
- Recall the appropriate VU meter readings for both an analog audio system and a digital audio system.

Professional Terms

audio mixer	mic level
automatic gain control (AGC)	mic mixer
background sound	microphone
boom	natural sound (nat sound)
boundary mic	off-camera narration
cardioid mic	omni-directional mic
condenser mic	on-camera narration
diaphragm	parabolic reflector mic
directional mic	pick-up pattern
dynamic mic	pop filter
electret condenser mic	potentiometer (pot)
feedback	power level
fishpole boom	ribbon mic
generating element	room tone
hand-held mic	shotgun mic
high impedance (HiZ)	stick mic
hypercardioid mic	supercardioid mic
lapel mic	uni-directional mic
lav mic	voiceover (VO)
line level	voice track
low impedance (LoZ)	VU meter
	wireless mic

Introduction

Audio is so important to the television production business that it is placed first in the phrase "Audio/Visual (A/V)." Watching television can be very frustrating if the audio is muted. A chase scene in a movie is nowhere near as entertaining without screeching tires and exciting music. The low, muffled sound of footsteps moving closer or a sudden, high-pitched scream adds to the suspense and shock of a horror movie. Audio is vitally important in getting a message across in television and the movies. Unfortunately, most students new to broadcasting tend to think primarily about the video aspect of a program and consider the audio only as an afterthought. This type of thinking is a recipe for disaster.

The Functions of Sound for Television

Without sound, programs become silent movies. All sounds on television serve one or more of the following four functions:

- Voice track
- Music and sound effects
- Environmental sound
- Room tone

voice track: The audio portion of a program created through dialogue or narration.

on-camera narration: Program narration provided by on-screen talent (seen by the camera).

off-camera narration: Program narration provided by talent that is heard, but not seen by the viewer. Also called *voiceover (VO).*

The *voice track* is usually the primary means of getting a message to the viewer's ears and may be considered the most necessary audio of a program. The voice track is the sound created through dialogue or narration. Narration takes two forms:

- On-camera narration
- Off-camera narration, also known as voiceover.

If the viewer sees the narrator speaking, this is typically called an *on-camera narration. Off-camera narration*, also called *voiceover (VO)*, is when viewers hear but do not see the narrator of a program (**Figure 6-1**). In broadcast journalism, the narrator may be the reporter or an on-the-street eyewitness describing the event or situation that the videographer is shooting. For example, a reporter asks one resident of a local neighborhood

Figure 6-1. The narrator often watches the video portion of a documentary and speaks the narration into a mic attached to an additional recorder. In the editing phase, the narration track is synchronized and recorded onto the same tape with the video.

to describe what the tornado sounded like as it bore down on his home. As the resident talks, the camera pans the destruction caused by the tornado.

PRODUCTION NOTE

On a remote shoot, sometimes the dialogue picked up by the mics is weaker than the background sounds. In a dramatic program, for example, the live background engine noise of race cars at a NASCAR race makes the dialogue between two characters in the scene very hard to hear. Later, the actors view the scene and re-record their dialogue in a sound booth—essentially lip-syncing with themselves. This process is called automatic (or automated) dialogue replacement (ADR), or may simply be called "re-recording." Look for the job title "ADR" or "Re-recording" in the end credits of a movie.

Music and sound effects help set the mood and enhance the action of a program. The emotion of a scene is dramatically enhanced with properly selected and timed background music. Sound effects can be created by the production team or be purchased on tape, DVD, CD, or downloaded from the Internet.

Complete silence is artificial and should rarely, if ever, be found in a television program of any type. Every location has a certain amount of normal sound associated with that particular location. This is the existing sound that a production crew may find in an environment when shooting occurs. For example, the sounds in a scene that takes place in a classroom may include papers rustling, pens tapping, the faint sound of hallway noise, and the hum of fluorescent lighting fixtures. If the expected environmental sounds are not present during the shoot, they may be added during post-production editing to help establish the setting in a dramatic production. Care must be taken to prevent the volume of environmental sound from becoming distracting.

There are two terms used to describe environmental sounds: background sound and natural (nat) sound. The difference between these environmental sounds depends on their importance in a shot.

Background sound is environmental sound that is not the most important sound in a shot. If it's in the background, some other sound is in the foreground—such as the voice of actors performing the dialogue in a dramatic production. Since background sound is not the most important sound in the shot, it must not overpower the foreground sound in a shot. The production crew must be aware of background sound and organize the shoot to make sure the background sound remains effectively in the background.

Natural sound, or *nat sound*, is environmental sound that is important to the topic of the story; it may often be *the* most important sound in the shot. For example, a feature story about a blacksmith should include some shots with natural sound of his hammer hitting the metal on an anvil. Nat sound is usually captured on a secondary recording called B-roll, either before or after the shot containing the voice track. During the editing phase of production, the nat sound is placed on a track underneath the primary audio of the program.

background sound: Type of environmental sound that is not the focus of or most important sound in a shot.

natural sound (nat sound): Environmental sound that enhances a story and is important to the shot.

PRODUCTION NOTE

In broadcast journalism, a reporter can record voiceover narration at the studio in a sound booth or conduct an interview in a quiet location (to ensure good sound quality) about a topic which includes action. In each of these situations, B-roll shots with nat sound become very important in the editing process to provide visuals and sounds of the action associated with a story. The B-roll video with nat sound audio of related action shots can be mixed with the voice track from the A-roll. As the clips with nat sound are inserted into the story during the editing process, the nat sound audio level can be controlled to ensure the volume doesn't overpower the reporter's voiceover. A talented and experienced reporter can plan/pace their voiceover to incorporate the nat sound during pauses in speech (at commas or periods in scripted narration). This way, the nat sound, which was recorded because it is an important part of the topic/event covered, becomes an important piece of raw material the reporter can use to build the story.

Nat sound is environmental sound that helps call attention to what a reporter is saying and entices the viewer to continue paying attention to the story. Nat sound is only the environmental sound that supports the story. Where things get tricky is the processing that must sometimes be done to make sure relevant nat sound is present in a recording, without the unwanted background sound.

VISUALIZE THIS

A busy roadway with holiday weekend traffic, horns honking, loud music playing, and motors revving are all part of the background sound to a traffic report. The reporter and photog must be very careful to make sure the reporter's audio is in the foreground and the traffic sounds are in the background. However, this same traffic scene can be shot without the reporter, taken back to the studio, and used as nat sound behind the reporter's voiceover. The traffic sounds can be coordinated with the reporter's voiceover, so that the honk of a car horn happens right when the reporter finishes a sentence and acts as an "exclamation point" for the story. When natural sound is used this way, it is placed with the visual that accompanied it in reality.

If nat sound is extracted from a video recording and manipulated, ethical issues arise about modifying reality. Using a shot and its accompanying sound to illustrate the reporter's narration is ethical. The only acceptable alteration of audio in news is to reduce the volume of nat sound to better hear the voices of those speaking.

room tone: The sound present in a room or at a location before human occupation.

Room tone is the sound present in a room, or at a location, before human occupation. Room tone is the "sound of silence" in the shooting environment. If shooting on location, it is important to clear the set for a few minutes after the equipment is set up. Once all the talent and crew have left the location set, turn on the recorder and record at least three minutes of the existing environmental sound. Having the environmental sound of each location recorded is useful when editing the program. The environmental sound may be used to cover unwanted sounds in the background of a scene that were not noticed while shooting. Using the environmental sound of a location creates a much less noticeable audio edit than if true silence were used.

ASSISTANT ACTIVITY

Listen to the silence of a room. Although you have probably never noticed it, the silence of different locations varies greatly. Go to your bedroom; remain perfectly still and listen. Do the same in the family room and in the backyard. You will soon realize that the "silence" is notably different in each location.

Place a good pair of stereo headphones with full earmuffs over your ears and do not turn on any sound. Listen to that kind of "silence." All of these "sounds of silence" are surprisingly unique and they help to define the video image's environment from an audio perspective.

Sound Frequency

Sounds are generally divided into three groups: low-frequency sounds, mid-range sounds, and high-frequency sounds. Most people are familiar with common band instruments and the sounds they create. Common instruments are used in the following examples of the three frequency categories:

*   Low-frequency instruments include the bass guitar, bass drum, and the tuba. A bass vocalist is also categorized in the low-frequency range.
*   Mid-range sound frequency instruments are trumpets, clarinets, and French horns. Alto and tenor vocalists fall within the mid-range. The human speaking voice is generally in the mid-range, as well.
*   High-frequency sound is created by flutes, piccolos, and soprano vocalists.

Types of Microphones

A *microphone* is the piece of equipment that picks up sounds in the air and sends them to the mixer or recorder. Fundamentally, all microphones work the same way. Sound waves in the air hit a thin surface inside the mic, a *diaphragm* or *generating element*, which then vibrates. In the most common types of mics, the vibration moves a tiny wire back and forth through a magnetic field creating an electrical signal, **Figure 6-2**. This electrical signal is sent through the mic cable to an amplifier or a recorder.

microphone (mic): The piece of equipment that picks up sounds in the air and sends them to the mixer or recorder.

generating element: A thin surface inside the mic that vibrates when hit by sound waves in the air and creates an electrical signal. Also called a *diaphragm*.

Talk the Talk

The term "microphone" is commonly abbreviated as "mic" or "mike."

Figure 6-2. This illustration provides a general representation of how microphones work.

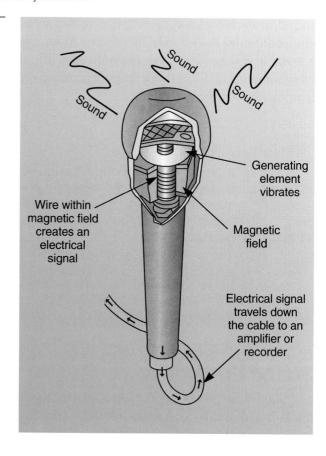

Microphones may be classified by the type of generating element (surrounded in a magnetic field) each uses:
- Diaphragm that vibrates a coil.
- Thin piece of coated film that vibrates a coil.
- Thin piece of metal foil.

Mics can be differentiated by examining the frequencies of sound each best captures. Some mics pick up certain frequencies of sound better than others. The audio engineer's goal is to match the right mic to the right frequencies of sound. All microphones are not created equal—you typically get what you pay for. Low-cost gear often yields results that are less than satisfactory.

Wired and Wireless Mics

Another classification of mics is "wired" or "wireless," which refers to how the signal gets from the mic to the recorder.

A wired mic is attached to the recorder by a cable. Cables can span a significant distance (at least 200 feet) without ill effects to the audio signal. With extremely long runs of cable, such as from the press box in a football stadium to a remote production truck in the parking lot, an amplifier may be put in the line to keep the signal strong. Wired mics and their cables are very reliable. However, the greatest disadvantage of wired mics is their cables. Performers using wired mics must be mindful of the mic cables and avoid tangling their feet in the cables. Additionally, when wires are run along the floor, they pose a tripping hazard and must be taped down securely. Removing the tape at the end of a shoot and re-coiling the cables neatly is also a substantial task that must be done carefully to avoid a huge spaghetti-like mess of wires, which would take a great deal of time to untangle.

Wireless mics have a practical advantage over wired mics in that there is no wire to run to the recording unit. A *wireless mic* has a short cable that runs from the mic to a radio transmitter with an antenna. The transmitter is sometimes built into the mic itself. The transmitter sends the audio signal through the air, via a radio wave, to a receiver that is on or near the recorder. The receiver picks up the transmitted signal from the air and sends it through a short cable to the recorder, **Figure 6-3**. A primary advantage of a wireless system is the freedom of movement it allows the performers—they do not need to be concerned with tripping over mic cables while performing. Additionally, the audio engineer does not need to lay many feet of mic cable, tape it to the floor for safety, and pull up the cable to re-coil it at the end of the shoot.

Some may consider wireless mics to be the best choice for all applications, but this is not the case. Wireless mics transmit and receive signals using a radio frequency, which is very effective as long as no one else in the vicinity uses the same radio frequency. Wireless mics are prone to interference from walkie-talkies, baby monitors, CB radios, heavy machinery, and other wireless mics operating at or near the same frequencies. Use wireless mics whenever appropriate and practical, but always keep a backup of wired mics and mic cable.

wireless mic: A mic that uses a short cable to connect the mic to a radio transmitter with an antenna, or the transmitter may be built into the mic itself. The transmitter wirelessly sends the signal to the receiver, which sends the mic signal through a short cable to the recorder.

PRODUCTION NOTE

When recording a theatrical event, there is an excellent chance that the stage performers or the theater's tech crew will be using their own wireless mics on the actors, as well as on the stage crew's headsets. Always check the frequencies used by the theater and compare them to the frequencies of your equipment. If the frequencies are close, use wired mics instead.

Figure 6-3. The wireless mic allows a performer to have freedom of movement, without the danger of tripping over a mic cable.

Dynamic Microphones

dynamic mic: A very rugged type of mic that has good sound reproduction ability. The generating element is a diaphragm that vibrates a small coil that is housed in a magnetic field.

The generating element in a *dynamic mic* is a diaphragm that vibrates a small coil that is housed in a magnetic field. It is a rugged mic with good sound reproduction ability, **Figure 6-4**. Dynamic mics are designed to "hear" different sound frequencies—the pitch of a sound, not its volume or strength. The dynamic mic most commonly found in a television studio is designed to pick up sounds best in normal speaking voice frequencies. They are not designed to mic musical instruments or accompanying vocals. The dynamic mics used in a studio setting do not pick up high- and low-frequency sounds as effectively as the mid-range sounds of speech. Using a mic that picks up mid-range sounds to mic musical instruments or singing stage performers would result in music that lacks good sound quality and reproduction.

Condenser Microphones

condenser mic: A type of mic that requires an external power supply (usually a battery) to operate. The generating element is a thin piece of metal foil or coated film. Also called an *electret condenser mic.*

The generating element used in *condenser mics* is a thin piece of metal foil or coated film. This type of mic requires an external power supply (usually a battery) in order to operate, **Figure 6-5**. Condenser microphones are also called *electret condenser mics*. They can pick up a greater range of sound frequencies than dynamic mics and good condenser mics are usually more expensive.

Figure 6-4. A dynamic mic is the perfect choice for most applications. It is extremely rugged in design.

Figure 6-5. The condenser mic requires an external power supply. In this example, the mic is battery-powered.

Ribbon Microphones

The *ribbon mic* is the most sensitive of all mic types used in television. A thin ribbon of metal surrounded by a magnetic field serves as the generating element in this type of mic. At one time, ribbon mics were the only type found in commercial radio stations. In television applications, a ribbon mic is most commonly placed on a talk show host's desk. These mics are now primarily used in music recording studios. Superb sensitivity is a great advantage of this type of mic. However, the fragility of the generating element is an expensive disadvantage of the ribbon mic. An accidental bump of the mic itself or the "pop" produced from the rush of air released when pronouncing a "p" sound could break the ribbon inside the microphone. In recording studios, a barrier made of shaped wire covered with a piece of nylon is placed between the ribbon mic and the talent. This *pop filter* protects the mics from explosive "t" and "p" sounds, and catches moisture and rushes of air before they hit and damage the diaphragm of the ribbon mic, **Figure 6-6**.

ribbon mic: The most sensitive type of mic used in television. A thin ribbon of metal surrounded by a magnetic field serves as the generating element.

pop filter: A barrier made of shaped wire covered with a piece of nylon that is placed between a sensitive mic and the talent to avoid damage to the diaphragm of the mic.

Talk the Talk

The pop filter is sometimes referred to as a "spit guard."

Non-Professional Microphones

The microphone built into low-end camcorders should not be used in professional recording scenarios. It has a very limited pick-up range and, when inside a room, produces audio that sounds like the person speaking has a bucket over his head. This microphone picks up the grinding sound of the zoom lens motor, the rubbing or knocking sounds of the operator's

Figure 6-6. Using a pop filter protects the generating element of a microphone. The generating element may be damaged by the sudden rush of air created when speaking or singing "t" and "p" sounds. (Popless Voice Screens)

fingers and hands operating the camera, and the sound of the operator breathing. None of these are components of quality audio.

Specialized Microphones

The *boundary mic* is most commonly a condenser type, previously described. Boundary mics are becoming the most common way to mic an entire stage or large room. These mics do not look like any others that most consumers commonly see. They work on the principle that sound is reflected off hard surfaces, and are usually placed on a table, floor, or wall, **Figure 6-7**. They have a very low profile, rising no more than an inch above the surface they are placed upon.

A *parabolic reflector mic* is a very sensitive mic that looks like a satellite dish with handles, **Figure 6-8**. This type of mic is designed to pick up sounds at a distance. The operator simply aims the mic at what he wants to hear and the sound is received very clearly. The pick-up range of a parabolic reflector microphone depends on the refinement of the electronics on the inside of the mic. The sensitivity of the electronics inside the mic is directly related to the cost—the more sensitive the electronics, the higher the purchase price.

Figure 6-7. The boundary mic is commonly used to mic a stage for a dramatic performance.

Figure 6-8. The parabolic reflector mic is capable of clearly picking up sounds from a significant distance.

The parabolic reflector microphone is often seen on the sidelines of professional football games and picks up the grunts and crashes of bodies slamming against each other, which add to the excitement of the game. When using these mics at a professional sporting event, an experienced operator knows when to turn off the audio feed from the mic. Some of the vocalizations it can pick up are not likely to be appropriate for prime-time television.

Pick-Up Pattern

Microphones are further classified by their pick-up pattern. *Pick-up pattern* refers to how well a mic "hears" sounds from various directions.

An *omni-directional mic* has a pick-up pattern that captures sound from nearly every (omni) direction equally well, **Figure 6-9**. The only weak area for this type of mic is the sound coming directly from the rear of the mic.

Mics with a *uni-directional* pick-up pattern pick up sound from primarily one (uni) direction, **Figure 6-10**. A uni-directional mic is also known as a *directional mic* or a *cardioid mic*. The term "cardioid" is derived from the shape of the pick-up pattern; it is shaped like a valentine heart. The point of the heart is aimed at the source of the sound. Sounds from the sides and rear of the mic are not heard as well, or not at all. The cost of a uni-directional mic is usually proportionately related to how far away it can pick up sounds,

pick-up pattern: A term that describes how well a mic hears sounds from various directions.

omni-directional mic: A mic with a pick-up pattern that captures sound from nearly every direction equally well.

cardioid mic: A mic with a pick-up pattern that captures sound from primarily one direction. Also called a *uni-directional mic* or *directional mic.*

Figure 6-9. The omni-directional mic picks up sounds from nearly all directions.

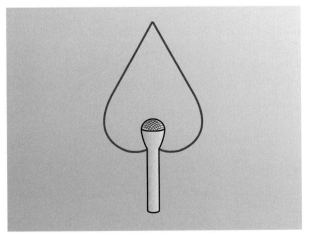

Figure 6-10. A uni-directional mic has a pick-up pattern in the shape of a heart, thus the name "cardioid."

and how much sound outside of the pick-up pattern is eliminated. The longer and narrower the pick-up pattern, the more expensive the microphone.

In a noisy environment, a directional mic is a better choice for a narrator or reporter than an omni-directional mic. If an omni-directional mic is used, the viewer may have a difficult time separating the talent's voice from the background sounds.

Directional mics are available in various degrees, or grades, of directionality, **Figure 6-11**. A *hypercardioid mic* has a narrower and longer pick-up pattern than a cardioid mic. A *supercardioid mic* has an even narrower pattern. More directional still is the *shotgun mic*. Sometimes, the shotgun mic is literally mounted on a rifle stock and even has a sight to help in aiming it at the sound source. The parabolic reflector mic is a version of a directional mic.

Directional mics are most important when recording music. If the performer uses an omni-directional mic and steps in front of the band's speakers, a high-pitched squeal is emitted. The squeal is called *feedback*. Feedback occurs when a microphone picks up the sound coming from a speaker that is carrying that microphone's signal, **Figure 6-12**.

A feedback loop is created when:

1. Sound enters the microphone.
2. The sound is transmitted to an amplifier.
3. The signal from the amplifier is sent to a speaker.
4. The sound from the speaker goes through the air and back into the microphone.

Each time this circle is made, the pitch gets higher and louder and more painful to the human ear. If the cycle is not stopped, the speakers could be

hypercardioid mic: A directional mic with a narrower and longer pick-up pattern than a cardioid mic.

supercardioid mic: A directional mic with a narrower pick-up pattern than a hypercardioid mic.

shotgun mic: A directional mic with an extremely narrow pick-up pattern.

feedback: A high-pitched squeal that occurs when a microphone picks up the sound coming from a speaker that is carrying that microphone's signal.

Figure 6-11. The cardioid mic can be purchased with varying degrees of narrowness in its directional pick-up pattern.

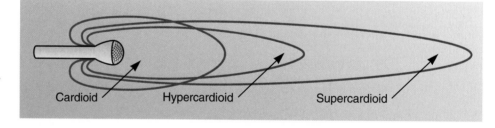

Figure 6-12. Feedback occurs when a mic "hears" its own signal.

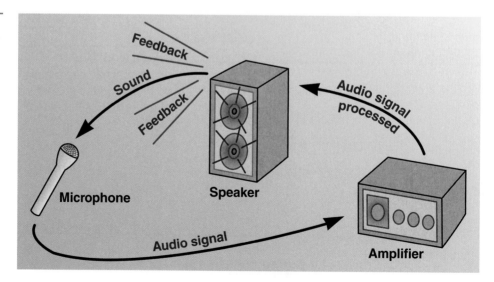

permanently damaged. To prevent feedback, the sound coming from the speakers needs to be blocked from hitting the mic. Using a directional mic decreases the likelihood of feedback because of its narrow pick-up pattern.

PRODUCTION NOTE

To stop the high-pitched squeal of feedback:
- Move the mic away from the speaker.
- Bury the mic in your armpit.
- Turn down the amplifier.
- Turn the speaker away from the mic.

Determine what caused the feedback and take precautions to ensure it does not happen again.

Mics on the Set

A mic stand is the most commonly known device designed to hold a microphone in place. In television, however, the mic stand is not often seen. A talk show host may have a microphone on a desk stand, but that mic is often only a prop.

A *hand-held microphone* is designed to be held in the hand, rather than placed on a stand or clipped to clothing, but can be placed on a stand or boom. Hand-held mics are sometimes referred to as a *stick mics*.

A *boom* used in television production is essentially a pole that is positioned over the set with a microphone attached to the end of the pole. The mic picks up the sound of the talent performing on the set. Any type of mic can be attached to the end of the boom as long as the connectors will mate. The goal of the boom operator is to get the mic as close to the talent as possible without dipping the mic into the top of the picture. A *fishpole boom* is a type of boom that must be physically held over the heads of talent, **Figure 6-13**. This requires that one or both arms be extended over the operator's head and held in position for the duration of the shot.

hand-held mic: A mic that is designed to be held in the hand, rather than placed on a boom or clipped to clothing. Also called a *stick mic*.

boom: A pole that is held over the set with a microphone attached to the end of the pole.

fishpole boom: Type of boom that must be physically held over the heads of talent.

PRODUCTION NOTE

If using a fishpole boom, most operator's arms tire quickly and the mic begins to dip into the frame of the picture. A simple solution is to obtain a microphone stand that can be raised to a height of 6–8 feet. Take the fishpole boom to an audio store and purchase a microphone stand mic clip that fits the shaft of the fishpole boom. Place the boom into the mic clip on the stand, positioned about 4 feet from the back end of the pole. Let the stand be a fulcrum to bear the weight of the pole, and you can adjust the height of the mic by swiveling and raising or lowering the mic-end of the boom. No more tired arms!

Large studios have the same type of stand for booms, except the stand is often more like a camera tripod on a dolly and can roll easily. If you have a spare tripod and dolly, you may be able to be creative and devise something to hold the fishpole boom securely onto the top of the tripod.

Figure 6-13. A mic can be positioned over a set using a fishpole boom. (K-TEK)

Talk the Talk

Many students ask about a "boom mic" during class discussion. No such piece of equipment exists. If you were to walk into a professional audio store and ask where the boom mics are located, the staff would have no idea what you are asking for. Any mic that is attached to a boom is a boom mic. Therefore, the phrase "boom mic" refers to *how* a mic is used and not a piece of equipment.

lapel mic: The smallest type of mic that can be worn by talent and is attached to clothing at or near the breastbone with a small clip or pin. Sometimes referred to as a *lav*.

Mics on Talent

The smallest mic worn by talent is the *lapel mic*, sometimes called a *lav*. It is attached at or near the breastbone of the talent with a small clip or pin, **Figure 6-14**. The cord is routed under the clothing to be less obvious. The most common lapel mic is about the size of a pencil eraser.

Figure 6-14. A lapel mic is quite small and can be attached to the talent's clothing. A—The mic cord is run under or behind a piece of the talent's clothing. B—The appearance of the mic on the front of the talent is discrete.

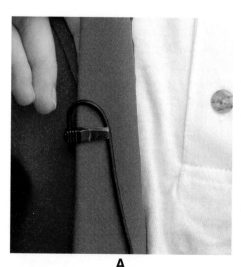

A

B

Handling and Care of Microphones

Proper microphone etiquette requires that reasonable judgment be used to ensure longevity of the audio equipment and the safety of staff and talent. This applies to everyday use and handling, and the storage of equipment. Microphones should be handled very carefully.

PRODUCTION NOTE

Any noise or action that can damage a human eardrum may also damage a microphone. Being slapped in the ear with a cupped hand is the equivalent of slapping the head of a microphone. This action could burst the eardrum. Treat a microphone the same way you would treat the ears of someone you care for.

Below are a few general guidelines:
- Never blow into a microphone to see if it is working. A strong burst of air can damage or tear the microphone's diaphragm, just as it can damage or permanently impair the human eardrum.
- Do not shout into a microphone. Extreme sound vibrations can stretch the diaphragm out of shape, just as these vibrations can stretch the tissue of the eardrum. This causes a mic to receive sound vibrations improperly, if at all. The tissue of the human eardrum can be stretched by the extreme sound vibrations at a loud concert, which results in temporary hearing difficulty for hours after the concert.
- Never let anyone put their lips directly on the mic. The saliva that enters the mic moistens and softens the diaphragm. This obstructs the microphone's ability to receive sound vibrations just as sound vibrations are muffled to the eardrum when water is trapped in the ear after swimming.
- Do not slap the head of the microphone to hear the muffled thump through the speakers. The increased air pressure can tear the microphone's diaphragm, just as being hit with a cupped hand over the ear can burst the eardrum.
- Do not exhale directly into or inhale through the microphone. Exhaling into the mic forces moisture in and that moisture softens the diaphragm. Inhaling through the mic transfers all the bacteria inside the mic to you through your mouth. Inhaling through the mic also creates a loud hiss in the sound reproduced.
- Never swing a mic by its cord. The centrifugal force created can easily separate wired connections inside the mic cable connector and prevent any sound from being reproduced.

Proper Use of Microphones

Many amateur bands use mics that have a silver or black ball of mesh on the tip, **Figure 6-15**. Beneath the wire mesh is usually foam that protects

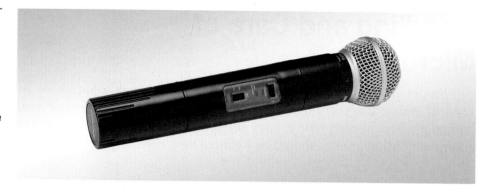

the diaphragm of the mic. The foam provides a barrier to moisture and rushing air when a performer places their lips directly on the mic. Speaking or singing with your lips directly on a mic is not necessary for sound reproduction, but it is a style commonly seen in popular music videos. Amateur bands, trying to emulate popular bands, imitate this style while performing. Singing enthusiastically with lips pressed on the mic pushes saliva into the black foam under the surface of the wire mesh mic tip. After a performance, all of the mics are typically packed up and stored until the next practice or show. The moisture on the foam surface is stored in the dark, at room temperature, and away from any airflow. This is the perfect environment for bacteria growth. From practice to practice or show to show, it is very unlikely that the same mic will be used consistently by the same person. The next band member to use the mic will place his lips directly on the bacteria-infested mic while singing. That person will very likely exhale and inhale right through the mic while performing. In an effort to keep the performers and equipment germ-free and healthy, do not allow talent to place their lips directly on the surface of a microphone.

When using a hand-held mic, hold the mic firmly in your fist and keep your hand and fingers still. Moving or adjusting your fingers produces a very distracting sound that is picked up by the mic. Place the knuckle of your thumb against the sternum of your chest to properly position the mic while reporting, **Figure 6-16**. A common mistake made by novice reporters when interviewing is to point the mic at the subject/guest when asking a question, and then point the mic at themselves as the guest answers. Even though this is backwards, it is a very easy mistake to make.

When interviewing children, do not stand over them. Being on their level creates a much more pleasing picture and children are less likely to be intimidated when an adult is physically at their level. A reporter should squat down so that their head is at or below the level of the child's head. The child can also be raised up to the reporter's level using a stool to achieve the same effect.

When running cables on the ground, never place an audio or mic cable beside an electrical cable. Electrical cables produce magnetic fields, which can cause interference in the audio signal. The interference may be detected as a persistent hum heard through the audio system. In some circumstances, it may be unavoidable for an audio cable to be near or have to cross a power cable. Keep the runs of cable apart for as much of the length as possible. Limit the portions of cabling that are close in proximity to as few as feasible. If the cables must intersect, make sure it is at a 90° angle.

Figure 6-16. This reporter demonstrates the proper position for a hand-held microphone when speaking.

Impedance

In television, a microphone's purpose is to create a signal that is sent through a cable to be recorded. There are two different kinds of signals that mics can send: high impedance and low impedance. A *high impedance (HiZ)* mic is typically inexpensive, low-quality, and cannot tolerate cable length much longer than 8′. For these reasons alone, high impedance mics are not usually found in TV studios. A *low impedance (LoZ)* mic is typically of high-quality, more costly than a HiZ mic, and can tolerate long cable lengths.

Levels

It is important to know the three levels of audio, **Figure 6-17**, because the output of one level cannot be connected into the input of another. The result is either massive distortion or no sound at all. The three levels of audio are:

- *Mic level*. The level of audio that comes from a microphone. It is designed to be sent to the "mic in" on a recorder or mixer.

high impedance (HiZ): A type of mic that is typically inexpensive, low-quality, and cannot tolerate cable lengths longer than 8′.

low impedance (LoZ): A type of mic that is costly, high-quality, and can tolerate long cable lengths.

mic level: The level of audio that comes from a microphone. It is designed to be sent to the "mic in" on a recorder or mixer.

Figure 6-17. The three levels of audio are mic level, line level, and power level.

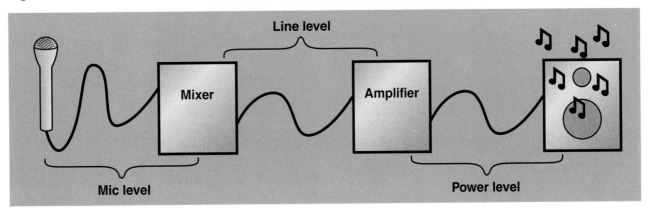

line level: The level of audio between pieces of audio equipment. For example, the level of audio going from the output of a CD player to the input on an amplifier.

power level: The audio level from the output on an amplifier to the speaker.

mic mixer: A piece of equipment that combines only the microphone signals into a single sound signal.

audio mixer: A piece of equipment that takes the sounds from a variety of sources, such as mics, a CD player, or tape player, and combines them into a single sound signal that is sent to the recorder.

- *Line level.* The level of audio between pieces of audio equipment. For example, the level of audio going from the output of a CD player to the input on an amplifier.
- *Power level.* The audio level from the output on an amplifier to the speaker.

Mixers

A *mic mixer* combines only the microphone signals into a single- or dual-channel sound signal. An *audio mixer* is designed to take the sounds from a variety of sources, such as mics, a CD player, or tape player, and combine them into a single sound signal that is sent to the recorder, **Figure 6-18**.

In both the mic mixer and audio mixer, each signal coming into the mixer can be controlled with a *potentiometer*, or *pot* for short. The operator can increase or decrease the strength of each signal, so each audio source is properly balanced in the output signal. A pot is usually a knob or a slider. As a knob, it functions like the volume knob on a stereo. The signal coming in gets stronger as the knob is turned to the right. If the control is a slider, the signal coming in gets stronger as the slider is moved farther away from the operator.

Figure 6-18. The audio mixer has a different pot for each audio input. The volume unit (VU) meters indicate the levels of the audio.

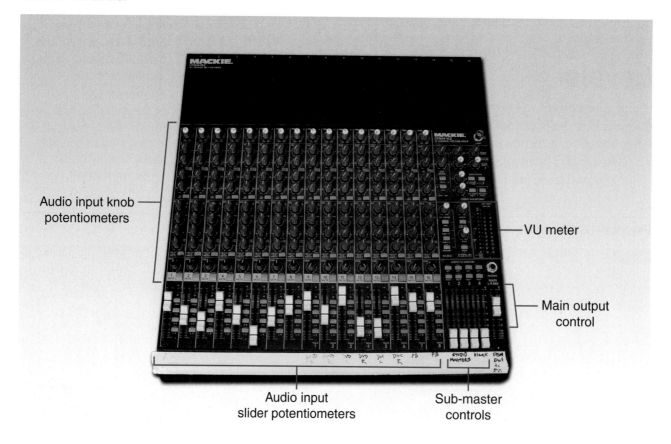

Audio input knob potentiometers

VU meter

Main output control

Audio input slider potentiometers

Sub-master controls

Figure 6-19. In analog recording, the audio meter should fluctuate between −3 and +3. In digital recording, the audio meter should hover in the area of −20.

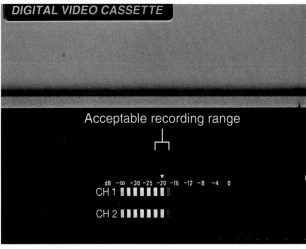

A   B

When using either of the mixers, the operator gauges the signal strength by watching a *VU (volume unit) meter*, **Figure 6-19**. VU meters take two forms:

- One type looks similar to a car's speedometer, with a scale and a needle to indicate the signal strength.
- Another type is a series of LEDs that light as the signal gets stronger.

potentiometer (pot): A knob or a slider control that regulates the strength of a signal.

volume unit meter (VU meter): A meter on either an audio or mic mixer that indicates signal strength.

PRODUCTION NOTE

Do not assume that red LEDs on a VU meter signify over-recorded sound. Different manufacturers use different colored LEDs to mean different things. Always check the operator's manual for your mixer.

To operate either mixer type:

1. Activate the sound source. Play a CD or instruct the talent to talk.
2. Bring up a single pot until the sound reaches a desired level on the VU meter.
3. Repeat for each individual microphone or sound source. All the pots are typically not adjusted identically. The pot running a naturally loud voice is set much lower than the pot for a soft-spoken person's microphone. All background music should be relatively low or it drowns out the dialogue in the scene.
4. Bring up the pot labeled "master" to send the mixed signal out of the mixer to the recorder.

By adjusting the potentiometer, the audio engineer makes certain that the audio signal is appropriately strong. If the system uses analog technology, the master VU meters should fluctuate between −3 and +3 dB. If the system is digital, the VU meters should hover near −20 dB.

PRODUCTION NOTE

When monitoring the master VU meters, the needle should never touch the far right edge of the meter. This is called "burying the needle," or recording "in the mud." Burying the needle means the audio is being over-recorded. Over-recorded audio is distorted. This *cannot be fixed* in post-production. Over-recording or under-recording the audio when shooting raw footage is an error that a professional usually makes only once. This mistake requires that everyone involved in the production reconvene to re-shoot the otherwise perfect footage. Calling everyone back for a re-shoot and gathering all the necessary equipment is so expensive that most people do not make this mistake a second time.

It is critically important for whoever is recording the audio to wear a good set of headphones that cover the entire ear, like earmuffs. This is the only way to accurately monitor the quality of the audio being recorded. The small foam "ear bud" style or collapsible earphones many people use with their personal audio players are totally inadequate for this purpose.

PRODUCTION NOTE

A confusing situation that a novice audio technician may encounter involves recording a stage performance. Do not take a direct feed off the audio board of the theater, if offered by the theater's audio engineer. Set up your own microphones on the stage. While connecting to the audio board is the easier option, you run the risk of recording poor quality audio. If recording the audio from the theater's audio board:

- The recording has only the sounds picked up by the audio mixer's microphones. Any performers who are not specifically speaking or singing into a microphone are not heard at all. The same holds true for any instruments not playing into a mic. The live audience hears all the instruments, but the audience of the recording hears only what is played into a microphone.

- The audio mixer mixes the signals from the mics and sends them to an amplifier to be sent out over the theater's speakers. The purpose of the theater's audio mixer is to reinforce and amplify the sounds, so the live audience hears them well. The live audience hears a blend of live, electronically mixed, and amplified sounds. When recording a performance, your purpose is to record all the sounds in the theater. This is not possible when the mixer sends only some of the sounds—just those picked up by microphones.

- Any mistake made by the audio mixer is clearly evident in your recording, but you will bear the responsibility and blame for the audio quality.

automatic gain control (AGC): A circuit found on most consumer video cameras that controls the audio level during the recording process.

Automatic Gain Control

The *automatic gain control (AGC)* is a circuit found on most consumer video cameras that controls the audio level during the recording process. If the sound is soft, the AGC turns the recording levels up, while also bringing up a noticeable tape hiss and background noise. If the sound is loud, the AGC turns the recording levels down. While this function sounds very helpful, the circuit is always about a second behind "real life." Using the AGC should be avoided in most analog recording situations.

The AGC circuit works quite well when it is part of a digital camcorder, whether consumer or professional. Because the AGC circuit used with digital camera recording technology operates much faster than with analog technology, the undesired tape hiss is not recorded. Therefore, disengaging the AGC circuit on a digital camcorder is not always necessary.

Wrapping Up

Microphones can be classified in many ways. Understanding the features of each classification is critical in choosing the right microphone for a specific application or shooting scenario. The following chart summarizes the mic characteristics and classifications discussed in this chapter.

Classifying Microphones	
Mic Characteristic	**Options**
Frequency	Low, Mid-range, High
Cabling	Wired, Wireless
Type	Dynamic, Condenser, Ribbon, Specialized (boundary and parabolic reflector)
Pick-up Pattern	Omni-directional, Directional/Uni-directional (cardioid, hypercardioid, supercardioid, and shotgun)
Mounting	Hand-held, Lapel (lav)

Review Questions

Please answer the following questions on a separate sheet of paper. Do not write in this book.

1. What is the difference between background sound and nat sound?
2. Explain how microphones work.
3. List five types of microphones available and the unique characteristics of each.
4. How does feedback occur? How can it be prevented?
5. What is the difference between high impedance and low impedance?
6. What are the acceptable VU meter readings for analog audio systems and for digital audio systems?

Activities

1. Create illustrations that demonstrate the pick-up patterns of omni-directional microphones, uni-directional microphones, and supercardioid microphones.

STEM and Academic Activities

 Science

1. Investigate how sound frequency is measured. What is the audible range of sound frequencies? Make a list of sounds common in your everyday environment. Categorize the sounds as low, medium, or high frequency.

 Mathematics

2. Research the prices for various studio-quality microphones and compare the costs of different types of microphones. Why are some microphones more expensive than others?

 Social Science

3. Find clips of comments made by celebrities or politicians when they thought their mics were turned off. Prepare a presentation of some of the embarrassing moments you discover.

4. Watch a scene from one of your favorite movies. Pay particular attention to any background sound, nat sound, or room tone in the scene. Have you ever noticed those sounds in the scene before? What kind of feelings do these sounds emphasize? How do you think the scene would be different if those sounds were not present?

 Language Arts

5. Record a few video clips of action only. Add voiceover narration to describe the action and events in the clips.

www.aes.org The Audio Engineering Society is a professional audio technology society whose membership includes audio engineers, creative artists, scientists, and students.

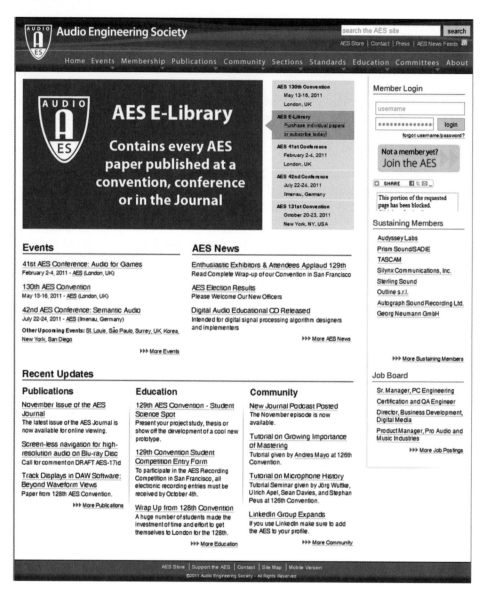

Chapter 7

Connectors

Objectives

After completing this chapter, you will be able to:

*   Explain how connectors and adapters are used in the broadcast industry.
*   Identify the types of connectors used in the broadcast industry.

Introduction

The names of the various connectors must be learned and used correctly. In the remaining chapters of this textbook, connectors are referenced by name. Having access to connectors, cables, and electronic devices will reinforce your understanding and recognition of the connectors and adapters used in the broadcast industry. Upon completion of this text, it is recommended that you review the connector names and descriptions in this chapter before beginning to work with production equipment in earnest.

Professional Terms

⅛″ connector
¼″ connector
adapter
barrel adapter
BNC connector
cable end connector
cannon connector
chassis mount connector
connectors
DIN connector
F-connector
female connector
FireWire
HDMI
jack

male connector
mini connector
phone connector
phono connector
PL259 connector
plug
RCA connector
S-VHS connector
T-connector
USB
XLR connector
Y/C connector
Y-connector

Figure 7-1. All connectors are either permanently attached to a piece of equipment (chassis mount), on the end of a cable (cable end), or designed to change one connector type into another type (adapter).

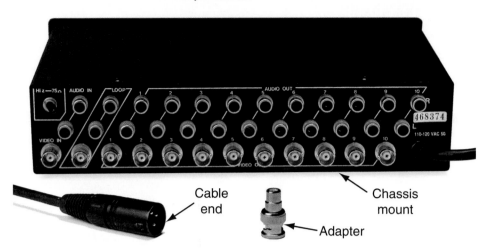

Cable end

Chassis mount

Adapter

Connectors

connectors: Metal devices that attach cables to equipment or to other cables.

chassis mount connector: A connector that is built into a piece of equipment.

cable end connector: A connector found on the end of a length of cable.

adapter: A connector that changes the type, or connector end, of an existing connector.

plug: A connector with one or more pins that are designed to fit into the holes of a jack (female connector). Also called a *male connector.*

jack: A connector with one or more holes designed to receive the pins of a male (plug) connector. Also called a *female connector.*

BNC connector: A type of connector commonly used in television production. The female and male versions lock together securely with a simple ¼-turn twist.

Connectors are metal or metal and plastic devices that attach cables to equipment or to other cables, **Figure 7-1**. Any cable can carry any video or audio signal, as long as the cable is adapted for the necessary connector. A connector can be classified into only one of the following categories:

- *Chassis mount connectors* are built into a piece of equipment.
- *Cable end connectors* are on the end of a length of cable.
- *Adapters* change the type, or connector end, of existing connectors.

One type of connector end is referred to as the *plug*, or the *male connector*. The other connector is a *jack*, or the *female connector*. Male connectors have one or more pins that are designed to fit into the holes of a female connector. Female connectors have one or more holes designed to receive the pins of a male connector.

Talk the Talk

When referring to these types of connectors aloud, only the connector name is used: BNC, DIN, ¼", mini, or Y/C. "Please bring a female RCA to male PL259." The words "connector" or "adapter" are understood and, therefore, not actually spoken when industry professionals use these connector terms.

BNC Connector

The female and male versions of a *BNC connector* lock together securely with a simple ¼-turn twist. A BNC is the most common connector used in television production, **Figure 7-2**. BNC actually stands for British Naval Connector, but this connector is simply referred to as a "BNC."

Figure 7-2. This barrel adapter consists of 2 female BNCs.

DIN Connector

"DIN" is a generic term that refers to any connector with four or more holes/pins. *DIN connectors* are also commonly found on computer equipment cables. In television production, use this term only if there is no other more exact term available.

DIN connector: A term that refers to any type of connector with four or more holes/pins.

F-Connector

The *F-connector* (**Figure 7-3**) comes in two styles: push-on and professional. With the push-on style connector, the male end simply pushes onto the female connector. The professional style F-connectors have a small nut that secures the male and female ends together. BNC connectors have almost completely replaced F-connectors in the industry. While an F-connector requires tedious manipulation of a tiny nut to secure it, a BNC requires only a ¼-turn. F-connectors are commonly found on the back of consumer VCRs and TVs. The female chassis mount connectors are marked "Ant. In" and "Ant. Out." The corresponding male F-connector is on the cable running from the VCR to the cable box or television.

F-connector: A type of connector that carries an RF signal and is commonly found on the back of consumer VCRs and televisions.

Figure 7-3. This cable end connector is a male F-connector.

Phone Connector

A *phone connector* is ¼" in diameter and single-pronged, with a little indentation near the end of the prong (**Figure 7-4**). It is also called a

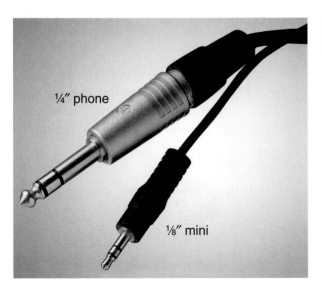

¼" phone

⅛" mini

Figure 7-4. 1/4" phone and 1/8" mini connectors.

phone connector: A connector that is ¼" in diameter and single-pronged, with a little indentation near the end of the prong. This type of connector is commonly found on the cord used with large stereo headphones. Also called a ¼" *connector.*

¼"*connector.* Most large stereo headphones have a phone connector at the end of the cord to connect with the stereo equipment.

Talk the Talk

When verbally referring to a phone connector, it is very common to drop both the words "inch" and "connector." For example, "Please bring me a male quarter to a male quarter 6 feet long."

Mini Connector

A *mini connector* is ¼" in diameter and is single-pronged, **Figure 7-4**. It is also often called a ⅛" *connector*. The mini looks similar to the ¼"connector, but is smaller in size. This type of connector is most commonly found on headsets for portable CD players, iPods, and MP3 players.

Phono Connector

Many of the components of a home entertainment system usually have *phono connectors*, or *RCA connectors*, on the back (**Figure 7-5**). They are labeled "audio in," "audio out," "left," and "right." The female phono connector is usually a chassis mount connector, with the male then being a cable end connector. The male has a single center prong surrounded by a shorter crown.

Figure 7-5. A female RCA cable end connector.

PL259 Connector

The *PL259 connector* is similar to the F-connector, but is much larger. A CB radio is very likely to have a PL259 connector for the antenna. The male end has a single prong with a nut to tighten, like the F-connector, **Figure 7-6**. The nut is larger and easier to handle, but must still be turned

Figure 7-6. A male PL259 to female BNC adapter.

many times to secure the connection. The PL259 has also been replaced, for the most part, by BNC connectors.

Y/C Connector

A *Y/C connector* has four tiny round pins and a rectangular plastic stabilizing pin. This connector may be nickel or gold plated. Because of the fine pins, it is a fragile connector for video inputs and outputs. The consumer term for Y/C connectors is *S-VHS connectors* or S-connector.

XLR Connector

An *XLR connector* is usually a 3-pin connector for microphones, but can be 4-pin, 5-pin, and other pin configurations, **Figure 7-7**. This connector may also be called a *cannon connector*. When it is a 3-pin connector, it is simply called an "XLR connector." If there are more than 3 pins, it is referred to by the number of pins. For example, "4-pin XLR" or "5-pin XLR." The advantage of an XLR connector is that the male and female ends fit together and a hook automatically locks the two together, **Figure 7-8**. The male and female ends do not separate once locked. Depressing a button on the connector disengages the hook, and the ends separate easily.

Y/C connector: A video input and output connector that is characterized by four tiny round pins and a rectangular plastic stabilizing pin.

S-VHS connector: The consumer term for a Y/C connector.

XLR connector: A connector for microphones that usually has 3-pins, but can have 4-pins, 5-pins, and other pin configurations. The male and female ends lock together with a hook. Also called a *cannon connector*.

Figure 7-7. A female XLR (circled) to male ¼″ phone adapter.

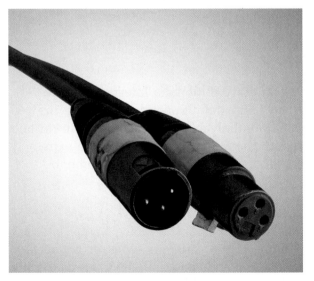

Figure 7-8. Male (left) and female (right) XLR connectors.

4-pin FireWire

6-pin FireWire

9-pin FireWire (FireWire 800)

FireWire Connector

FireWire connector: A type of connector designed to carry digital signals and available with 4-pin (audio and video only) and 6-pin (audio, video, and power) connections. This connector is also known as *IEEE 1394.*

FireWire is made of copper cable and is available with 4-pin (audio and video only), 6-pin (audio, video, and power), and 9-pin (audio, video, and power) connections, **Figure 7-9**. This connector is also known as IEEE 1394. FireWire connectors and cables are designed to carry digital signals.

USB Connector

USB connectors: A durable digital connector for video, audio, and power. The electrical contacts are enclosed in a metal housing and buffered by a plastic plate.

A *USB* is a digital connector for video, audio, and power, **Figure 7-10**. This type of connector is most often found on computers. The USB is durable connector, with the electrical contacts enclosed in a metal housing and buffered by a plastic plate.

Figure 7-10. A USB connector.

HDMI Connector

HDMI connector: This connector is designed to carry high definition video and audio, as well as power. HDMI stands for High Definition Multimedia Interface.

HDMI stands for High Definition Multimedia Interface, **Figure 7-11**. This connector is designed to carry high definition video and audio, as well as power. HDMI connectors are commonly found on consumer Blu-ray and high definition DVD players, but the HDMI connector is also extremely common on all professional, high definition production equipment.

Adapters

An adapter is used to connect two different types of connectors, **Figure 7-12**. For example, a cable that ends in a male BNC needs to plug

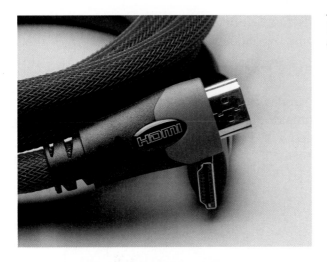

Figure 7-11. HDMI cable end connectors.

Figure 7-12. Several types of adapters.

Female BNC male RCA   **Female PL259 to male BNC**   **Female RCA to male BNC**

into a female RCA chassis mount connector. Using a female BNC to male RCA adapter, the two connectors can be joined. Most studios have a wide variety of adapters, with almost every conceivable combination available.

PRODUCTION NOTE

It is best to have as few adapters as possible in runs of cable. A little bit of the signal is lost at every adapter connection in the run. Think of it as a leaky hose—each adapter in the run is like poking another hole in the hose.

T-Connector

A *T-connector* is a special kind of connector that takes its name from its shape; it looks like the capital letter *T*, **Figure 7-13**. It is made of metal and

Figure 7-13. This arrangement of this T-connector is two female BNCs to male BNC.

T-connector: A connector that is shaped like the capital letter *T* and is made entirely of metal. The three ends of a T-connector are used to split one signal into two signals, or to combine two signals into one.

does not flex. The three ends of a T-connector are used to split one signal into two signals, or to combine two signals into one.

Y-Connector

A *Y-connector* is very similar to the T-connector, except it has three connectors separated by wires or all molded together. In either case, the device looks like a capital letter *Y*, **Figure 7-14**. A Y-connector serves the same function as the T-connector.

Y-connector: A connector that has three wires with a connector on the end of each. All the wires are tied together in the middle. The three ends of a Y-connector are used to split one signal into two signals, or to combine two signals into one.

Figure 7-14. A two female RCAs to male RCA Y-connector.

Barrel

A *barrel adapter* has the same type of connector and connector end on both sides, **Figure 7-15**. The barrel has a specific purpose—it allows two identical short cables to be connected together and make one long cable. For example, suppose you need a cable with a male RCA on each end to connect two pieces of gear together. The problem is the two pieces of gear are 10 feet apart and you have two 6-foot cables with male RCAs on each end. If you have a female RCA barrel adapter, you can connect the two cables to each other creating a 12-foot cable with a male RCA on either end. Problem solved!

barrel adapter: A type of adapter that has the same type of connector and connector end on both sides.

Figure 7-15. This barrel adapter has 2 female RCAs.

Wrapping Up

Knowing the appropriate use of and name given to each of the connectors is an important step in learning how to record quality audio. The importance of memorizing the names of all the connectors cannot be stressed too much. If a supervisor asks a P.A. to obtain a certain cable or adapter from the storage area, the P.A.'s job may actually depend on bringing back the correct item.

Review Questions

Please answer the following questions on a separate sheet of paper. Do not write in this book.

1. List the three categories of connectors.
2. What is the difference between male and female connector ends?
3. Which connectors combine multiple signals into one?
4. Which connectors are specifically designed to carry high definition and digital signals?
5. How are adapters different from other connectors?

Activities

1. Inspect the connectors on various pieces of electronic equipment in your home. List several of the items and identify the type of connector(s) used with each.

STEM and Academic Activities

 1. Connectors are made of or contain metal to conduct signals between the cables they connect. Research the conductivity of various metals. Which metals are the best conductors? Which metals are commonly used in A/V cables and connectors?

 2. What do you think the next generation of A/V connectors will look like? How will they operate? Write a summary of your vision for the next generation of A/V connectors and sketch a prototype.

 3. For one day, make note of every time you use a connector of some type (cell phone charger, mp3 player dock or headphones, portable DVD player, gaming system controller, etc.). You may be surprised at the number of connectors involved in your daily life!

www.studenttelevision.com The Student Television Network aims to support and enhance broadcasting and video production education in schools by providing a network of students and instructor resources and activities.

Chapter 8

Scriptwriting

Objectives

After completing this chapter, you will be able to:

- Identify each of the program formats presented and summarize the unique characteristics of each.
- Identify the expected components of a program proposal.
- Explain the format of a program treatment.
- Recall the elements in each type of script used in television production.

Introduction

Too many students experience anxiety when they hear the word "writing." One of the best things that can be said about television scriptwriting is that it bears little resemblance to the writing style required for academic courses. Although scriptwriting is relatively simple to do, *good* scriptwriting takes talent and skill.

Professional Terms

actors
big talking face (BTF)
concert style music video
documentary
drama
format script
interview
lecture
lecture/demonstration
magazine
montage
music video
newscast

nod shots
outline script
panel discussion
program proposal
public service
  announcement (PSA)
script
story style music video
storyboards
talking head
treatment
visualization
word-for-word script

Program Formats

A *script* is an entire program committed to paper. It includes dialog, music, camera angles, stage direction, camera direction, computer graphics (CG) notations, and all other items that the director or scriptwriter feels should be noted. There are many different kinds of television programs, each with unique requirements of the script. Most programs fit into one of the following categories: lecture, lecture/demonstration, panel discussion, interview, documentary, newscast, magazine, drama, public service announcement/ad, and music video.

Lecture

The *lecture* program format is the easiest format to shoot—the talent speaks and the camera shoots almost entirely in a medium close-up. All that is needed for this format is the talent, a camera, and perhaps a desk or podium for the talent to sit or stand behind, **Figure 8-1**. Other names for the lecture format are *BTF (big talking face)* or *talking head*. The lack of either camera movement or talent action creates a very dull and uninteresting program. This format has the lowest viewer retention of information and is often the mark of an amateur production team.

Lecture/Demonstration

The *lecture/demonstration* format lends itself to the numerous cooking shows, how-to shows, and infomercials seen on television today. This format is more interesting to watch than a lecture alone because of the action and many props used by the performers, **Figure 8-2**.

---

script: An entire program committed to paper, including dialog, music, camera angles, stage direction, camera direction, and computer graphics (CG) notations.

lecture: A program format in which the talent speaks and the camera shoots almost entirely in a medium close-up. Also known as *big talking face (BTF)* and *talking head*.

lecture/demonstration: A program format that provides action and makes use of props in addition to lecture. Examples of this format include cooking shows, how-to shows, and infomercials.

Figure 8-1. A single individual speaking from behind a podium provides little visual interest or action. Because of this, the lecture format has the lowest viewer retention rate of all the programming formats.

Figure 8-2. The lecture/demonstration format adds action that corresponds to the lecture and is more interesting for viewers to watch.

Panel Discussion

The many Sunday morning network programs that bring a group of professionals together to discuss current news and political topics are examples of the *panel discussion* format. Also included are the popular daytime talk shows. These programs are not difficult to produce, as long as there are a limited number of people on the panel, **Figure 8-3**. Panel discussions are driven by the program's content, not action. As more people

panel discussion: A program format that presents a group of people gathered to discuss topics of interest. Daytime talk shows are an example of this format.

Figure 8-3. The panel discussion format is relatively easy to shoot and provides viewers with interesting information, depending on the talent and the topic.

are added to a panel discussion, the group shot to include all members gets rather wide. A wide shot is also a tall shot, which increases the risk of shooting off the top of the set. To keep the top of the set in the shot, the camera may need to tilt down and inadvertently make the studio floor the most prominent item in the picture. As the industry moves more and more to shooting in 16:9 screen format, one or two people can be added to the ends of the panel. The 16:9 screen shape provides more width and less height than the 4:3 screen format.

Interview

On location or in the studio, the two-person interview can be electrifying. People like Barbara Walters and Oprah Winfrey have built entire careers on making a simple conversation a compelling program for the audience. The *interview* format is often shot with only one camera. To get various camera angle cuts between the interviewer and the interviewee, the interviewee is shot for the entire duration of the interview. The audio picks up the questions asked by the interviewer, but the camera only shoots the interviewee's face. After the interviewee has left the set, the camera shoots the interviewer asking the same questions a second time and records some *nod shots*. Nod shots are a special kind of cutaway (discussed in Chapter 19, *Production Staging and Interacting with Talent*). The interviewer does not say anything, but simply "nods" naturally as if listening to the answer to a question. When collecting nod shots, the interviewer faces the direction where the interviewee was positioned during the interview. Nod shots are critical to the editing process when an interview, which may have originally taken 30 minutes, must be cut to 12 minutes in order to fit into a time slot between spots. In the editing room, the angles and nod shots are cut together to create what looks like a conversation between the interviewer and interviewee.

Documentary

A *documentary* program is essentially a research paper for television. The program topic is researched, the information is outlined, and the script is written. See **Figure 8-4**. The audio in a documentary may be either off-camera narration, on-camera narration, or a combination of both. A documentary may also contain interviews. The audio portion of the script should be roughly written out before any shooting begins. In the process of writing, a shot sheet is developed. For a documentary program, a shot sheet is like a grocery list of shots needed to support the audio portion of the script. In addition to capturing the shots on the list while shooting, the director watches for other shots that include specific items, people, or anything that adds to the program's content and would be interesting to the viewers. Shot sheets are only a guide and are rarely long enough to provide enough footage to assemble an entire program. Always shoot more footage than is listed on the shot sheet.

Newscast

By definition, a *newscast* program is a collection of individual news stories. Each story within the program may be developed with a different

Figure 8-4. A documentary script combines research information on a topic and shots that support the information presented.

Documentary Script	
Video	**Audio**
592a Channel 9 sign	Television
592b Channel 7 sign	Production trains
592c Channel 4 sign	students for
593 DW TV studio sign on door. Door opens on studio in production	entry-level positions in television studios as production assistants.
594 Shot of SEG, tilt up to monitors	It also provides students with a greater
595 Waveform adjustment	hands-on background than most colleges
596 Shoot studio camera viewfinder. ZO rack focus to interview in studio set	offer. Students write, direct, shoot, edit,
597 Operate editors	and deliver their own programs.
598 Focal Point title on CG, run title program	A 30-minute program is produced by students
599 Music video clip	each week for the Fox Cable
600 Passive switcher	System.
601 Dark studio, switch on lights, light board in foreground	We produced the program you are watching right now.
602 A crew shooting a program on location	Location shooting with portable equipment is
603 Loading a car with equipment	a favorite of the students.
604 Drum solo tape	So are music videos.
605 Rayburn music video	Students may work for
606 Channel 10 control room	Channel 10 while taking the class.
607 Wedding	We frequently accept jobs working for the
608 Floor manager gives cue	community as fund-raisers. The students
609 Open barn doors	even earn a salary.
610 Operate audio mixer	The class is run like a real video production
611 Director talks into headset. Shot from studio into control room. ZO to see studio camera perform pan to aim at "us"	company, so student responsibility and dependability are strongly emphasized.
613 AFI book	Students in this class are considered to be college-bound.
614 College survey form	The instructor provides considerable help
615 Place lapel mic on student	in matching student interest
616 Move platform	with schools of communications.
617 Hall of fame plaque	If you are interested in the lucrative,
618 "Digital Wave Productions" rolls up on screen. Student stops tape.	glamorous, and demanding field of
619 CU hands taking tape out of machine. Slow ZO. Hands place into case. MS of person smiling at camera and walking out of control room	Television Production, check us out.

Shot number

Camera direction

magazine: A program format comprised of feature packages, each addressing a different story for seven to eleven minutes.

drama: A program format that includes both dramas and comedies and requires actors to portray someone or something other than themselves.

actors: Individuals who participate in a drama or comedy program, performing as someone or something other than themselves.

public service announcement (PSA): A program that is 30 or 60 seconds in length and aims to inform the public or to convince the public to do (or not to do) something in the interest of common good.

script style, but the overall program has its own script. As a script style, news is practically in a category all of its own. News scripts are discussed in Chapter 9, *Broadcast Journalism*.

Magazine

The *magazine* format originated from programs like "60 Minutes," but has become more than news-oriented programming. A regular news broadcast presents each story in two minutes or less. A magazine format program is comprised of feature packages and each package addresses a different topic. This allows more interesting detail to be included about each story, but fewer stories to be included in each program.

Drama

This term includes both drama and comedy programming, **Figure 8-5**. The *drama* format requires a different kind of talent—actors. *Actors* take on a role in a program and perform as someone or something other than themselves.

Public Service Announcement (PSA)/Ad

Generally, *public service announcements*/ads are 30 or 60 seconds in length. The purpose of a PSA is to inform the public or to convince the public to do (or not to do) something in the interest of common good, **Figure 8-6**. A typical television ad, on the other hand, attempts to convince the public to purchase goods or services. Examples of some PSA themes include "Just Say No" (anti-drug), "Friends Don't Let Friends Drive Drunk," "Keep America Beautiful" (litter prevention and waste reduction), "Rock the Vote" (voter registration), "Get Caught Reading," "Buckle Up

Figure 8-5. A drama requires that talent with acting ability be used in the program.

Figure 8-6. A PSA provides the public with information or tries to persuade the public to do or not to do something.

America," and "Change Your Clock, Change Your Battery" (changing batteries in smoke alarms).

Music Video

The *music video* has become a common and influential force in our culture. Items such as clothing, shoes, fashion accessories, and hairstyles gain popularity when seen in a music video. Music videos also serve to promote a band or a new song or album, in the hopes of increasing sales of CDs and concert tickets. Most music videos are one of three types:

- *Concert Style Music Video.* The audience sees the band perform the music that is heard. A concert style music video may include a compilation of different concerts the band has performed, a studio performance, or various locations.
- *Story Style Music Video.* The audience hears the music, but never sees the band. Instead, actors act out a story line that is supported by the lyrics of the song.
- A hybrid of a concert style and a story style music video.

music video: A program format in which all or most of the audio is a song.

concert style music video: A type of music video in which the audience sees the band perform the music that is heard.

story style music video: A type of music video in which the audience hears the music, but does not see the band perform. Instead, actors act out a story line that is supported by the lyrics of the song.

PRODUCTION NOTE

When producing a music video, copyright permission is the first and foremost consideration. Do not break the law! More information about music copyright is presented in Chapter 12, *Legalities: Releases, Copyright, and Forums* and Chapter 13, *Music.*

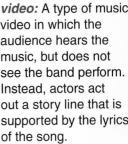

Visualization

visualization: The ability to mentally picture the finished program.

Visualization is the ability to mentally picture the finished program. Visualizing a program is similar to daydreaming. The visualized details of a program should be put on paper, so that others can share the vision. Only when everyone—the crew and cast included—shares the vision for the program can it become a reality. George Lucas waited to make *The Phantom Menace*, the fourth *Star Wars* film, until computer graphics technology was sophisticated enough to realistically reproduce onto the screen the creatures and worlds he visualized in his mind.

The Program Proposal

program proposal: A document created by the scriptwriter that contains general information about the program, including the basic idea, applicable format, message to be imparted to the audience, intended audience, budget considerations, shooting location considerations, and rough shooting schedule used to present the program to the executive producer to obtain permission and funding for the production.

The *program proposal* is created by the scriptwriter and provides general information about the program, including:

- The basic idea of the program.
- The applicable program format.
- The message to be imparted to the audience.
- The program's intended audience.
- Budget considerations.
- Shooting location considerations.
- A rough program shooting schedule.

The program proposal is presented to an executive producer for approval, either in written form or orally in a meeting. A program proposal is presented before writing a full script, to avoid wasting time and expense on a script that may be completely rejected by the executive producer. The program proposal allows for an initial "green light" on the project.

It is important to think through a script idea during the initial proposal stage. Using visualization, the scriptwriter can get a feel for the program and determine the direction of the script. The executive producer may reject the proposal, make suggestions, ask for further details, or accept it. Depending on the selected program format, the next step may vary.

Research

Both documentaries and interviews require that the program topic be researched. When interviewing someone, it is important to be proficient enough on the topic to hold a conversation that is interesting and informative. When developing a research paper, the research information is often organized on note cards. The notes are then turned into individual paragraphs of the paper. In television, the individual paragraphs become scenes.

Storyboards

storyboards: Sketches that portray the way the image on television should look in the finished program.

Some professionals use storyboards to help with visualization, **Figure 8-7**. *Storyboards* resemble comic books, in that they present a sketch of the way the image on television should look. Storyboards aid a director in communicating his vision to everyone on the production staff who sees them. They also help the director and camera operators to plan intricate camera moves. The disadvantage in using storyboards is the considerable

Figure 8-7. Storyboards assist the crew in creating the director's vision of the program. *(Courtesy PowerProduction Software)*

time and talent required to hand draw each scene. However, storyboarding computer software is also available, which draws storyboards using templates and "click and drag" elements. Several storyboarding software products are now available at prices that are reasonable for both professional and academic environments.

The Outline

If a program proposal is accepted, creating an outline is usually the next step. All dramas, lectures, lecture/demonstrations, and documentaries use the same kind of basic outline. The outline is very brief, not like the outline written for a research paper. An outline for a program includes comments that note the direction of the program.

ASSISTANT ACTIVITY

Find the last research paper you wrote. Reduce the major theme of each paragraph to a single, brief sentence. In doing this, you would create something very similar to the outline for a documentary on the topic of that research paper.

Basic Outline

A basic outline breaks each major event in a program into the fewest number of words possible, and places each on a different line. Each line begins with one or two words that identify the shooting location. **Figure 8-8** is an example of an outline for a drama called "Little Red Riding Hood." It is a brief, chronological listing of the program's progression. The dialog is either nonexistent or minimal—just enough to relay the main point of each scene. The normal progression for lecture, lecture/demonstration, and documentary program outlines follows the outline of most research papers—introduction, body, conclusion. A standard way of starting is with the "Tell 'em what you're going to tell 'em" type of introduction. Then, "Tell 'em" in the body of the program outline. Complete the outline with a "Tell 'em what you told 'em" conclusion.

Panel Discussion or Interview Outline

The outline for either an interview or panel discussion does not list major events or show progression. In these outlines, the only necessity is a list of at least twenty questions to ask the interviewee. "Who," "what," "where," "when," "how" (not "how long"), and "why" are the best kinds of question-starters to use. Think of questions that will get the talent to start talking, instead of just answering a question. The questions should spark and guide the conversation. Any questions that can be answered in ten words or less, with a number, or with a "yes/no" response should not be counted in the twenty question minimum. Short answers make for an uninteresting program. For example, the question "How long have you been…?" is widely overused in student-produced programming, but is almost unheard of in professional programming. Unless the answer to "How long have you been…" is unusual and sparks interest, viewers often do not pay attention to the answer.

Figure 8-8. The outline script for a drama is a brief, chronological presentation of a program.

House: Mom gives basket to LRR.
Warns not to stray from path.
Doorstep: Kiss goodbye, wave.
Path: LRR walking.
Path: Wolf sees LRR.
Path: LRR walking.
Path: Wolf running ahead to GM's house.
Path: LRR walking.
GM's house: Wolf breaks in and eats GM.
GM's house: LRR arrives and goes into bedroom.
Bedroom: LRR and Wolf conversation "what big…."
Bedroom: Wolf jumps up and chases LRR.
Bedroom: Woodsman bursts in and kills wolf.
Bedroom: Out pops GM.
The end.

PRODUCTION NOTE

A 7-year-old child who has just played a piano concerto at a major concert hall is asked how long he has been playing the piano. If the answer is "5 years," the question is worthwhile. However, if a 50-year-old man is asked the same question and responds "30 years," it is not particularly interesting.

Every question listed may not be asked in the course of the interview or panel discussion program. A particularly interesting answer to a question may lead to one or more impromptu follow-up questions. However, if the conversation lags, standby questions can jump-start the conversation.

Music Video Outline

Concert style music videos do not require an outline. However, story style music videos do require an outline. The second step in producing a music video may be to obtain copyright permissions, then begin outlining and scripting. (Permissions are discussed in Chapter 12, *Legalities: Releases, Copyrights, and Forums* and Chapter 13, *Music*.) By this point, the executive producer should have approved the quality and suitability of the lyrics and music. The lyrics of some songs are wholly inappropriate for broadcasting to the general public. The Federal Communications Commission has some detailed regulations regarding obscenity and decency on the public airwaves. Moreover, the school administration probably has regulations governing acceptable language for broadcasting student produced programming over in-house cable systems.

Expanding an Outline

Once completed, it may be necessary to expand your outline to include more detail about the program. To do this, review each line of the outline and list details related to that line. Much like the outline created for a research paper, list sub-topics and supporting details for each main topic line of the outline, **Figure 8-9**. Provide five to seven lines of detail for each main topic. When the outline has been sufficiently expanded, it is ready to be developed into a script.

Treatment

A *treatment* must be created for some types of programs, particularly dramas and long documentaries, before going to a full script. A treatment is, essentially, a narrative written from the outline that tells the story in paragraph form, **Figure 8-10**. Dialog is not included in a treatment. Each scene listed in the outline is expanded to an entire paragraph that details what happens in the scene. Creating a well thought out treatment makes script development much easier.

treatment: A narrative written from a program outline that tells the program's story in paragraph form.

Writing the Script

It is recommended that all scripts be written using a computer word processing program. If written with a word processing program and saved,

Figure 8-9. Use Roman Numerals, letters, and numbers to create an expanded outline.

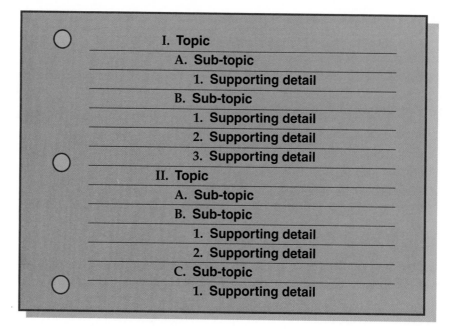

I. **Topic**
   A. **Sub-topic**
      1. **Supporting detail**
   B. **Sub-topic**
      1. **Supporting detail**
      2. **Supporting detail**
      3. **Supporting detail**
II. **Topic**
   A. **Sub-topic**
   B. **Sub-topic**
      1. **Supporting detail**
      2. **Supporting detail**
   C. **Sub-topic**
      1. **Supporting detail**

Figure 8-10. This is a treatment for scene 22 in the drama script featured in Figure 8-13.

Scene 22—Interior apartment, late night: Lenny and Christine are snuggled on the couch, watching a movie on TV. Evan comes home from the theatre. Greetings and small talk. Evan asks how the day went. Not understanding the question, Lenny and Christine launch into a litany of the frivolous things they did all day until they mention lunch. Evan interrupts. They didn't mention something Lenny was supposed to do. It becomes apparent that Lenny forgot to go to an audition Evan set up for him. Lenny and Christine sit in stunned embarrassed silence. There is a long pause as the camera lingers on Evan's face, which is full of fury. Composing himself, Evan asks Christine to leave. She does. Evan and Lenny are alone. Evan switches the TV to the channel of the security camera feed from the lobby of their apartment building. Lenny tries to apologize several times. Evan will not listen.

any alterations and changes requested by the executive producer or client are easily made without rewriting multiple pages of script.

PRODUCTION NOTE

Make sure to keep a copy of each revised version of scripts. After your revisions are complete, do not simply press "Save." Choose "Save As" and rename the file to reflect the revision sequence, such as "Scene 4 revision 3." Otherwise, the previous version cannot be retrieved for future review or if a previous version is preferred later in the process.

Unlike film scripts, television scripts are always written in two columns. The left column is reserved for the video and technical information. The right column holds the audio and stage direction. The information in the right-hand column of a television script is exactly what is contained in a "play-style" script for a theatrical performance.

The right and left columns are not the same size. The video column is narrower than the audio column, taking only 1/3 of the page width, **Figure 8-11**. The audio column (right column) occupies 2/3 of the page, because there is always more audio detail to include than video information.

Each line of the video column lines up horizontally with the corresponding line in the audio column. The result may appear to be a lot of wasted space on a page, but the empty space makes the page very easy and clear to read. When video and audio events occur simultaneously, they line up together on the script page.

The video portion of the script can make use of many abbreviations, as long as the director and crew all understand the meanings. All camera movements on the script should be abbreviated. There is not enough time to speak full-sentence directions over the headsets. For example, "ZO-2S Brian/Mike" is the equivalent of "zoom out to a two shot of Brian and Mike."

Types of Scripts

In television production, there are three types of scripts:
- Word-for-Word
- Outline
- Format

Word-for-Word Script

In a *word-for-word script*, every word spoken by the talent is written out, **Figure 8-12** and **Figure 8-13**. This type of script is used in dramas, music videos, lectures, and documentaries.

When writing a word-for-word script, write the right-hand column material first (audio and stage direction for performers). While writing the audio, visualize how the program will look. When you imagine a camera angle switch, move to the left column of the next line in the script and note "switch" in the video box. A change in camera angle can even occur in the

word-for-word script: A program script in which every word spoken by the talent is written out.

Figure 8-11. A television script is always written in two columns, with the video in the left column and the audio and stage direction in the right column.

PSA Word-for-Word Script	
Stay In School!	
Video	Audio
	Music Note: Rowdy rock instrumental at significant volume in beginning. Slowly lower the volume of the music track to zero when the narrative indicates the band broke up.
WS amateur band performing	**I WAS GONNA BE A ROCK STAR.**
CU wall of beige tile. Very slow pan/tilt. It's ok if the audience is not quite sure what the image is.	
CU of Neil (lead singer) performing	**I HAD MY ACT DOWN.**
CU wall of beige tile. Very slow pan/tilt	**I WROTE A LOT OF STUFF.**
CU Neil writing at a keyboard	**ME AND MY BUDDIES WERE NEGOTIATING A RECORD DEAL.**
CU wall of beige tile. Very slow pan/tilt	**WE HAD A LOT OF IDEAS OF WHAT WE'RE GONNA DO AND WHO WE'RE GONNA BE.**
LS Neil walking out of a school directly toward the camera, ending with a MCU of his smiling face.	(Neil walking out of school in direction of camera) **AND I FIGURED, "HOW IS HISTORY OR SCIENCE GONNA HELP ME IN THE MUSIC INDUSTRY?" SO, I DROPPED OUT OF HIGH SCHOOL.** (Tosses notebook in trash can, looks at camera and gives "thumbs up" sign)
CU wall of beige tile. Very slow pan/tilt. Cut music.	**THEN THE RECORD DEAL FELL THROUGH AND THE BAND BROKE UP** (Music ends)
MLS as Neil watches fellow band member drive off	**CAUSE EVERYONE WENT OFF TO COLLEGE.**
CU wall of beige tile. Very slow pan/tilt.	**THAT'S OKAY, WHAT'S RIGHT FOR THEM ISN'T NECESSARILY WHAT'S RIGHT FOR ME.**
Continue CU wall of beige tile. Very slow pan/tilt bringing ECU of 3/4 side view of Neil's face into the picture. He speaks to the camera	(Music Note: Bring in "elevator music" noticeable) **THAT WAS FOUR YEARS AGO, EVERYONE HAS A JOB NOW. ME, I FINALLY FOUND ONE, TOO.** (Audio: loud Beep)
ZO to MS as Neil turns profile to face a computer screen revealing that he is wearing an intercom headset on the previously hidden side of his face	(startled by the beep) (To camera) **OH!** (Neil turns to computer screen and speaks to the microphone of the headset) **I'M SORRY, DID YOU WANT FRIES WITH THAT? OKAY, PULL UP TO THE SECOND WINDOW.**
Graphic	**STAY IN SCHOOL. GIVE YOURSELF SOME CHOICES.**

Figure 8-12. A word-for-word music video script with time code.

Word-for-Word Music Video Script		
Empire of the Sun		
Seconds	Video	Audio
1. 0.00-0.05	Intro shot beats 1, 2, 3, 4 PC Walking beats 5, 6 Blaine walking beats 7, 8	Opening music
2. 0.05-0.10	Andy walking beats 9, 10 Phil walking beats 11, 12 PC MCU pose beat 13 Blaine MCU pose beat 14 Andy MCU pose beat 15 Phil MCU pose beat 16	Music
3. 0.11-0.16	Phil hits cymbal-beat 17 shot starts zoomed in on cymbal and zooms out for the rest of bar ends in a LS of whole band	Music
4. 0.16-0.21	New LS of whole band	Music
5. 0.22-0.26	MLS of Chapin and Rachel on bench from front	Now you've left me to die in this forgotten cell
6. 0.26-0.31	LS of Austin and Matt throwing football	You've left me a bitter man or can't you tell
7. 0.31-0.36	LS of Matt throwing ball too high over Austin's hands	Well I'm here now girl and from grace I have fell
8. 0.36-0.41	LS of Chapin and Rachel as ball lands next to them	To you I'd have given up my soul to sell
9. 0.42-0.46	LS of Rachel getting off bench to grab ball	But you rejected my love, told me to stay away
10. 0.47-0.51	Quick shot of Austin walking toward her as she picks ball up and turns into a subjective shot	Well I'm back to offer my love for just one day
11. 0.51-0.55	CU of Rachel's eyes, and then Austin's eyes	So you got this last chance, think about it please
12. 0.56-1.00	MLS Rachel walks over to Austin and they hold hands, quick CU of Chapin looking mad	Don't waste your time on that guy, besides I've heard he's a tease.
13. 1.01-1.04	LS of whole band from front	Music
14. 1.05-1.09	MS of Blaine	Music
15. 1.10-1.14	MS of Andy	Music
16. 1.15-1.18	MS of Phil	Music

Figure 8-13. A word-for-word drama script based on the treatment in Figure 8-10.

Word-for-Word Drama Script	
Scene 22	
Video	Audio
Fade in Cam 2, 2S, ZO for 3S	*(We are inside the apartment. Christine is snuggled up against Lenny. They are watching a movie on TV. A Christmas tree is in the background. Evan comes into the apartment from the theater.)*
	Evan: *(singing)* Hello young lovers, wherever you
Cam 1, 2S Christine and Lenny	Evan: are. Lenny and Christine: Hi, Evan! Lenny: How'd it go tonight?
Cam 3, MS of Evan	Evan: Knocked'em dead. Watcha watchin'?
Cam 1, 2S C & L	Christine: A Summer Place Lenny: At Christmas! Can you believe it?
Cam 3, MS of Evan	Evan: *(laughs)* Easter Parade would be worse. Hey, *(interrupting as they turn back to watch the movie)* tell me about today.
Cam 2, 2S of C & L Cam 1, move to CU of Lenny	Lenny: Well, we started out by pretending we were rich. Christine: Yeah, dressed up in our finest and walked into Saks. You should have seen the saleslady when Lenny told her he didn't like the $25,000 fur coat I had been fawning over. Lenny: I thought she was going to have heart failure. *(laughs)* Then we came back to the apartment for lunch and --
Cam 3, MS Evan	Evan: *(interrupting)* Lunch!
Cam 1, CU Lenny	Lenny: Yeah. And then we --
Cam 3, Begin ZI to MCU Evan	Evan: You didn't do anything else this morning?
Cam 2, 2S L & C	Lenny and Christine: No.
Cam 3, MCU Evan	Evan: Do you know what day this is?
Cam 2, 2S L & C	Lenny: Sure, it's Tuesday.
Cam 3, MCU Evan	Evan: *(evenly)* The audition.
Cam 2, 2S L & C	*(Lenny freezes. Christine slowly looks from Evan to Lenny. Silence. Lenny looks frightened.)*
Cam 3, MCU Evan	*(Evan looks from one to the other.)* Evan: Well, what happened at the audition?
Cam 1, CU Lenny	Lenny: Evan, I...it was an accident. I mean --
Cam 3, CU Evan	Evan: You mean what? What about the audition I set up for you?
Cam 1, CU Lenny	Lenny: *(Unable to face Evan)* I forgot about it. *(silence)*
Cam 3, CU of Evan	Evan: *(Calmly enraged)* Christine, would you excuse us please?
Cam 2, 2S L & C Cam 3, Move to 2S of L & E	Christine: Well, it is getting kinda late. Call me tomorrow? *(Lenny nods and helps her on with her coat.)* *(In a whisper to Lenny)* Are you sure I should leave? *(Lenny nods; Christine exits.)*
Cam 3, 2S L & E	Lenny: Evan, I– Evan: I don't want to hear it, Lenny.
Cam 1, CU Lenny	Lenny: But, I–
Cam 2, CU of Evan	Evan: Lenny, I don't want to hear it now.
Cam 3, 2S E & L	Lenny: But Evan, I want to ex–
	Evan *(With quiet fury. Turning to Lenny)* Lenny, no.
Fade out	

middle of a sentence. Be sure to make a quick note of the visualized switch before moving on with the audio column.

Once the entire right column is complete, allow the content specialist to review the script, if applicable. With the content verified, begin determining the shots needed for each audio box in the script. In the left column, describe each shot needed including the size of the shot, subject of the shot, the camera movements, and all other information pertinent to the video. Remember that a box in the script only contains the video or audio for a single shot. For example, one sentence of audio may include five shots. Therefore, that one sentence should span five boxes in both the audio and video columns.

Outline Script

The *outline script* usually has a word-for-word introduction and conclusion, but an outline for the body of the script. For example, some interviews may not be news-oriented and may be completely pre-planned. For a pre-planned interview, the questions may all be scripted, **Figure 8-14**. In the initial draft, the scriptwriter does not know how the interviewee will respond and the answers cannot be scripted. The interviewee's response is noted in the audio column of the script as "the talent answers," "talent response," or a similar phrase.

For an interview program, consider holding an informal rehearsal of the questions with the interviewee. Record the audio of the conversation to use as a reference when writing the script. Keep in mind that the interviewee probably will not give exactly the same responses during the actual interview. But, this prepares you for the type of responses to expect and helps you to better understand how the program will flow. It may also prompt some additional follow-up questions, or may lead to an entirely new direction for questions. This information is important in developing the video column of the script. Plan to cut to a different image about every seven seconds. This requires a variety of shots planned in the video column of the script to obtain many cutaways.

A cooking show is another program that uses an outline script. When writing the script for a cooking show, each step involved in the preparation of a dish is detailed, including the exact measurement of each ingredient, in the right-hand column of the script. Each step should be placed in a separate box of the audio column. When the script is complete, the chef should review it to ensure every step is included and is accurate. After the audio is verified, determine the shots needed for each step. The credit roll for cooking programs should include the recipe(s) featured and corresponding ingredients for each dish prepared.

Format Script

The *format script* is very brief and is used for panel discussions, talk shows, game shows, and other programs where the format does not change from episode to episode. See **Figure 8-15**. The on-screen talent and lines may change, but the shots are predictable from a production point of view. The order of events in programs of this type is predetermined and the sequence of every episode is consistent.

outline script: A program script that usually has a word-for-word introduction and conclusion, but an outline for the body of the script.

format script: A program script that is very brief and used for programs in which the order of events is predetermined and the sequence of each episode is consistent.

Figure 8-14. An outline script for a television interview with notations for interviewee responses.

Outline Format Television Interview Script	
Ttile: Movie Theaters in the 21st Century	
Scene 23	
Video	Audio
Host walks past the camera. Pan right as he walks into the stairwell and up the stairs.	Host: **Now that you've had some insight on running a theater, I'm going to show you where all the magic happens,**
Host walking from the stairwell into the projection room.	**the projection room. We'll talk to the projectionist.**
Pan right and zoom to MS of Host walking up to the projectionist.	Host: **Hello, (**projectionist's name**). Can we hang with you for a while and see how you do your job?**
Cut ELS left side angle shot of projectionist and host.	Projectionist: **Sure.**
Cut between host and interviewee every time a question is asked.	Host: **What kind of training does a projectionist need?**
	Projectionist answers.
Get many cutaways for editing variety.	Host: **What kind of training does a projectionist get?**
	Answer.
	Host: **How long is the film for most movies?**
	Answer.
	Host: **Do you need to clean the film before loading it?**
	Answer.
	Host: **What can you do if the film is damaged?**
	Answer.

Assistant Activity

Write a format script as you are watching a late night talk show. If everyone in the class writes one for a different episode this week, you will discover that all the scripts are nearly identical. The only variations are the faces on the screen and the dialog. The format/order of events remains constant from episode to episode.

Format Script	
Late Night Talk Show	
Video	Audio
Segment 1	*Intro*
LS Walk on MS Host during monologue Cutaways of audience reactions	Host welcomes audience to show Lists guests Opening monologue
Segment 2	*Guest 1*
Intercuts between MCU of host, MCU of guest, 2S of both, and cutaways of audience	Host and guest chat/interview
Segment 3	*Guest 2*
Intercuts between MCU of host, MCU of guest, 2S of both, and cutaways of audience	Host and guest chat/interview
Segment 4	*Musical Group*
Variety of MLS, MS, MCU, and CU of performers	Band plays
Segment 4	*Guest 3*
Intercuts between MCU of host, MCU of guest, 2S of both, and cutaways of audience	Host and guest chat/interview
Segment 5	*Wrap-up*
MCU host then WS to include all guests	Host thanks guests and audience all guests

Figure 8-15. A format script for a late night talk show program.

Writing Style

In most academic writing situations, students are encouraged to carefully choose their words, be mindful of the rules of sentence structure, and abide by the rules of composition. In general, students are expected to follow the commonly accepted grammar and usage rules. This type of writing is called "formal." Formal writing is difficult for some people. Nearly all school textbooks are written with a formal writing style. Formal writing is not used in script writing. Scripts are written the way people talk, using contractions and slang. Sometimes scripts do not even contain complete sentences.

The television script is written in an informal style to aid in easy understanding. For example, if you are reading a book and find a passage that you do not understand, you go back and reread it. This cannot happen on television. On television, if a concept or sentence is missed, it is gone. Therefore, on television, sentences are short, simple, and easily understood.

Those who have anxiety about writing can try dictating scripts into a voice recorder. After dictation is complete, the recording can be transcribed.

Another option is "voice-to-text" computer programs, which automatically types what is spoken into a microphone connected to the computer.

PRODUCTION NOTE

Do not waste words! The audience only sees what you show them. If showing a close-up of a rose, do not waste words by having a narrator state the obvious: "As you can see, here we have a rose." It is not necessary to describe what the audience is seeing, unless providing information they may not be able to acquire with their own eyes. With complex visuals, on the other hand, some explanation may be necessary for the audience to understand what they are seeing.

Montages

montage: A production device that allows a gradual change in a relationship or a lengthy time passage to occur in a very short amount of screen time by showing a series of silent shots accompanied by music.

A *montage* is a script/production device that establishes a setting, allows a gradual change in a relationship, or depicts a lengthy time passage in a very short amount of screen time. Montages are usually set to music and do not include any dialogue. The following is an example of shots in a montage, presented in shot sheet format:

Shots of:

A couple having dinner.

The couple going to a museum.

The couple playing in the park.

The couple coming out of a movie theater.

The couple swimming at a public pool.

The couple raking leaves.

The couple at a Halloween costume party.

The couple shoveling snow.

The couple decorating their home for the winter holidays.

The couple assembling kites in the park for their children.

A love song accompanies the series of shots. As a result, two minutes of real time shows that a year has passed in the couple's lives and depicts how their relationship has grown.

Wrapping Up

Organizing your ideas and developing a script, however brief, helps to focus your thoughts. Never shoot a program without a script of some kind. When this rule is broken, the crew inevitably ends up reshooting on location because the first shoot lacked a plan. Few people would attempt a cross-country auto trip without planning the trip on a map ahead of time. At the same time, people don't often strictly adhere to the original plan. Traffic backups, taking side trips on a whim, and road construction are just some of the things that may sidetrack a journey. The same is true for a script. Few scripts are shot exactly the way they are written. They do, however, provide the backbone structure to hold the director's creative vision together. Deviations from the script are common during the shooting process, but the basic structure of the program is constant because a script exists.

Review Questions

Please answer the following questions on a separate sheet of paper. Do not write in this book.

1. What are *nod shots*? How are they used?
2. What items are included in a program proposal?
3. What is a script outline?
4. How is a program treatment developed?
5. List the three types of scripts used in television production and the unique characteristics of each.
6. Why are television scripts written using informal language?
7. What is a *montage*?

Activities

1. For each of the program formats listed below, name a television show currently on the air that serves as a format example:
   - Lecture
   - Lecture/Demonstration
   - Panel Discussion
   - Interview
   - Newscast
   - Magazine
   - Drama
   - Music Video
   - PSA

   Be prepared to explain the characteristics of the selected television show that qualify it as an example of the corresponding program format.

2. Record an episode of your favorite sitcom and create an outline for the program. Remember that an outline for this type of program breaks each major event in the story into the fewest number of words possible and progresses chronologically.

STEM and Academic Activities

1. Explain how word processing programs have impacted the process of writing scripts.

2. When expanding an outline for a program, you should provide five to seven lines of detail for each main topic. For an outline with 15 main topics, how many lines of detail will you need to provide?

3. Choose a 12-hour block of programming for a network or cable channel. On the programming schedule, categorize each program as lecture, lecture/demonstration, panel discussion, interview, documentary, newscast, magazine, drama, or public service announcement/ad. What is the overall percentage of programs in each category?

4. Watch a 30-minute television program and record it for reference. Write a treatment for the program that lists each scene and tells the story of the program in paragraph form.

5. Choose a current event topic and select a person to interview who is involved in or related to the topic. Write relevant and interesting "who," "what," "where," "when," "how," and "why" questions for the interview.

Chapter 9

Broadcast Journalism

Objectives

After completing this chapter, you will be able to:

- Explain the responsibility broadcast journalists have to the viewing public.
- Identify news programs as mainstream, non-mainstream, or tabloid.
- Recall the news elements used to judge the newsworthiness of a story.
- Recognize the different story types broadcast during a newscast.
- Explain the elements of a package.
- Identify the various abbreviations used on a newscast script.
- Recall the workflow and responsibilities involved in a typical day in a newsroom.

Introduction

The general public turns on the television and watches the news, typically without considering everything involved in creating that news program. The audience merely sees and hears the news program and usually accepts what they see and hear as *fact*. News programs have an awesome responsibility to the public.

Broadcast journalism is the profession that brings television news to the public. "Broadcast" refers to the television production necessary to technically bring the video and audio to the viewers' television screens. "Journalism" refers to the careful determination of facts included in the stories presented during the newscast.

This chapter introduces the broad area of journalism in "broadcast journalism." Many schools that offer courses in broadcast journalism also assign that class the task of producing a news-type program. This program may then be sent throughout the school building on a regular basis—as often as once a day, several times a week, or several times a month. The student-produced news program should not only be a presentation of school appropriate news, but also a practical demonstration of the skills students learn in the broadcast journalism course. For this reason, many broadcast journalism courses are modeled after a newsroom in the real world of broadcast journalism. In mirroring the operation of a professional newsroom, many students in broadcast journalism classes are actually participating in career training to enter the field of broadcast journalism.

Professional Terms

beat
evergreen
extended package
feature
feature package
hard news
IFB
live shot
mainstream media
news
news feature package
news package
non-mainstream media
outro

package
patter
personality feature
reader
rundown
soft news
SOT
sound bite
stand-up
tabloid media
TRT
VO
VO-SOT

The News Media

news: Information people want to know, information they should know, or information they need to know.

News is information people want to know, information they should know, or information they need to know. Television news may be classified as one of three basic types: mainstream media, non-mainstream media, and tabloid media. There are subdivisions of these categories, but this section concentrates on the three basic areas.

In carefully examining the characteristics of the three types of news media, the gray nature of some television "news" programs is easily recognized. In the past, it was easy for the public to identify news that was biased and that clearly presented a particular point of view. These opinion pieces represent only the speaker's point of view and were clearly labeled and announced as "commentary." Unfortunately, the practice of announcing "The following is commentary" before airing an opinion piece has become inconsistent or nonexistent. As a result, the public may be unaware when a news story is biased or unbiased, unless they switch between various news programs to see how several different reporters report on the same story.

Good journalism requires that every effort be made to present stories factually and allow the audience to form their own opinions based on the facts they are given. Ethically, any commentary should be labeled or noted as such.

Mainstream Media

mainstream media: Television news programming that is expected to provide a fair and unbiased presentation of facts, without any particular viewpoint.

Mainstream media is programming that is expected to provide a fair and unbiased presentation of facts, without any particular viewpoint. Mainstream media is the most highly respected form of broadcast journalism. This type of news programming includes 24-hour cable news networks, the network-level news programs broadcast on major television networks, and local news programs broadcast by network affiliates. Therefore, the early morning news, news at noon, evening news (airing between 5 and 7 p.m.), and the 11 p.m. news are all considered mainstream media.

PRODUCTION NOTE

The public has two absolute expectations of the news media:

- The public expects the news media to report on what is happening in the world around them.

- The public expects that they will not be told what to think about what is happening in the world around them. If the audience is told what to think, or facts are presented in a way intended to influence the opinions of viewers, it is no longer news—it is propaganda. Always be wary when television news begins to tell viewers what they should think.

24-hour cable news networks, such as FoxNews, CNN, and MSNBC, do not always provide news 24 hours a day. Scheduled hard news broadcasts are separated by extended commentary and news/talk programming, which may be described as current event discussion and opinion. In times of crisis or breaking news, these networks interrupt the regularly scheduled programming to provide extended news programming. It is important for viewers to recognize that opinion and discussion programming is not news programming. Hard facts and truth are often diminished when opinion is the focus of a program.

Non-Mainstream Media

Non-mainstream media is programming that is expected to report news from a particular point of view. For example, the news presented on a religious-oriented cable station is expected to examine the news from a religious perspective. Additionally, news on a sports channel is expected to provide sports-oriented news programming.

Tabloid Media

Tabloid media stretches and exaggerates facts by dealing with sensational stories. The news stories presented on tabloid media programs are often so far removed from unbiased truth, that they are nearly fiction stories using real people's names. These programs can be found on fringe cable stations and even on local broadcast stations during non-network programming hours. However, tabloid programs are usually not network-provided programming. Tabloid media is generally considered more entertainment than news. Tabloid media is sometimes derogatorily called "gotcha journalism." Print versions of tabloid media can often be found near grocery store checkout counters and often include sightings of UFOs and Elvis, and stories about celebrities who have gained or lost weight.

Ethics and News Judgment

The First Amendment of the United States Constitution guarantees the freedom of the press, **Figure 9-1**. At first glance, many falsely interpret this to mean that journalists can do anything they want to do. In reality, all rights come with responsibilities.

Ethics in Journalism

While the law covers many situations, other content and production decisions are guided by ethics. Technology provides the incredible ability to capture reality, and modern editing equipment gives journalists tools to alter reality. Recording in almost any environment (openly or secretly) is quite easy, as is sharing the recording with a worldwide audience.

non-mainstream media: Television news programming that is expected to express a particular point of view.

tabloid media: Television news programming that stretches and exaggerates facts by dealing with sensational stories; generally considered more entertainment than news.

> *Congress shall make no law respecting an establishment of religion, or prohibiting the free exercise thereof; or abridging the freedom of speech, or of the press; or the right of the people peaceably to assemble; and to petition the Government for a redress of grievances.*
>
> *The Constitution of The United States of America,*
> *Bill of Rights—Amendment I*

Figure 9-1. The First Amendment to the Constitution of the United States of America provides for five freedoms: speech, press, religion, assembly, and petitioning the government.

PRODUCTION NOTE

Just because you have the *right* to do a story does not mean you *should* do the story.

Journalists, both professional and student, capture reality, process information, and prepare a valid, edited story for their viewers. The viewing public demonstrates trust in broadcast media and expects truth and reliability. Journalists must meet viewers' expectations and serve the viewers by making the "right" decision in tough call situations. Sometimes the "right" thing to do is complicated, given the legal freedom journalists have and all the technological tools at their disposal. Even student journalists must make judgment calls in deciding *what* to cover and *how* to cover it. Following established guidelines for ethical journalism is a basic step in learning the skills to become a credible journalist, **Figure 9-2**. Following a strong industry code of ethics early in your education sets a beneficial pattern of behavior for your professional career.

VISUALIZE THIS

The negative ramifications of making the story public may far outweigh the "glory" a reporter would receive for getting the "scoop." In a time of war, for example, a reporter discovers a story that would reveal classified information if broadcast. The classified information would be harmful to national security and helpful to the enemy. Should the reporter do the story? The line between what news organizations have the right to do and what is right for them to do has become blurred in recent years. A good reporter is willing and able to make ethical decisions based on the "greater good."

News Judgment

News programs have a finite and unmovable amount of time to report on the most important stories of the day. To conform to the time frame of a newscast, some stories do not make it to broadcast. Sometimes, a story idea may be rejected or postponed because it is deemed not as newsworthy as other available stories. Other well-produced stories do not make the newscast simply due to the run time constraints of a newscast. Unlike a newspaper, which may add additional pages or use a condensed font to squeeze a story in, minutes cannot be added to a television news program.

In a professional broadcast journalism environment, the producer and news director make decisions about which stories will be covered by journalists. The producer and news director positions are typically held by long-time industry professionals with many years of experience in reporting. Most professional journalists are required by their superiors to produce a mandatory number of sources to verify a story before it can be broadcast. Appropriate sources must be trustworthy and knowledgeable about the story topic. Multiple sources (at least two) should be required in an academic broadcast journalism environment, as well.

Figure 9-2. The Student Television Network adopted a code of ethics appropriate for use in school broadcast journalism programs; it is reprinted here with their permission. The STN code of ethics was adapted from the Radio and Television Digital News Association and the Student Press Law Center's codes of ethics.

Student Television Network

Student Television Network Code of Ethics

Preamble

Members of the Student Television Network believe in the importance of a free news media in a democratic society, including the academic/school setting. We believe the duty of student journalists is to seek truth and report it fairly and thoroughly to their audience, both in school and in the community. We understand that journalistic integrity is the basis of public trust and is developed and maintained by adherence to ethical principles. As members of STN, we adopt this code as the statement and evidence of our principles and our commitment to them.

Seek Truth and Report It

STN journalists should search for the truth. Student Journalists should:
- Be accurate in reporting information.
- Make every effort to avoid distortion of information.
- Identify sources of information as accurately as possible.
- Be objective in reporting both sides of a story.
- Avoid stereotyping.
- Never plagiarize.

Minimize Harm

STN journalists should value news subjects and news as topics and people who deserve respect. Student journalists should:
- Be compassionate when covering stories that may be painful to subjects.
- Be sensitive to those in grief or affected with tragedy.
- Be cautious and aware of the law in identifying minors involved in crime or victims of crime.
- Recognize the privacy of individuals.
- Avoid reporting that is libelous or slanderous.
- Avoid reporting that may endanger students or school personnel.
- Show good taste in reporting, avoiding vulgarity or obscenity.

Act Independently

STN journalists should aspire at all times to fulfill their obligation to their audience and the public, and maintain their independence of any attempts to obstruct or hinder these efforts. Student journalists should:
- Avoid conflicts of interest, and disclose any unavoidable conflicts.
- Hold those in power accountable.
- Resist pressure from individuals or groups to influence coverage.
- Seek to maintain the integrity and credibility of themselves and the organizations they represent.

Be Accountable

STN journalists are accountable to their audience and to their peers. Student journalists should:
- Understand they are a supervised group, just as their counterparts in the business world are.
- Seek and encourage open dialogue with the public about their coverage and conduct.
- Admit and correct any mistakes, promptly and publicly.
- Follow the highest standards of integrity and ethical behavior.

Adapted from the Radio and Television News Directors Foundation (RTNDF) and SPLC codes of ethics.
Adopted June 2006

In an academic broadcast journalism environment, the instructor often fills the role of producer and/or news director in the classroom. As the producer or news director, the instructor makes programming decisions and helps students understand the basis for these decisions, which assists students in developing a sense of news judgment. Modeling and teaching these skills may lead to increased student involvement in decisions, but a teacher functioning as the news director maintains authority for final story approval.

In judging the newsworthiness of a story, various news elements must be considered. It is critical that a reporter recognize these news elements to effectively develop a news story.

- **Proximity.** Is the story important to viewers because it concerns their immediate environment? Building a new freeway in Washington, DC might be very important to residents of Washington, DC and its suburbs, but the story means very little to residents of Phoenix, AZ.
- **Timeliness.** Does the story report an event that just happened, which viewers need to know about right now? Such as, the "just-in" election results of the town's mayoral race or a breaking news report of an Amber alert for a missing child.
- **Prominence.** Is the principal character in the story a well-known (prominent) individual, making the story newsworthy? The town mayor has decided to run for governor. A celebrity quits Hollywood to enlist in military service.
- **Consequences.** Does the story directly affect a significant number of viewers? If taxes are not raised by the city council, the salaries of all public workers (including firefighters, police officers, teachers, and city employees) will be reduced next year. Do the consequences of the story require the public to act or react in a specific way? Tornados have been sighted five miles outside of town. Schools are closing three hours early due to inclement weather.
- **Conflict.** Does the story contain a controversy, struggle, or issue with two or more sides? Is the final outcome of interest to the public? For example, political campaigns, crime stories, governmental votes, and sporting events.
- **Unusualness.** Is there a particular aspect that makes the entire story unusual? A four-year-old boy is a piano prodigy and has been asked to perform at Carnegie Hall. Someone in the community celebrating a birthday is ordinary, but someone celebrating her 105th birthday is unusual and interesting.
- **Emotion.** Will the story "pull at the heartstrings" of the public? A young soldier endured a two-year tour of duty in a war-ravaged land and finally returned to his hometown. He was crossing the street at the bus station to greet his wife and infant son, and was killed by a drunk driver who ran a red light.
- **Achievement.** Does the story involve an amazing effort that leads to an outstanding achievement? The story of a young athlete who suffered a devastating injury and was told he would never be able to play his sport again. Sheer determination and training brought him back from his injuries to win a spot on the US Olympic Team.
- **Contrast.** Does the contrast of two elements in the story create general interest? A story of two very different families celebrating the

same religious holiday. A story detailing the life of a carnival worker 30 years ago, in contrast with the life of a carnival worker today.

Determining which of these news elements is the strongest helps establish the angle needed for the story. (A story's "angle" is discussed in Chapter 10, *Newswriting for Broadcast*.) When more news elements are included in a single story, the bigger the story becomes. The first five news elements listed (proximity, timeliness, prominence, consequences, and conflict) are *usually* associated with hard news. The last four news elements (unusualness, emotion, achievement, and contrast) are *usually* associated with soft news. However, there are no firm and fast rules dictating which news elements define hard and soft news. Hard news stories and soft news stories are two classifications of content in a newscast.

Hard news is characterized by seriousness and timeliness. These stories may address politics, economics, war, crime, health crises, weather crises, and governmental messages to the public. Hard news stories contain information that viewers need to have immediately. "Breaking news" bulletins and stories are examples of hard news.

Soft news is characterized by information that may be interesting, but is not necessarily something viewers *need* to know—these stories may focus less on timeliness. Soft news often consists of human interest stories and may include sports, updates on celebrities, entertainment, consumer tips, and gardening hints. In many cases, these stories may be appropriate to be told any time there is room in the news program. A story that may be broadcast at any time is called an *evergreen* story. However, soft news stories may also have a degree of timeliness. For example, a story about a friendly competition between two neighbors for the most extravagant holiday lights display is appropriate for broadcasting only in December.

Ethically Funding the News

The news is not a fund-raiser for a television station. A news operation requires many paid employees, whose salaries and equipment must be funded by some revenue source at the station (**Figure 9-3**). Most television stations have a studio in the building. The sole purpose of the studio is often to present the news. Revenue earned by airing local ads throughout the entire day must fund the operation of the studio. Most ads for local car dealerships or restaurants are produced by advertising production companies or by a television station's production department. The ad is then aired by the station according to the contract held with the advertiser. Sometimes, the television station's studio may produce ads for clients as a way of increasing revenue. Each time the ad airs on the station, even more funds are generated.

To ethically fund a news operation, the news cannot have any relationship to the station's advertisers and cannot be influenced by advertisers. Advertisers cannot be given special treatment merely because they have purchased a substantial amount of ad time on the station. If a scandal was uncovered, for example, involving the owner of a local restaurant bribing a health inspector, the newsroom must not be unduly influenced to avoid covering the story because the restaurant advertises on the television station. The news operation must be kept separate from the advertising department to maintain the untarnished appearance of unbiased news.

hard news: Type of news story that contains information that viewers need to have immediately; characterized by seriousness and timeliness.

soft news: Type of news story that contains information viewers may find interesting, but not necessarily information they *need* to know.

evergreen: A story that is appropriate to be broadcast at any time, regardless of season or time of day.

Figure 9-3. From the control room to the studio floor, all the staff and equipment require funding to operate. *(Countryside High School, Clearwater, FL)*

Airing Stories

A topic can be covered, from a technical standpoint, in a variety of ways. The type of coverage often depends on how quickly a story needs to be aired, how big the story is, and how much information/footage is available to work with.

IFB: Interrupted feedback; a line of communication between the anchors and the producer in the control room. An earpiece worn by the anchor is connected to the producer's headset, allowing the producer to speak directly to an anchor while the anchor is on the air live.

The fastest way to get a story to the public is through an *IFB* (interrupted feedback), **Figure 9-4**. The anchors on a news broadcast wear an earpiece with a wire that runs behind their ear and down their back (typically under the anchor's shirt or jacket). This earpiece is particularly noticeable if a camera shoots the anchor from a side angle. In the news industry, the earpiece worn by an anchor is the IFB. The IFB is connected to the headset worn by the producer in the control room of the studio. Under normal conditions, the anchor's earpiece carries the audio of the news program as

Figure 9-4. An IFB fits snuggly in the ear and provides a direct line of communication from the producer to the talent/anchor.

heard by the television audience. However, the producer can break into the IFB feed and speak directly to the anchor who is on the air live. The anchor immediately repeats what the producer is saying without embellishment. It may be challenging to maintain composure on camera when "channeling" the words of the producer. The producer must speak clearly and concisely because there is no filter between the producer's words and the words the anchor broadcasts.

PRODUCTION NOTE

Footage of newscasts from any of the major networks on September 11, 2001 and the following three days provide excellent examples of news covered as it happens and anchors relying solely on the words fed to them by producers through the IFB.

Assistant Activity

Sit in front of a mirror and call a friend on the phone. Ask your friend to read a newspaper or magazine article aloud to you. Try to repeat what your friend is saying to you *while* they are saying it. Maintain a normal facial expression and repeat your friend's words accurately.

Types of Stories

Many different types of story formats are used in a single newscast. Each type has unique characteristics and complexities, which allows stories to be told in different ways and with varying depth.

Reader

A *reader* is a story that an anchor simply reads aloud from the teleprompter for the viewing audience to hear, **Figure 9-5**. A reader does not include video to support the story.

VO

A *VO* (voiceover) is a type of story that incorporates B-roll video rolled-in from the control room, in addition to the script read by the anchor. The audience hears the nat sound on the B-roll behind the anchor's voice, **Figure 9-6**. A VO takes the reader story one step further with the addition of supporting video.

Talk the Talk

When speaking the term *VO* aloud, simply say the letters "V-O," just as you might say "OK" in response to the question, "How are you?"

reader: A story, written by a reporter or anchor, that does not have video to accompany the story. The anchor simply reads the text on the teleprompter aloud for the viewing audience to hear.

VO: Voiceover; a type of story that incorporates B-roll video rolled-in from the control room, in addition to the script read by the anchor.

Figure 9-5. A—An example of a reader script in two-column format. B—The audio portion of the reader script is uploaded to the teleprompter and read by the anchor during the newscast. *(South County Secondary School, Lorton, VA)*

Video	Audio
Cam 2—Med shot Anchor 1 CG—Lower third super identifying anchor 1	Anchor 1: The Randolph Community Theater is now in rehearsal for their spring musical. This year's production is "Murder by Lottery." It is a dark comedy and with a cast of 23 locals. Karen Telesco will play the lead—a 70-year-old eccentric millionaire. The director is Jim Smythers. Opening night is set for April 15 in the Randolph Auditorium.

A

B

Figure 9-6. A sample VO script that an anchor reads as the audience sees B-roll footage.

Video	Audio
Med shot Anchor 1	Anchor 1: The Randolph Community Theater is now in rehearsal for their spring musical.
Wide shot of several actors on stage in rehearsal with nat sound in background	This year's production is "Murder by Lottery." The three-act play is a dark comedy with a touch of suspense. All of the 23 cast members are local—some with experience and some totally new to the stage.
Full shot of Karen Telesco on stage with nat sound	Karen Telesco will play the lead—a 70-year-old eccentric millionaire. In reality, Karen is a much younger lady who will have a complete makeover to play the role.
Med shot of Jim Smythers sitting with director's notebook in lap	The director is Jim Smythers. Smythers has directed several community theater productions, but this will be his first with The Randolph Group.
Med shot Anchor 1	Opening night is set for April 15 in the Randolph Auditorium.

VO-SOT

VO-SOT (voiceover–sound on tape) is a type of story that is one step higher in complexity than a VO. The audience sees B-roll video and hears the anchor reading from the teleprompter, followed by footage of a comment from a principal player in the story. The B-roll is one file and the comment (SOT) is another. *SOT*, also called a *sound bite*, is footage of a principal player connected to the story and includes voice that supports the reporter's story. This footage is often the answer to a reporter's question and should be a reliable source that is connected to the story in some way, such as an official person from the event or an eyewitness to the event. SOT footage is usually between 5 and 10 seconds in length, but rarely more than 15 seconds. The B-roll is seen by the viewers as the anchor reads and, at the appropriate time, the switch is made to the comment footage, **Figure 9-7**.

VO-SOT: Voiceover-sound on tape; a type of story in which the audience sees B-roll video and hears both the anchor reading from the teleprompter and footage of a comment from a principal player in the story.

SOT: Sound on tape; footage of a principal player connected to a story, which includes voice/audio that supports the story. Also called *sound bite*.

Talk the Talk

When speaking the term *SOT*, pronounce the letters as a word ("sot")—rhymes with "got" and "not."

When speaking the term *VO-SOT*, pronounce the letters "VO" as a word (rhymes with "toe") and "SOT" as described above—"voe-sot."

Video	Audio
Med shot Anchor 1	Anchor 1: The Randolph Community Theater is now in rehearsal for their spring musical.
Wide shot of several actors on stage in rehearsal with nat sound in background	This year's production is "Murder by Lottery." The three-act play is a dark comedy with a touch of suspense. All of the 23 cast members are local—some with experience and some totally new to the stage.
Full shot of Karen Telesco on stage with nat sound	Karen Telesco will play the lead—a 70-year-old eccentric millionaire. In reality, Karen is a much younger lady who will have a complete makeover to play the role.
Med shot of Jim Smythers sitting with director's notebook in lap	The director is Jim Smythers. Smythers has directed several community theater productions, but this will be his first with The Randolph Group.
Med shot of Jim Smythers (talking head) CG: Lower third super identifying Jim Smythers as director	SOT Jim Smythers: "The challenge of this production is that three different locations must be used, which calls for creative set building and quick changes. I guarantee the audience will enjoy the fast pace of this show."
Med shot Anchor 1	Anchor 1: Opening night is set for April 15 in the Randolph Auditorium.

Figure 9-7. This sample VO-SOT script includes scripted anchor lines, B-roll footage, and recorded comment footage.

package: A story that is about 1 1/2–2 minutes in length, contains its own intro and outro, is edited, and can be inserted into a live program at any time the producer chooses.

Packages

If a story is shot and edited prior to the newscast, the story is called a *package*. A package is a complete unit that can be inserted into a live program at any time the producer chooses; it is simply rolled-in after the anchor introduces it. A package is fully thought-through, usually 1 1/2–2 minutes in length, contains its own intro and outro, and is edited, **Figure 9-8**. It is called a "package" because the beginning, middle, and end of the story are neatly tied together to create a complete packet—the story can stand alone. The topic and content of packages range from in-depth news stories to human interest

Figure 9-8. A package script contains an intro, body of the story, and the outro, in addition to all of the shots, lines, and footage.

Video	Audio
Wide shot of two actors on stage choreographing a fight with nat sound	Reporter VO: The Randolph Community Theater is now in rehearsal for their spring musical.
Wide shot of several actors on stage in rehearsal with nat sound in background	This year's production is "Murder by Lottery." The three-act play is a dark comedy with a touch of suspense. All of the 23 cast members are local—some with experience and some totally new to the stage.
Full shot of Karen Telesco on stage with nat sound	Karen Telesco will play the lead Karen Telesco: "I'll be playing a 70-year-old eccentric millionaire.
Med Shot Karen Telesco (talking head from interview) CG: Lower third identifying Karen Telesco as female lead.	Just getting the makeover will be a challenge, but I also have to become arthritic and grumpy. I'm sure my high school English students will be glad I'm not REALLY like that when they see me."
Med shot of Jim Smythers sitting with director's notebook in lap	Reporter: The director is Jim Smythers. Smythers has directed several community theater productions, but this will be his first with The Randolph Group.
Med shot of Jim Smythers (talking head) CG: Lower third super identifying Jim Smythers as director	Jim Smythers: "The challenge of this production is that three different locations must be used, which calls for creative set building and quick changes. I guarantee the audience will enjoy the fast pace of this show. AND I guarantee most of them won't identify the murderer until the very end of the show."
Full shot of reporter on set holding a prop gun and a heavy candlestick	Reporter Stand-up: Was THIS the murder weapon? Or was it THIS? Only the cast knows, and they're not telling. If you want to know, you'll have to buy a ticket. Opening night is set for April 15 here in the Randolph Auditorium.

to sports. A package includes a reporter's audio track, one or more sound bites, and may have a stand-up by a reporter. An *extended package* may be 2–4 minutes in length and typically provides more in-depth coverage of a specific story. A documentary, 6–10 minutes in length, may also be considered a type of package.

A package that covers hard news/current events is often called a *news package*. For example, a local reporter might produce a package about a fire that occurred this morning in the town. The package shows the damage, includes comments from owners of some damaged buildings, and comments from firefighters on the scene. This type of package is produced very quickly—the reporter and camera operator get to the scene, shoot, write, and edit the package so it can air while the event is still current. Another example is a recall issued by a toy manufacturing company. The news package shows the toy, a demonstration of the danger, and tells the audience how to return it for refund.

A *news feature package*, also called a *feature package* or *feature*, covers soft news stories that are connected to current events. For example, a news feature on the celebration events taking place for the fifth anniversary of a local food bank may be included in the day's newscast. The story may include comments on the center's growth from the director of the food bank, as well as comments from people who have donated to the food bank and people who have benefitted from the food bank. Another example may be the rebuilding efforts of a family business destroyed by a fire last spring. This would be a follow-up type of story. After being in operation for three generations, the family is rebuilding—bigger and better. The news feature has the "now factor" because rebuilding is in progress, but also addresses the decision to rebuild, changes being made, and the emotions associated with starting over.

Other packages may not be related to current events and may not have any relevant consequences for the viewer. These are human interest stories, which may simply be interesting and entertaining. News elements typical of human interest stories include unusualness, emotion, achievement, or contrast. One type of human interest story is a *personality feature*, which focuses on one person. A personality feature introduces viewers to a person and explains why that person is newsworthy. Stories about people being honored for service, accomplishment, overcoming adversity, or having an unusual job or home are examples of personality features.

Packages are often roughly outlined before the crew arrives at a location to begin shooting based on research and previous knowledge of the story topic. The outline provides a list of the video and audio to obtain at the location, and may include some preplanned interview questions. With this rough outline, the crew is more likely to return to the studio with usable footage for the story. Once the required footage is obtained, the crew may gather additional footage and record other things of interest at the location. In post-production all the footage is examined. It may be determined that the footage supporting the original story outline is not very compelling, but some of the other footage reveals a different and interesting angle on the story. In this case, a new story is written to match the new footage.

A *stand-up* is footage in a package that depicts a reporter standing in front of the camera, speaking directly to the viewers from the location of a

extended package: A 2–4 minute story that is shot and edited before a newscast and typically provides more in-depth coverage of a specific story.

news package: A package that covers hard news/current events.

news feature package: A package covering soft news stories that are connected to current events. Also called a *feature package* or *feature*.

personality feature: Type of human interest story that focuses on one person and why that person is newsworthy.

stand-up: Footage in a package that depicts a reporter standing in front of the camera, speaking directly to the viewers from the location of a story.

story, **Figure 9-9**. The stand-up is shot at a location connected to the story topic, and may be used at the beginning, in the middle, or at the end of a package story. The purpose of the stand-up is to establish for the audience that the news team was actually at the location to cover the story or to demonstrate action relevant to the story. A stand-up is a story-telling tool.

The stand-up is a very common element of a package and allows the audience to see what is happening with their own eyes. For example, a story on severe winter weather has greater impact with a shot of a snow plow truck stuck in the snow. In some soft news stories, the reporter actually takes part in the story, which effectively allows the audience to take part in the story. A reporter that takes a test-drive in a NASCAR racing vehicle, for example, can take the audience on the ride by shooting from inside the vehicle while driving. Another use of a stand-up is to demonstrate a particular aspect of the story or to show the inner workings of an object or event.

If a reporter includes a stand-up in the report, there must be a reason for the shot. It should not be used simply because there is not enough B-roll to fill the package. Never do a stand-up from a location that is not linked to the story. Including footage from an unrelated location will confuse the audience or raise doubts about the reporter's integrity. A reporter's credibility will disappear in an instant if the audience realizes the reporter tried to "fool" them.

Editing Packages. A reporter needs a strong ethical standard when editing a package that contains sound bites. In a package story about a campaign speech made by a political candidate, for example, the sound bites must be edited extensively to complete the story and keep it to a *TRT* (total running time) of 2 minutes. The context of the candidate's speech must not be altered in the process of editing, regardless of the reporter's personal political views. The news media must remain truthful in all the stories reported.

TRT: Total running time; industry abbreviation.

Figure 9-9. A stand-up places the reporter at a location related to the story.

VISUALIZE THIS

A reporter is assigned to cover the speech of a political candidate. The candidate says, "I do not support giving consideration of any kind to drug dealers. These parasites feed on the innocence of our youth and we should place them behind bars and throw the key away" (the audience erupts in thunderous applause). An unscrupulous reporter could edit the footage so the candidate says, "I [edit here] support giving consideration [edit here] to drug dealers [edit out the rest of the speech]" and cut to audio of thunderous applause and shots of young people shouting approval. In distorting what the candidate said and broadcasting what the reporter knows is blatantly untrue, the reporter has breached ethical standards. That breach of ethics may ultimately cause the defeat of the candidate.

Live Shot

A *live shot* is a story that is introduced by the anchor and delivered through a live feed by a reporter on location. Typically, the word "Live" is displayed somewhere on the screen or someone will mention that the reporter is "Live from the scene." The reporter tells the story and delivers a standard *outro*, the closing at the end of a story. Lines such as "Back to you, Jim" or "This is Lisa Thompson, EyeWitness News," are commonly used to send the viewers back to the anchor in the studio. The anchor in the studio and the reporter in the field may have a live conversation on the air before the anchor continues to the next story. This conversation between anchor and reporter is usually set up in advance of the live shot and is often scripted. The live shot is more complex than other story types because the reporter must deliver a report live using notes on the spot, usually without a teleprompter. Sometimes, live shots include live interviews and may even include action happening as the report is delivered. An extreme example of a live shot is war correspondents giving live reports from the battle front.

live shot: A news story that is introduced by an anchor and delivered through a live feed by a reporter on location.

outro: The salutation at the end of a story; opposite of an intro.

PRODUCTION NOTE

High school journalists rarely broadcast a live shot story due to lack of technical ability. However, students can produce a live shot style story. This type of story is called "look-live." The footage is taped on location and a reporter speaks from notes. The footage is then edited to look like it is a live feed. When this type of story is included in a newscast, the audience should not be led to believe that it is an actual live feed.

Investigative Reporting

Investigative reporting is a difficult and complex type of reporting. It often involves a reporter digging into a topic, searching for wrongdoing by an individual or organization. This type of reporting may be viewed as exciting, particularly by students, but is typically both physically and legally dangerous. Investigative reporting (and the defamation of character that possibly results) is the foundation for many lawsuits. The National

Television Academy recommends there be executive-level approval (above the level of news director) before a reporter pursues an investigative story. The primary concern is to determine if the story will expose something of significant public concern, reveal a wrongdoing by a head official, or if the investigative report will profoundly harm the reputation of an innocent individual or group. These are very serious and significant issues.

Investigative reporting is sometimes associated with hidden-camera footage. To be justified, a compelling case must be made that there is no other possible means of acquiring the necessary video and that no laws will be broken in obtaining the video. Hidden-camera journalism typically involves many legal issues, the most notable being privacy rights. (See Chapter 12, *Legalities: Releases, Copyright, and Forums.*) This is not to say that journalists cannot cover controversial topics, analyze statistics to draw conclusions, and gain access to information of public record in order to develop stories. The level of investigation should be commensurate with the reporter's experience, skill, and position.

VISUALIZE THIS

You have come to the conclusion that local police drive too fast for no apparent reason and set out to prove your theory. You cruise the streets of your town with a friend in the backseat of your car. Your friend has a video camera and is ready to start shooting. You find a police car on the road, pull behind it, and follow, maintaining the same speed. Your friend shoots video of your speedometer in the foreground, and the police car you're following in the background. You feel this is your "proof" that the local police regularly speed in non-emergency, normal driving. The problem with your investigation is that you have broken the law by speeding yourself and have made a video of yourself breaking the law! If you use that video in your story, you publicly admit to the community and local law enforcement officials that you broke the law. Your story about local police driving too fast will be lost in the uproar caused by a reporter who put the community in danger by driving recklessly to get a story. Also, you can certainly expect a visit from the local police concerning your illegal activity.

The Newscast Script

Most television stations use scripting software that prints the script in different formats for different members of the production team. For example, the lighting director may receive a script that contains only lighting cues instead of the all-encompassing script generated for the technical director. Some newscasts also include portions that are not fully scripted. The *patter*, spontaneous conversation or small talk, between an anchor and a reporter on location is an example of unscripted dialog in a newscast.

For student newscasts, a two-column script with all the directions noted helps students see the "big picture" and understand how all actions are synchronized, **Figure 9-10**. The text in the right column of the script is displayed on the teleprompter. When a script is set up in a two-column table, each row indicates a change in video source—either a different camera or switching to a pre-recorded piece. A change in audio source may or

patter: The spontaneous on-air conversation or small talk between anchors or anchors and reporters.

Figure 9-10. A script sample of a fully-scripted newscast.

Video	Audio
VPB—Show Open	SOT
Cam 2: 2 shot	SOT soft light in background Anchor 1: Good Morning! Thanks for joining us. I'm Andrew Kendall. Anchor 2: And I'm Sandra Bailey. One Roane County football player has an artistic side and we've found a student that YOU might want to hire. Anchor 1: Raider Television starts right NOW.
VPB Show Open	SOT
Cam 1: MED shot Anchor 2 CG: Lower third super with name of Anchor 2	SOT fade Anchor 2: There have been some changes made to the Writing Assessment. The Writing Assessment will take place March 31st through April 10th. The BIG change this year is that Juniors and Freshmen will be taking the test, in addition to Sophomores. Yes, you heard right. Grades 9, 10, and 11 will take the state Writing Assessment right after spring break. The test will be administered online in the library computer lab. So, the lab will be closed to other classes during testing time.
Cam 3: MED shot Anchor 1 CG: Lower third super with name of Anchor 1	Anchor 1: While we're on the topic of writing, we have the winners of this year's Young Writers' competition.
FSG (still pictures and text) Amanda Jackson Miriam Hottle	VO Anchor 1 Freshman Amanda Jackson took first place for Grades 9 and 10. The title of her short story is "Lady". The story is about an older woman who is dying, but has a great impact on her granddaughter. Senior Miriam Hottle took first place for grades 11 and 12 with her story about a middle-aged man who has lost his factory job and faces traumatic changes.
Cam 2: 2 shot	Anchor 1: Congratulations and good luck to the winners. They will represent Roane County at the state competition in mid-May.
Cam 1: MED shot Anchor 2 (left of center) CG: OSG (FFA logo)	Anchor 2: While some students are competing indoors, others are gearing up for a national competition outside. The FFA land judging team has qualified for the national competition in Oklahoma. Students on the team are Justin Braddon, Chad Macklin, and Logan Philips.
Cam 3: 2 shot	Anchor 2: Another group of students has been preparing to entertain an audience.
Cam 3: Med shot Anchor 2	Anchor 2: The Roane Arts and Humanities Council is sponsoring a comedy play titled "Marriage by Indecision." The play will take place April 3rd through the 6th at the Spencer Middle School auditorium. The play has 28 cast members, including some high school students that you'll recognize. Rick Bradley has the story.
VPB—Community Theater	SOT Outcue—"...director says no one will want to miss "Marriage by Indecision" coming to this stage in April. This is Rick Bradley reporting for Raider Television."

may not accompany the change in video source. Cues, like SOT, VO, and Outcue, may or may not be included, depending on the preference of the crew. There are many abbreviations that may be used in a newscast script.

Newscast Script Abbreviations	
Abbreviation	**Meaning**
VPB	Video playback. Indicates that a pre-recorded, edited piece should be inserted into the show. The words following "VPB" on a script identify the filename of the piece to be played at that time.
SOT	Sound on tape. Instructs the audio technician to get audio feed from the pre-recorded, edited piece. "Tape" is still used in the term even though the material is on a computer—this has carried over from the days of tape technology.
CG	Character generator. Directions to the person responsible for displaying graphics at the appropriate time.
FSG	Full screen graphic. Directions to the person responsible for displaying graphics. FSGs may be a colored background with still pictures, text, graphs, maps, or diagrams.
VO	Voiceover. Lets the audio technician know to keep anchor mics open for audio, even though the anchors are not seen on-screen.
OSG	Over the shoulder graphic. Directions to the person responsible for displaying graphics. An OSG may be a box or a design with text overlaying about 1/3 of the screen. The shot of the anchor is moved to the side of the screen to allow room for the OSG.
Outcue	When audio comes from a pre-recorded piece, the last few words of the piece are noted in the script so the director can give a stand-by to the anchors and crew.

On-Air Appearance

The familiar phrase, "You can't judge a book by its cover" is generally a true statement. However, a newscaster's appearance and behavior directly affect their credibility, as perceived by the viewing public. Therefore, management judges these qualities quite critically. Newscasters' ability to speak correctly, clearly, and intelligently also affects their credibility.

Assistant Activity

Flip around from one newscast to another on both broadcast and cable network channels. What do you notice about the appearance of newscasters and reporters?

Newscasters diligently work to make themselves as visually attractive as possible by remaining physically fit and maintaining a mainstream appearance, which includes makeup, hair style, and clothing. A mainstream appearance does not include visible body piercings, tattoos, and radical hair styles and colors. Nothing about the on-air talent should distract from the news being reported. It is important that on-air talent does not alienate any segment of the viewing audience by appearing extreme in any direction. Alienated viewers will tune into a rival news program, which results in lost revenue for the entire television station. Presenting an appropriate and acceptable image is so important that some stations provide consultants and expense accounts for makeup and clothing for the on-air news talent.

Newscasters must dress professionally when on camera. Mainstream business attire that is neat, clean, and pressed is generally appropriate. A coat and tie is common for male newscasters. Conservative business attire is acceptable for women, which does not include plunging necklines, short skirts, or tight-fitting clothing. A short skirt, gym shorts, or torn jeans worn by talent seated behind an anchor desk is never seen on camera, **Figure 9-11**. However, skirt length on female talent becomes very critical if she is sitting on a chair, sofa, or stool for an interview with knees and legs included in the shot. In addition to a professional appearance, business attire offers many options for unobtrusive placement of a lapel-style microphone.

There are some situations when the requirement for professional attire may be relaxed slightly. A reporter interviewing a champion swimmer poolside in the heat of summer, for example, may dress more casually than when reporting from the studio set. Appropriate clothing is still required—the reporter would not wear swim apparel to conduct the interview. Brief, playful moments may also be reason for more casual dress. For example, an anchor may wear gag glasses with long, springy eyes to introduce a light, human interest story about the Halloween festival sponsored by local merchants.

Figure 9-11. When seated behind a desk, the anchor's waist and legs are not visible to viewers. Whether the anchor is wearing swim trunks or pajama pants, it won't be seen on camera. *(South County Secondary School, Lorton, VA)*

A Day in a Television Newsroom

The assignment editor arrives early in the morning to review potential stories. These stories come from several sources—wire feeds from national news organizations, stories the graveyard shift reporters have been working throughout the night, press releases for events happening during the day or in the near future, listening to police radio scanners, reputable Internet sites, and other sources.

The morning meeting is held with all early evening anchors, reporters, producers, news directors, and, often, photogs are also present. At this meeting, everyone offers story ideas and participates in the discussions. Prior to the morning meeting, reporters do their own research and know what is happening in their own beats. A *beat* is an area that a reporter is assigned to cover regularly, and may include a police beat, a city council beat, an education beat, etc. A beat may also be a specific geographic section of the viewing area. At the end of the meeting, the news directors and producers make decisions on which stories will be covered for the newscast. The assignment editor hands out assignments to the reporters and, if necessary, pairs them up with photogs.

beat: A specific area (topics or geographic location) regularly covered by a reporter.

PRODUCTION NOTE

In an academic broadcast journalism class, typical beats might include the English department, Guidance department, Student Government Association, sports, student activities, theater, co-op, cafeteria, school administration, music, etc.

After the morning meeting, reporters usually begin to "work" their stories by making phone calls to arrange interviews. They complete the research necessary to effectively interact with their interviewees. Meanwhile, the producer begins to organize the newscast with the assumption that the assigned stories will be complete before air time. The organization of the newscast script is called the *rundown*, and is extremely general in its first draft. The rundown is a constantly changing outline of time slots in the news program. A common, but not universal, sequence for a local news program is:

rundown: The organization of stories and sequence of a newscast in written form.

1. Hard local news
2. Hard national news
3. Lighter news
4. Sports
5. Weather
6. Arts, entertainment, and evergreen filler

As reporters head into the field to shoot their stories, they keep in communication with producers at the station. The producers continuously update the rundowns and begin to determine the TRT necessary for each story. They must also consider the amount of time consumed by ads that run during the newscast. The early evening anchors work with the rundowns and constantly update and revise the script for their part of the early evening newscast. The entire newscast must fit into the allotted block of time.

Reporters commonly work at least two stories each day. This is a critical issue. If reporters are required to produce more stories each day, the amount of time available to work carefully and accurately is directly affected. Even so, attention to detail and accuracy must be consistently represented in a reporter's work in order to remain employed.

During the afternoon, late evening anchors arrive at the studio and begin planning for the late evening newscast. At this time, reporters from the morning meeting are coming back into the station to write up their stories, record narration, and begin editing their packages. As the afternoon progresses, the packages are viewed and approved or tweaked, as necessary, to fit into the story's allotted time. The anchors and producers pull the script together for the early evening telecast, and the teleprompter is loaded with necessary text (**Figure 9-12**). The anchors rehearse with the teleprompter, if there is time.

The early evening newscast is broadcast while the daytime newsroom staff ends their workday, and the late evening shift takes over to prepare for the late night news. The same process begins during the late evening news broadcast, as the overnight shift comes in to prepare for the next day's early morning newscast.

As a result of the media convergence taking place in the broadcast news industry, most newsroom staff have the additional responsibility to place news content on the station's website. This means that a reporter must learn to write audio for TV, as well as text for the Web. The station's website may include graphics, maps, or footage related to a story, but not used in the original newscast. The reporter may even blog on the website about the development of the story. Many consumers have their personal electronic devices (cell phones, computers, etc.) set up to receive news

Figure 9-12. The scripted lines for the anchor(s) are entered into a computer to be displayed on a teleprompter. *(Countryside High School, Clearwater, FL)*

alerts and updates 24 hours a day. While the day in a newsroom is spent preparing for the next newscast, the website is updated continuously.

The previous information is a general plan for a normal day, but days in the television news business are rarely normal. Breaking news stories can throw any normal schedule into chaos. The television news business constantly adapts to the news of the world. This continuous state of change is part of the allure and excitement that surrounds the television news business, but is also the cause of high stress levels. Working in the broadcast journalism business is not a 9-to-5 job. The hours industry professionals work are the hours necessary to get the job done, regardless of when or how many hours that may be.

Wrapping Up

Broadcast journalism professionals have an awesome responsibility. Most of the viewing public believes that if they see something on the news, it must be true. Every member of the television newscast team must contribute to telling the truth in its most unbiased form. They must determine what the audience needs to know and wants to know, without expressing what the audience should think about any topic. A good reporter can tell a story without revealing how he personally feels about the story or individuals in the story. This unbiased approach extends even further to the entire news program. Fair coverage of a wide variety of stories and providing balanced coverage (presenting different sides of a story) is important to maintain credibility. The news media has the public's trust and must do everything in its power to maintain that trust, because once lost, it is nearly impossible to regain.

Review Questions

Please answer the following questions on a separate sheet of paper. Do not write in this book.

1. What is *broadcast journalism*?
2. Define *mainstream media*. Give examples of mainstream media programming.
3. What are characteristics of an appropriate story source?
4. Explain how "consequences" factors into judging the newsworthiness of a story.
5. Which news elements are usually associated with soft news?
6. What is an *evergreen* story?
7. What is the function of an IFB?
8. How is a reader different from a VO story?
9. What is an *outro*? Give examples of typical outro lines.
10. What is the purpose of a stand-up?
11. What are the characteristics of a package?
12. Identify the challenges in investigative reporting.
13. List and define some of the common abbreviations used in newscast scripts.
14. What is a *beat*?

Activities

1. View several different types of news programs and identify which of the three basic categories (mainstream media, non-mainstream media, and tabloid media) the program falls into. List at least one program for each of the categories and be prepared to share your list with the class.

STEM and Academic Activities

1. Explain how satellite communication has changed how the news media covers stories.

2. Research the salaries of the various broadcast journalists, both cable and network. Compare the salaries to determine if there is a connection between the journalist's salary and the program's rating.

3. If reporters commonly work two stories a day and work six days per week, how many stories will a reporter have worked on after a year on the job?

4. Record three local news programs and compare the number of hard news stories to the number of soft news stories each program airs. What is the average number of hard news stories aired? What is the average number of soft news stories aired?

5. Discuss how the Internet has affected broadcast journalism on a local and national level.

Chapter 10

Newswriting for Broadcast

Objectives

After completing this chapter, you will be able to:

- Identify ways to find newsworthy stories.
- Explain how the angle of a story affects how the story is written.
- Summarize the concept of "writing for the ear."
- Apply the guidelines for good news story writing.

Introduction

Whether a news story appears on a newscast as a reader or as a fully prepared and edited package, the story first needs to be written. This chapter presents many topics a reporter must consider in writing a story for broadcast, aside from the video footage (**Figure 10-1**).

Professional Terms

attribution
angle
close
hard lead

lead
reporter track
soft lead

Figure 10-1. A reporter must manage the entire process of writing a news story.

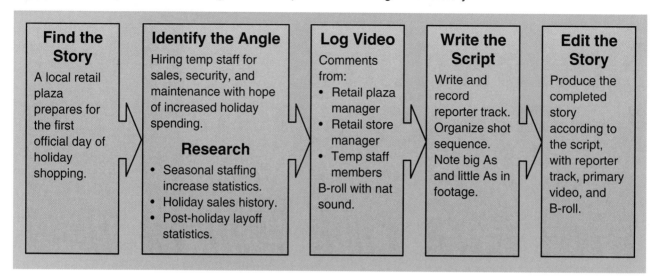

Find the Story	Identify the Angle	Log Video	Write the Script	Edit the Story
A local retail plaza prepares for the first official day of holiday shopping.	Hiring temp staff for sales, security, and maintenance with hope of increased holiday spending. **Research** • Seasonal staffing increase statistics. • Holiday sales history. • Post-holiday layoff statistics.	Comments from: • Retail plaza manager • Retail store manager • Temp staff members B-roll with nat sound.	Write and record reporter track. Organize shot sequence. Note big As and little As in footage.	Produce the completed story according to the script, with reporter track, primary video, and B-roll.

Finding Stories

It may be difficult to imagine how reporters can find stories to write about day after day. Whether in a small town of 1000 people or a metropolitan area of 2 million people, local television stations manage to run news programs with new stories several times a day. Stories are out there for a reporter to find. In determining if a story is newsworthy, ask yourself:

• Is there some conflict in the story to sustain viewer interest?
• Is the story unusual?
• Is someone well-known involved in the story?
• Is there a segment of the audience that will be impacted by the story?
• Can the story be brought "home" to the local audience?
• Does the story include emotion or human interest aspects?

Reporters have a well-rounded base of general knowledge, are particularly aware of their immediate surrounding environment (local government and politics, locations and geography, various agencies, current issues affecting the local population), and understand which topics of public interest motivate, excite, worry, and concern the audience. Reporters listen, read, watch, and ask questions. Remember: Who? What? When? Where? How? A reporter should always be thinking, "What is the story?" Anytime there is a lively conversation in the reporter's vicinity, he should be alert to the topic and recognize it as a possible story idea. It doesn't matter if people involved in the conversation are arguing, laughing, sharing, or discussing; if those people are interested in talking about something, it is a potential story. Also, any story or program on television is a possible springboard for a local story.

PRODUCTION NOTE

People are often somewhat self-centered—interested only in what *they are* interested in. An effective reporter must be interested in what other people are interested in, and be able to recognize and develop those topics.

In developing a human interest story, for example, consider that nearly everyone has something that is particularly interesting to them. This interest may be a hobby, craft, leisure activity, relative, living environment, physical location, memory, or an object (such as a special antique, heirloom, car, recording, or collection.). The reporter's job is to find the "thing" about a person, ask questions to get the most interesting information, and turn it into a story.

Assistant Activity

Begin a conversation with a classmate and do not end the conversation until you have discovered five new and interesting things about that person. The longer you talk, trying to find something to say, the more both of you will open up and share interesting information.

Finding Stories in an Educational Environment

Reporters watch what is happening around them—in school, in the car, on the bus, in the cafeteria, at the mall, on television, and on the Internet. Any event, visitor, poster, bulletin board, or classroom assignment is a potential story. Student reporters should tour the school and try to view it as someone who has never been there before, **Figure 10-2**. Story ideas may come simply from the surroundings. For example, passing the classroom where the yearbook is produced might spark a story about the new yearbook staff and this year's yearbook theme. Walking past a student wearing a trendy, branded t-shirt may inspire a story about brand status or the shopping habits of students. The activities director or master calendar in the school is a great source of information on upcoming events. Knowing about events that are scheduled provides a direction or source to gather more information, such as the organizations sponsoring an event, and write a story.

Figure 10-2. Pay attention to activity in the school hallways. From activity banners to the students themselves, potential stories may be right in front of you!

VISUALIZE THIS

You see that there is a shoe drive scheduled for next month, but you've never heard of a shoe drive. Bingo! If you've never heard of it, many other students at your school probably have never heard of it either. Just the name, "shoe drive," sounds odd enough to make you curious to find out more. You note the event information and seek out the sponsors.

It turns out that the shoe drive is a request for students to bring in old shoes they no longer wear. The shoes are collected and donated to a charity that gives them to people who need shoes.

You could write a story that merely announces the upcoming drive. Or, you could use your curiosity to create an interesting package that might actually increase the number of shoes collected in the drive. Perhaps you can go to the charity shoot on-camera interviews with those involved in the drive. Maybe you can contact some people who have been helped by the charity—they may not be willing to have their faces on camera, but the story is about donating shoes, anyway. Get footage of feet of all sizes, both with and without shoes.

You can either simply make an announcement, or you can develop a thought-provoking feature story. And remember, it all started with two words on an activity calendar in an office: "shoe drive."

Go into the community outside the school building to discover people and events in the local area. Visit craft fairs where artists and crafters sell their work. Read the local newspaper to find stories about local people, events, and retailers. A story from the newspaper may be further researched and enhanced with video to make a compelling package.

There are innumerable stories that can be written based on interviews and activities surrounding the sports, music, theater, and art programs at your school. Competitions are almost always news story topics because they contain the element of conflict. Competition stories may include Mathletes, science fair and social studies projects (may also showcase students for achievement), or the band preparing for a marching competition, and a follow-up story with the results of the competition. Any course that involves visual classroom or laboratory activities, such as career and technology classes, can provide compelling video.

As course registration time comes around, small features may be written about the guidance department and the various elective classes offered. Additionally, consider an in-depth interview with a teacher who has an interesting, but little-known characteristic, pastime, or life experience.

- The English teacher who is a weekend paintball aficionado.
- The drama teacher who was once in the New York cast of *CATS*.
- The math teacher who just returned from a tour of duty in the Middle East.

Student reporters may also be assigned different beats that cover all areas of the school. Covering a beat involves developing a relationship with people in that area of the school, knowing the purpose and responsibilities of the group or department, and knowing their calendar of regular and special events. The reporter should check-in on a regular basis to keep up-to-date with any new or unusual changes or events. For example, a new science credit is required for graduation. The student assigned to the science department beat should be on top of the story. Which class is now required? Who will be teaching it? What will the class cover? The student assigned to the guidance department beat should also be involved in the story. When will the requirement go into effect? Which students will be affected? When can students begin registering for the class? Ideally, both student reporters would get this information by diligently covering their beats and would bring up the topic in the pre-production morning meeting. The producer or assignment editor decides if it is a story, who will do the story, and from what angle the story should be developed.

Researching Stories

Once a story is determined, research begins. The purpose of research is to gather all the information necessary to frame a story responsibly, fairly, accurately, and completely for viewers. There are usually several ways to get information, and the reporter should be persistent in finding out everything possible about the story and getting the facts straight. The information and details of a story should be double-checked to ensure that every word is verifiably truthful and factual. Hearsay is as unacceptable in reporting as it is in courtrooms; hearsay is gossip, not reporting.

The amount of research necessary can vary a great deal depending on the story type and approach to a story. In the previous "Shoe Drive" example, the research could be as brief as getting the details of the drive in a quick conversation with the event sponsor, or may be as involved as researching the charity, its operation, and its clients. A story may cover an accident on the main highway through town, which caused the road to be closed in both directions and will affect hundreds of local residents. Researching the exact road location and other details with the police department may be as simple as listening to a police scanner and making a follow-up phone call. In this case, the story can be written and put on the air as soon as possible. Or, the news director may choose to send a camera crew out to the location and do a live feed or produce a package of the story for a later newscast. The actual research necessary for this story is very minimal, but is absolutely necessary to verify the facts of the story.

Deadlines are a constant concern for reporters trying to be diligent about responsible research. An approaching deadline should not compromise a reporter into writing and airing a story that has not been fully researched. Once a story is aired, incorrect information cannot be taken back without embarrassing the reporter, news program, and higher administration. Airing incorrect information may also have legal consequences for the station and can endanger viewers. Reporters should always work under the assumption that they will have to prove everything they say or write. *Attribution* (crediting the source of information) should always be given for the quotes, information, and facts of a story.

attribution: Crediting the source of information used in a story.

Two important concepts in developing a news story are "KISS" (keep it simple, silly) and "be complete." These seem to be in opposition with each other, so reporters must carefully balance both concepts at all times. To keep a story simple, write with simple sentence structure using simple language. A story should not be cluttered with insignificant or irrelevant details. A complete story leaves viewers with no unanswered questions.

A reporter needs to fully understand the story and find an angle before writing the story. An *angle* is the approach used to tell a story, which helps the viewer understand why this story is important, why the viewer should

angle: The approach or point of view used to tell a story.

care, and what makes this story unusual or different. In the previous "Shoe Drive" example, using feet and shoes as the main images in the story could be an interesting approach to the story for both the reporter and the viewer. The language used to write the story should allow the audience to under-stand and care about the story, as well.

One angle that is frequently used is to tell a story through a character. The reporter chooses a person who is part of the event or affected by the topic of the story and uses that person as the "face" of the story. Taking this angle personal-izes events, issues, and conflicts, and allows viewers to identify with the char-acter. For example, a story about minimum wage jobs could focus on statistics (such as income, living expenses, and number of people with minimum wage jobs). However, when the topic is examined through a day in the life of a single mother with a minimum wage job, statistics become personal and the story has a face and significance with the viewing public. The character can tell the story from firsthand experience, which the reporter cannot do.

Newswriting Fundamentals

Chapter 8, *Scriptwriting* addressed the type of writing necessary for var-ious kinds of non-news programs. Newswriting is different from the writ-ing style used for other program types. However, one concept that applies to all types of scriptwriting is the kind of language used. In scriptwriting, informal language is used to write the way people speak. This also applies to writing stories for news programs and is called "writing for the ear."

Reporters should have command of language, sentence structure, grammar, and vocabulary, and should actively search for the precise right words to use in a story. The language used in a news story needs to be simple and direct, so the meaning is understood the first time it is heard. Sentences should be short and should not contain long clauses. Remember, not all television viewers can rewind the program and re-play something they did not understand the first time. While some viewers *may* use their DVR to rewind and replay something they want to hear again, viewers should not need to listen twice to understand the information.

For newswriting, use simple sentences written in active voice, rather than passive voice. Also, use simple subject-verb-object sentence construction.

- Active: "The stunt car hit the ramp, flew through the air, and landed in the pile of hay bales. The driver climbed out and waved to the crowd." "The mayor called a city council meeting."
- Passive: "After going up the ramp, through the air, and landing in hay bales, the driver climbed out and waved to the crowd." "A city council meeting has been called by the mayor."

Try to avoid using forms of the verb "to be" coupled with a past participle, such as "has been called." These phrases typically make a sentence passive.

Use present tense as much as possible. The very nature of news does not always lend itself to telling a story in present tense, but using the pres-ent tense engages the interest of the audience.

- "The police are investigating last night's accident on Route 13, which resulted in one fatality."
- "Governor Jones says that…"
- "The Health Department urges consumers to …"

Assistant Activity

Write a news story. Keep "writing for the ear" in mind while writing and revising your story. Read the story out loud to a friend or family member. At the end of the story, simply stop talking and wait. If your listener asks a single question about the topic, that question needs to be addressed in the story.

Perhaps you need to word something differently or you accidentally left something out. Fix your story and read it to another person—do not read it to the same person again. Continue to read aloud and revise your story until your listener has no questions at the end. Television does not allow conversation between the reporter and viewers; the audience cannot ask the reporter questions. Television is similar to a lecture format, without the opportunity for questions and answers at the end.

With the very first line of a story, viewers decide if they will continue paying attention. Phrasing is crucial in delivering your message to the audience. To effectively communicate with viewers, the content of a story should be stated as clearly and accurately as possible. The following are suggestions for good news story writing.

- Never start a story with a participle or word ending in -ing: "Saving the resources of the Chesapeake Bay was always in the thoughts of the conservation group." The listener must unscramble the sentence to make sense of it—it's just not the way people talk.

- Avoid introducing a story by asking viewers a question: "How do you feel when you receive a speeding ticket in the mail?" Instead, make an attention-grabbing statement: "Drivers caught by traffic cameras are speaking out."

- Do not begin a story with a quote read by the reporter.

- Do not scare the audience with your words. Say: "Officials urge you to go into your basement and move near a masonry wall until the tornado passes." Don't say: "The tornado will destroy your house and everything in it. Hide in your basement until the danger has passed."

- Give suggestions that repeat the message of officials; do not give orders. If an order needs to be communicated, turn the mic over to an official to state the order.

Visualize This

A snowstorm has started and the reporter goes on air with a story about the local transportation officials mobilizing the snowplows, drivers, and salt spreaders. Viewers are usually interested in stories about preparations for weather events that may affect them personally. In the newscast, the reporter passes along a request from the head of transportation, "Transportation officials request that citizens stay off the roads during the snow clean-up efforts." The reporter is not actually telling the public to do or not to do something, but is passing a message along from the officials. The reporter attributes the action to the official who gave the recommendation, which takes the reporter out of the story. It would not be acceptable for a reporter to tell the audience to do something on his own authority.

- Use action verbs, when possible.
- Do not offer your opinion by commenting that something is bad, good, interesting, or shocking.
- A person's name should never be used at the beginning of a story, unless the person is well-known. When a person's name is used in a story, always provide an identifying title or the reason the person is in the news story. Mention the person's title or reason for involvement *before* stating their name, so viewers understand the importance or context of the person.

VISUALIZE THIS

You are a single male at a party. Many of your friends have been trying to "fix you up" with potential prom dates. You've already been introduced to four ladies at this party. Your friend Chuck arrives and introduces another woman, Christine. Christine is the most attractive woman you've met so far. It would certainly be important to mention that Christine is Chuck's new girlfriend, right at the beginning of the introduction. Knowing Christine's title or connection to the event/person (Chuck) is a great deal more important than knowing her name, considering all the attempts at fixing you up at this party. Having this information immediately would avoid a very embarrassing situation.

A reporter has a finite amount of time to tell a story and relay all the information viewers need to know. It is important to purposefully choose the most effective words to tell the story.

- Do not use a long word when a short one will do. Say: "The colors on a plasma television match the colors of things in real life." Don't say: "The chrominance and luminance on a plasma television are reproduced accurately."
- Do not start a story with trite and cliché phrases that do not provide any useful information, such as "Once again," "In the news," "A new development," "As expected," and "In a surprise move."
- Mention a person's age only when it is relevant to the newsworthiness of the story. For example, "A ten year old graduates from Harvard."
- When footage or images are included in a news story, the reporter should not waste words by narrating with information the viewer can plainly see, or stating the obvious. Phrases like, "As you can see," "Here is a," and "This is an" typically describe what the viewer can see for themselves. A picture is worth a thousand words. The reporter's time and words are better spent in providing viewers with more relevant information that may not be obvious in the image on screen.

Preparing a News Package

The fundamentals of newswriting, including simple sentences and language, present tense, and active voice, apply to any type of news story. A package story that incorporates interview footage with narration by the reporter poses additional writing challenges. For example, a reporter extensively researched the designer of a new high school theater and contacted the designer to schedule an in-person interview. After conducting the interview, the reporter and photog return to the studio with interview footage and notes. The reporter must now put the story together.

The first step is to log the video footage. Logging the footage is necessary so the reporter can quickly find each statement on the recording during the editing process. The reporter first views the recorded interview footage and notes the time code (specific location address code) for each question, each answer, and, if necessary, the main point of each answer. Also, all the comments made by the interviewee should be transcribed with the corresponding time code noted. By reviewing a written copy of the interviewee's comments, the reporter can easily decide which comments to use as sound bites and where to find the video and audio for the comments. After logging the interview, the reporter logs the B-roll footage to review other footage that may be inserted to support the story. Nat sound is also logged at this time for use later, as necessary.

Everything spoken by the reporter in a package is the *reporter track*, **Figure 10-3**. The reporter track connects all the interview sound bites used in the story and provides viewers with additional information not contained in the sound bites. A package rarely includes audio of the reporter's original question. A novice reporter may be tempted to ask interviewees to restate the question in their answer, but this is not recommended. This technique results in unnatural and awkward responses from interviewees. A good reporter can write the reporter track and cut sound bites together so that the meaning of the interviewee's statement is clear without hearing the original question.

Some of the interviewee's comments recorded on the footage may phrase things better than the reporter can. These are noted as "big A," or "big answer," comments. "Big A" comments may be emotional statements made by interviewees, narration, or simply a good turn of a phrase. Other recorded comments may be more efficiently and clearly summarized by the reporter. These are noted as "little A" comments, and they become the basis for the reporter track.

> **reporter track:** Everything spoken by the reporter in a package.

PRODUCTION NOTE
Experienced reporters can actually begin writing a story in their mind while on location or even while an interview is in progress. They envision the presentation sequence for information and recognize which comments from an interview should be used as sound bites in the package. Experienced reporters can often write and shoot stand-ups while on location for a story. However, beginning reporters need to work through the process of logging the tape and studying the information and footage to determine the best way to present a story to the viewer.

After the video footage has been logged, the reporter can begin writing the story. The very first sentence of a story is the *lead*. A *hard lead* begins the story abruptly and does not waste words. It contains a straightforward

> **lead:** The very first sentence of a story.
>
> **hard lead:** The first line of a story that begins the story abruptly and immediately presents the most important information.

Figure 10-3. The reporter track is presented in the audio column of a package script. In this package example, the reporter track is highlighted in three cells of the audio column.

Video	Audio
Wide shot of reporter Stephanie Carter standing on stage in new auditorium Stephanie Carter Raider Television Super	Stephanie Carter/Stand-up The programs are printed. The tickets are on sale. Tonight is dress rehearsal for the first performance in the new Roane County High School auditorium. This new facility is equipped with features that will enhance any production. Senior Doug Miller is the sound engineer. While others have studied their lines, Doug has been busy learning everything he will have at his fingertips to make those lines sound perfect.
Doug Miller Sound Engineer B-roll of sound board	SOT Doug Every cast member will have wireless microphones like these placed at their temples, sort of like this. They will be almost invisible. I have a volume control here on the sound board for every person, and I've had to learn the settings for each to make them all even. Then, I also have to deal with the music and sound effects. This new sound system has the potential to be perfect, but only if I'm perfect at MY job. I'm pretty nervous.
B-roll of rehearsal	Reporter VO Doug has been at every rehearsal, fine-tuning those settings. He's not the only one who has to learn some new technology. When the curtains open and the lights come up...
Su Kiki Lighting Technician	SOT Su Kiki That will be me, I'm running the lights. We're using a total of 75 different lights for this production. This is my light board and I have a play script with all my lighting cues marked. Some of them are programmed in for different combinations for certain scenes, but I still need to have my timing just right to make everything happen.
	Stephanie Carter/Stand-up Making everything happen—that's what it's all about. Tomorrow night, we'll meet the cast of "The Execs" as they get ready for opening night on Saturday, right here at Roane County High School. This is Stephanie Carter for Raider Television.

soft lead: The first line of a story that communicates the general idea of a story, but does not offer any facts.

action verb and is active, not passive. The most important information is presented immediately. For example, "A bomb threat caused the evacuation of City Hall today." A *soft lead* communicates the general idea of the story, but does not offer any facts. It often sets the scene or introduces the characters. For example, "It's noon. It's quiet. That's about to change. In less than three hours, the Cowboys will take the field in front of thousands of fans and the quarterback decision will be history. The controversy started last week when..."

The reporter scripts a package by writing the reporter track to connect the "big A" comments. A good reporter does not write, "We asked Joe about the new theater and this is what he said" as a lead in to Joe's answer (big A). A more eloquent and interesting lead in may be, "Visitors to the new theater at Roane County High School find several features especially nice." The script then cuts to the "big A" of Joe talking about the surround sound system of the theater, with a lower third key identifying him by name and title. The B-roll footage is reviewed to determine which images may be inserted into the script to make it stronger and provide visuals during the audio of the reporter track. The time code of the B-roll is entered into the story script, as well.

The ending of a story, or the *close*, may look to the future—what will happen next, who will be called to testify next, or when is the next game? Sometimes the close may be a "punch line" that sums up the story.

close: The conclusion of a story.

Once the story is written, the reporter usually records the reporter track or VO. The written story, reporter track, primary video, and B-roll tapes are then sent to the editor to put the story together. In smaller studios, the reporter may be responsible for editing the video and story together. The package is given to the producer when complete, and the reporter moves on to the next story.

Reporting the News

Viewers choose a preferred news program for a variety of reasons. Aside from the availability of channels, some viewers may choose a news program because they like the "look" of the set (**Figure 10-4**), the personalities of the anchors and reporters, or the physical appearance of the on-air personnel. Some choose to get the news from websites for convenience and may access additional information and features not included in a regular on-air newscast. News professionals hope viewers choose their news program because the content of their news show is the best produced in that time slot—excellent video and audio, near perfect performances by reporters and anchors, and the most pertinent, complete, and accurate news. No matter the viewers' reasons for watching, reporters have an obligation to the audience—report the news truthfully. Reporters obtain information,

Figure 10-4. An attractive set may attract loyal viewers. *(Countryside High School, Clearwater, FL)*

Television Production & Broadcast Journalism

process and organize the information, and give facts to viewers in the most understandable way. Viewers then make their own decisions and form their own opinions.

VISUALIZE THIS

Today is April 15—Federal Tax Day. A reporter does a story about people who wait until the last minute to file their taxes. Video for the story includes a long line of cars waiting to get into the parking lot of the post office, which is completely full. The reporter simply reports that taxes are due and does a human interest piece with video of people who waited until the last minute to file. Included may be man-on-the-street interviews with a few drivers commenting on why they waited until the last minute to file.

The reporter does not launch into an opinion piece about taxes being too high, blaming all the ills of the country on the current administration, and suggesting that citizens rebel by not paying their taxes at all. It is unacceptable and unprofessional for reporters to present their own opinion in a news story.

Wrapping Up

Finding stories in the world around you is easy once you realize that a story can be anything that keeps people, including yourself, engaged. If something is interesting to one person, it is likely interesting to others, either as participants or observers. The reporter is responsible for bringing topics of interest to the viewers. Reporters diligently research subjects and double-check facts before passing information along to ensure earnest reporting, not gossiping. To write a story, reporters find just the right angle to keep viewers interested in the story and choose words purposefully to avoid interjecting the their own opinions. A story that is told truthfully and well, informs the public and supports the reporter's professional reputation.

Review Questions

Please answer the following questions on a separate sheet of paper. Do not write in this book.

1. What questions should you ask yourself to determine if a story is newsworthy?
2. What are some story sources in an educational environment?
3. What is the purpose of researching a story?
4. What is *attribution*?
5. Explain "writing for the ear."
6. What is the *reporter track*?
7. What is a *hard lead*? Give an example.

Activities

1. Watch several news programs and choose five stories that you find interesting. For each story, identify what makes the story newsworthy, what the angle of the story is, and note any attribution given during the story. Be prepared to share your findings in class.

STEM and Academic Activities

1. Identify the technological advancements that have made the process of researching stories easier. Explain how each advancement is used to research stories.

2. Create a list of possible news stories that can be written about your school. Of the topics on your list, what percentage of the stories are sports topics? What percentage are academic topics? What percentage are entertainment topics?

3. Choose three current event news stories and write a soft lead for each story.

Social Science

4. Watch three local news programs on different stations. Which of the three news programs do you prefer? Why do you prefer one program over the others? Compare and contrast the news set designs and on-air personalities when explaining your preference.

Chapter 11

Interviews

Professional Terms

background
B-roll
lead

Objectives

After completing this chapter, you will be able to:

- Explain the purpose of gathering background before an interview.
- Create interview questions and topics based on background research.
- Identify the differences between shooting an interview that is aired live and shooting an interview that will be edited into a package story.
- Explain the function of B-roll.
- Recognize effective techniques for conducting an interview.

Introduction

The interview is the most common element of television news. Nearly every story involves either an on-camera or off-camera interview with someone involved in the story (a major participant or person affected by events). Light-hearted interviews may be simple and require little preparation, such as man-on-the-street interviews asking people what they bought their significant other for Valentine's Day. Interviews with reputable individuals that address serious topics require considerable preparation and should provide viewers with in-depth information. Reporters who competently conduct substantial interviews find their credibility with peers and viewers increases with each successful interview.

This chapter addresses how a reporter should prepare for and conduct a successful interview, the journalistic skills involved in a productive on-camera conversation, and technical aspects of recording an interview.

Preparing for an Interview

Once the assignment editor or news director assigns an interview to a reporter, the reporter must become acquainted with the topic and the interviewee. A topic that is relatively unknown to the public is probably also unfamiliar to the reporter. Properly preparing for an interview involves thorough research and development of informed and well-crafted questions.

Research

As with a regular news story, research is the first step in preparing for an interview. All the information gathered through research prior to conducting an interview is called *background*. Thorough research demonstrates to the interviewee that the reporter put forth effort to obtain knowledge about the topic. Sufficient background allows the reporter to hold up his end of the conversation with an interviewee, rather than absently asking questions without interest in the answers.

background: All the information gathered through research prior to conducting an interview.

The sources available for research depend on how well-known the interviewee is. While much of the research for some interviews may be accomplished by talking to a few people (**Figure 11-1**), other sources for background research may include residential, business, and government agency listings in the telephone directory, the library, the Internet, newspapers, and magazines. Additionally, the reporter should always research what has already been reported by other media outlets.

Research for a story about a successful gymnast named Michael Christopher, for example, who sustained a serious injury at the last gymnastics meet may involve finding the answers to the following questions:

- What is Michael's past gymnastics record?
- Which event Michael was participating in when the injury occurred?
- How common is this injury?
- How did Michael's injury occur?
- What is Michael's prognosis?

Figure 11-1. Speaking directly with people who are knowledgeable or involved in a topic is a common research resource for interviews.

This information could be gathered during the interview with Michael, but having thorough background information allows the reporter to better formulate the interview questions. The reporter may find this information in previous media reports, on Web sites, earlier newscasts, or in sports stories of newspapers. Michael's family or coach may also be helpful sources while researching this story. In this example, the reporter should also get information on the type of injury Michael sustained and the coach's opinion of how Michael's gymnastics future may be affected by the injury. If information about Michael's injury is available to the reporter, talking to a sports medicine professional may provide details about the nature of the injury and the approach doctors typically take for treatment. However, the reporter is unlikely to get information from Michael's personal doctors due to privacy issues.

While researching a person or topic, a reporter may encounter technical or topic-specific jargon. When speaking to the gymnastics coach, for example, words or phrases specific to the sport of gymnastics may be used. The general public is probably not familiar with these phrases. The reporter must become acquainted with this jargon in order to research and conduct an effective interview. In researching the gymnast story, the reporter might visit a gymnastics practice to observe the particular gymnastic event and the environment in general.

Taking the time to gather appropriate background for a story prevents insulting the interviewee and sends the message that the interview is important. By properly researching the topic or person, the reporter can avoid asking questions the interviewee may interpret as uninformed or offensive. For example, asking Michael, the gymnast, about executing backward flips on the balance beam is not appropriate because men do not perform on the balance beam—it is a women's event. Intelligent questions will flatter the interviewee and contribute to building rapport, which helps the person *want* to carry on the conversation. If the reporter has not properly prepared for the interview, the interviewee will sense the reporter's lack of interest and knowledge; the reporter will find it very difficult to get candid, in-depth answers.

Preparing Interview Questions

A good reporter does not let an interviewee control the direction of the interview. Developing a list of interview questions based on the background obtained helps the reporter remain in control during the interview. Given the background research, the reporter should also be able to anticipate some of the interviewee's answers to the questions listed. This allows follow-up questions to be developed before conducting the interview. Questions that can be answered in just a few words should be avoided. With proper preparation, a reporter can formulate questions that are more likely to provide good material for sound bites, such as

- "What will probably be the next step?" (prediction question)
- "How do you feel about…?" (opinion question)
- "Tell me just how this happened." (narration question)

Some of the planned questions may not be asked as written or may not be asked at all during the interview. However, having a list of interview questions can certainly save a stalled interview, **Figure 11-2**. Some may

Figure 11-2. Reporters may refer to written questions during an interview, but should not read questions directly from the card or list.

say that a reporter should never go into an interview with a list of questions, but even Barbara Walters has a few index cards visible in many of her interviews. An interview may go in unexpected directions that can be more interesting than the direction originally planned. Reporters are free to follow the unplanned direction with questions. During Presidential press conferences, White House press corps reporters can be seen actually reading their questions to the President off of cards. These questions are carefully prepared and worded to get the most information possible from the President's answer. For student journalists, writing the questions before an interview helps to mentally prepare the student interviewer.

After questions are written and thoroughly reviewed by the reporter, the questions can be shortened to just a few words that represent specific topics or categories the reporter can explore with questions. For example, a reporter is preparing for an interview with a broadcast attorney about issues related to music copyright releases for television. One of the items on the list of interview questions is, "If I produce a video yearbook, is it legal to include popular music in the audio track?" The abbreviated topic for this question may simply be "video yearbook." Since the entire interview is about music releases, the two-word topic is enough to remind the reporter to ask the question during the interview. The reporter can formulate questions on the spot, rather than appearing to read questions off of a page. Reading questions word-for-word from a list is uninteresting and does not engage the interviewee or the viewers—the interview will quickly fall flat.

Scheduling an Interview

Once background work is complete and the interview questions are formulated, the reporter must contact the interviewee and schedule the interview.

Professional phone etiquette should always be observed. During the phone conversation, the reporter should speak confidently about the interview topic, having the knowledge gained through background research. How to approach an interviewee varies from topic to topic, person to person, and reporter to reporter. Experience is the best teacher in this area. The following are some possible approaches for capturing an interviewee's interest and cooperation.

- "I am doing a story on <the subject> and have information about <be specific>. I'd like to hear your side of the story."
- "A friend told me that you have a technique for saving all the small slivers of soap that remain when a bar of soap is almost used up. As a consumer reporter, I'd like to let the public know…"
- "I'd like to do a human interest story that focuses on your gymnastics career and recovery from your injury."
- "I just saw a story in the local paper about the art award you won. I'd like to do a feature about you and show how you create your stained glass pieces. I'm sure our viewers have no idea exactly how these works of art are made."
- "I was looking through our old yearbooks for a possible story about how sports have changed, and I noticed that your father was on the state championship football team in 1985. Now, *you're* our quarterback! Can I do an interview with both of you about playing football at this school?"
- "We all know you as one of our math teachers, but I heard you were on the winning paintball team last weekend! Most of the students here would be surprised to find out that you play paintball, and play it quite well. I'd like to do a story about you. Could we set up an interview and maybe go out to the paintball field to talk about your strategies?"

Not all interviews are completely pre-planned. For example, a reporter might do a story on a local band playing a popular venue. After the reporter completes the stand-up, he may decide to approach a few patrons to ask their opinions on the band or the performance. The reporter creates questions on the spot for these interviews.

Shooting an Interview

Sometimes, an interview happens on location with little preparation time. Even on short notice, the goal remains to provide quality video and audio signals. The reporter and camera operator should arrive at the location prior to the scheduled interview time to allow for proper setup, **Figure 11-3**. For example, lighting instruments must be in place and turned on, and the camera operator must have the camera set up and white-balanced before the interview begins.

In general, journalists conduct two types of interviews for broadcast:

- An interview that is either aired live or recorded to be shown in its entirety, without editing.
- An interview that is designed to be edited into a package story.

An interview that is either aired live or recorded to be shown in its entirety may be very formal and lengthy. This type of interview takes place

on a set in a television studio or at a location related to the topic of the interview. To shoot an interview on location, the crew arrives in advance and arranges portable lighting instruments to ensure the lighting is even in the interview area. Each person involved in the interview, including the reporter, is outfitted with a microphone (probably a wireless lapel mic) that likely goes into an audio mixer. In some cases, more than one microphone can be fed into a camera to provide even audio. More than one camera may be used for this type of interview, and the shots switch between cameras in the final, broadcast interview. The reporter is treated as on-camera talent during the interview, and must look and act accordingly. This type of interview may also be conducted as a shorter, live-feed interview from a remote location that is broadcast during a newscast. In this situation, the reporter and interviewee stand together in front of one camera, with the reporter's handheld microphone shared with the interviewee—the reporter alternately points the mic at himself and the interviewee.

In an interview designed to be edited into a package story, the reporter is not usually seen asking questions. Instead, the interviewee's answers are recorded so the responses can be used as sound bites in the package. This type of interview is short in length and is shot with one camera. The interviewee is the only person wearing a lapel microphone. If the reporter uses a handheld microphone, the shot should be framed to show very little, if any, of the microphone. Raw footage of the interview includes the reporter's voice asking questions, as picked up by the interviewee's mic. However, the reporter's voice will be edited out, so sound quality of the reporter's question is not important. These interviews are usually shot using only natural lighting. However, a reflector or on-camera light may be used to fill shadows on the interviewee's face.

Reporters and photographers position themselves differently when interviewing one person to obtain sound bites for a package. The reporter faces the interviewee with his back to the camera, so that the reporter and interviewee can make eye contact, **Figure 11-4**. The photographer stands behind and slightly to the side of the reporter, and shoots the entire interview as an over-the-shoulder shot. The photographer can zoom in slightly to frame the interviewee's face so it fills most of the frame, while leaving enough room below the interviewee's chin to add a lower third graphic. The shot should leave only a little head room and some nose room on the side of the screen that the interviewee is facing. Remember: the interviewee should be making eye contact with the reporter, not the camera. The resulting visual effect is that the viewer is a spectator to the conversation between the reporter and the interviewee—the viewer is not addressed directly. Photographers may get creative with the shots for this type of interview, such as including related items in the foreground or background of a shot for impact or clarification. However, what the interviewee says is the most important part of the shot for a sound bite. If the camerawork is too creative, the visual image will override the verbal message.

Interview Audio

Depending on the type of interview conducted, one or two lapel mics or only one hand-held mic may be used. Regardless of the interview type,

Figure 11-4. An interview for a package story is typically shot over the shoulder of the reporter.

the mics must be cabled, attached to the clothing of both the reporter and interviewee, and batteries checked (if applicable).

At the beginning of the interview, the reporter prompts the interviewee to provide *lead* information that is recorded with the interview footage. The reporter asks the interviewee to state his name and spell it, state his title (if pertinent to the story), and state contact information for a possible future follow-up. Starting the recording with lead information should become a routine habit for reporters. This information should always be placed at the beginning of the recording for every interview so that the reporter and editor know where to find the information, if it is needed. For example, a lower third graphic of the interviewee's name and title should appear the first time the person is seen on-screen in the final, edited version. With the lead information at the beginning of the recording, it is easy for the CG title to be created in post-production.

lead: Basic information provided by the interviewee that is recorded at the beginning of every interview. The lead typically includes the interviewee's name and proper spelling, title (if pertinent and applicable), and contact information.

PRODUCTION NOTE

An additional benefit to hearing an interviewee speak his name for the lead information is that the reporter gets a refresher on the exact pronunciation of the name from the person who knows it best!

Recording the lead is also an effective method to get a reading on the interviewee's normal speaking voice, so the audio levels can be properly set. When interviewees are asked to, "Give me an audio level," "Count to ten," or "Say something" to get an audio reading, they often speak louder than normal. When providing information as common as their name and address, interviewees will likely speak in their normal tone of voice.

Interview B-Roll

B-roll should be shot immediately after shooting the interview, while the interview information is still fresh in the minds of the camera operator and the reporter. The *B-roll* should include shots of anything visual that was mentioned during the interview and any natural sound associated with the story. The importance of recording B-roll cannot be overstated.

Representational video should also be shot as part of the B-roll. For example, a reporter is covering the story of drastic changes in local funding for schools in the community. The reporter will detail the changes in funding allocation and how the changes impact local schools, while viewers see shots of local school buildings, school busses entering the school parking lots (**Figure 11-5**), and playground equipment. Representational shots help to visually communicate the meaning and focus of the story. Footage of the flashing blue lights on a police car is an example of representational video for a story about a traffic accident on a major expressway.

B-roll: Footage that includes shots of anything visual mentioned during the interview or that is related to the topic, and any natural sound associated with the story.

The more variety in B-roll shots, the better the finished product will be. The camera operator should shoot a wide, medium, and tight shot of every B-roll shot to triple the variety of shots. Nod shots should be shot while on location, as well. (B-roll is further addressed in Chapter 19, *Production Staging and Interacting with Talent.*)

Figure 11-5. Representational video of school busses in front of a school provides a visual that is related to the school funding news story.

In the editing room, the interview is cut into pieces of the most relevant information that will fit into the allotted time, and shots of the interviewee speaking will be full of jump cuts. These jump cuts can be removed and covered with B-roll, only if there are appropriate B-roll shots available. Using the same piece of video twice or more in a story because inadequate video was recorded while on location is considered highly unprofessional. There is usually no time to revisit the location and shoot more B-roll, so the final product will suffer without sufficient footage.

Conducting an Interview

Reporters have many options for conducting the interview itself. Each interview requires different methods depending on the reporter's relationship with the interviewee, the topic, and the personalities of the reporter and the interviewee. The following are some suggestions for conducting a successful interview. As reporters gain experience, they also develop their own collection of effective interview techniques.

Putting Interviewees at Ease

If an interviewee is nervous, make small talk while the camera operator is setting up the equipment. This allows the interviewee to see the reporter as a person instead of a threat, and keeps the interviewee's attention off of the camera. A good reporter can perceptively choose small talk topics—the interviewee's car or job, an unrelated news item of interest, or things the reporter and interviewee have in common.

Begin the interview with "easy" factual questions that confirm the background research. The reporter should not put the interviewee on-the-spot at the beginning of the interview. If the interviewee feels that he is being attacked, he may end the interview abruptly or become very guarded and defensive. Tips for combating talent nervousness are further addressed in Chapter 19, *Production Staging and Interacting with Talent.*

Asking and Listening

Reporters should word all questions neutrally and state all questions in an even, objective tone of voice. Neither the interviewee nor viewers should be able to detect a reporter's personal feelings about the topic of discussion. Questions such as "You didn't really do that did you?," "How could you act so irresponsibly?," or "How could the Republicans ever vote against such a wonderful Democratic piece of legislation?," reveal the reporter's personal opinions and feelings. Also, interview questions should not contain words or phrases that imply a value judgment ("irresponsibly" and "wonderful"). A reporter should report, not judge.

The reporter should be engaged in conversation at all times with the interviewee. Under no circumstances should the reporter look at the index card of abbreviated question topics while the interviewee is answering a question. If the reporter looks at the cards while the interviewee is speaking, both the interviewee and the viewers are made aware that the reporter is not listening and is being rude.

PRODUCTION NOTE

Some beginning reporters rely too heavily on their notes and are so concerned with asking the next question on their list that they don't *listen* to the answer being given. Looking at your notes while the interviewee is giving an answer sends the message: "I'm not listening to you because I'm thinking about something else." The lack of eye contact immediately dampens the interviewee's enthusiasm and he will probably cut his answer short. Relying too heavily on notes may also cause you to miss information in an answer. Perhaps the interviewee is giving information *now* that you planned to ask in another question *later.* Because you are looking at your notes, you aren't listening. When you later ask the planned question that has already been answered, your credibility as an interviewer is ruined.

The interview should "feel" like a natural conversation that viewers are allowed to listen to. The reporter should listen carefully to the interviewee in order to take advantage of opportunities for pertinent follow-up questions. An interview that is purely question and answer is often deadly boring and, in some cases, can feel like an interrogation to both the interviewee and the viewers. Never interrupt an interviewee's answer. For interviews that will go through post-production, the editing process can interrupt the interviewee, if necessary. If conducting an interview live, the reporter must exert more control to keep the interview within the allotted time.

Allow a short pause between the end of the interviewee's answer and the beginning of the reporter's next question or comment. In doing this, the reporter does not "step on" the words of the interviewee and avoids ruining the recorded answer—making editing much easier. Every noise the reporter makes while the interviewee is speaking will be heard during a live interview or will create a headache while editing the piece. Any sounds made by the reporter may cause the entire SOT to be unusable in the package.

An interviewee may say something that is not very clear during an interview, and asking him to say it again often results in the exact same, unclear words being repeated. Instead of repeating the question or asking another question, the reporter may try to remain silent and seem puzzled by what the interviewee just said. When interviewees "feel" silence, they typically react by talking more, which may serve to explain their point more clearly. The expanded explanation may provide a much better sound bite than the interviewee's original response.

Interviewees may try to deflect questions they do not want to answer by talking around the answer and trying to lead the interview in another direction. By simply repeating the question, the reporter can remain in control of the interview and note that the interviewee did not answer the question. If the question is avoided a second time, the viewers will realize the interviewee is dodging the question. To avoid this, the interviewee will likely answer the question when repeated. This technique is particularly effective when interviewing politicians.

Near the end of an interview, the reporter should confirm any questionable impressions picked up during the interview and ask for clarification on points, as necessary. When the interview concludes, the reporter should ask the interviewee if there is anything he would like to add. The reporter might say, "What is the most important point you'd like to make for our audience?" Quite often, the interviewee will concisely sum up the main point of the interview, which may provide the reporter with the best sound bite of the day. If the interview has been cordial and polite, this may give the interviewee a chance to address something the reporter had not thought to ask about. If the interviewee's response is valuable, it can be used. If the response is not valuable, it can be removed during the editing phase of production and it cost the reporter only a few minutes of time.

Body Language

The reporter's body language must communicate interest in the interviewee's answers, **Figure 11-6**. Eye contact is one of the most important ways to convey interest. The reporter should maintain eye contact with the interviewee as much as possible. The reporter can keep eye contact with the interviewee while he's talking, react to what the interviewee has just said with a follow-up question or comment (if appropriate), and thoughtfully glance down at a notes page for a prompt to the next question, if necessary. A glance down at the notes page is a logical pause that indicates a shift in the train of thought and does not unduly slow the pace of the interview. If the reporter must look away, the only direction to look is toward the page or index card of questions. If the reporter looks in any other direction, the interviewee will want to see what the reporter is looking at and turn his head in that direction, as well.

With experience, a reporter may be able to look directly at the camera (and viewers) during an interview. However, it is tricky to do this without seeming awkward. Also, looking at the camera is not appropriate for all types of interviews. Looking at the audience implies to the interviewee and the viewer that everyone should be in on this conversation, like family. Some interview topics do not lend themselves to this type of informality and may seem inappropriate or even offensive.

The reporter should give positive nonverbal feedback, such as nodding, smiling, and maintaining eye contact, while the interviewee is speaking. This feedback often fuels the conversation and keeps the interviewee talking.

Wrapping Up

The interview is a primary element of many news stories. Before an interview takes place, the reporter needs to know enough about the interview topic to hold up his end of the conversation/interview. The quality of answers given in an interview is directly related to the reporter's ability to prepare, establish rapport, and ask the right questions. It is incredibly important that a viewer watching an interview feels that the interviewee is treated professionally, fairly, and politely. Reporters should always try to maintain a good relationship with their interviewees, as they may need to contact the interviewee for follow-up or another interview.

Review Questions

1. What is the purpose of gathering background for an interview?
2. List some examples of technical or topic-specific jargon that a reporter may encounter when researching for an interview.
3. Identify the benefits of preparing a list of questions in preparation for an interview.
4. What are the two types of interviews that journalists conduct for broadcast?
5. What is included in the lead information recorded at the beginning of an interview?
6. How is B-roll footage used in a news story?
7. Explain how a reporter should phrase interview questions.
8. What is an effective method in getting interviewees to clarify a response without repeating themselves?
9. What is one of the most important ways a reporter can communicate interest in an interviewee's answers?

Activities

1. Watch a local newscast and make note of each interview included in the news program. For each interview noted, indicate whether the interview was aired live or edited into a package. Was there a sound bite from the interview used in the newscast or in teasers for the newscast? Be prepared to share your findings with the class.

STEM and Academic Activities

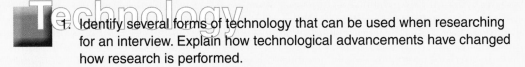

1. Identify several forms of technology that can be used when researching for an interview. Explain how technological advancements have changed how research is performed.

2. Keep a log of newsworthy and interesting information you hear of or read about over the course of three days. Determine what percentage of news and other information you receive comes from television, from print media, and from the Internet.

3. Create a list of interview questions for an on-the-street (or in-the-hallway) interview with another student about an upcoming event at your school. Phrase your questions so that the interviewee responds with more than just a few words.

4. View a recorded two-person, student interview. Did the interview feel like a natural conversation? How often did the interviewer look at notes? Did the interviewer remain objective throughout the interview? Did the interviewee appear to be at ease? Were there any moments when the body language of the interviewer or interviewee communicated more than their words?

Chapter 12

Legalities: Releases, Copyright, and Forums

Professional Terms

Copyright Law
Fair Use
limited public forum
non-public forum
passive talent release
public forum
private property

property release
public domain
public property
release
talent release
Trademark Law
transformative use

Objectives

After completing this chapter, you will be able to:

- Identify the different types of releases used in broadcast journalism and television production, and explain purpose of each.
- Recognize the differences between public and private property.
- Explain how Copyright Law applies in broadcast productions.
- Recognize how educational Fair Use applies in the classroom.
- Illustrate transformative use of material.
- Identify the criteria for public domain status.
- Summarize the characteristics of each type of public forum.
- Explain how the First Amendment applies in the organization and operation of a broadcast journalism course.

Introduction

This text does not offer legal advice; none of the information in this chapter should be construed as legal advice. Most of the legal information contained in this chapter was obtained during an extensive interview with an attorney at the Student Press Law Center, and is offered as general guidelines. In this country, a lawsuit can be mounted against an individual or a school for any reason—with or without merit. Once a lawsuit is filed, legal counsel must be retained, which costs money. Even defendants who are successfully cleared of wrong-doing must pay legal fees for their attorney's time, work, and expenses.

This chapter addresses many legal topics that affect broadcast journalism and television production, and includes:

- Talent and property releases
- How copyright applies in broadcasting
- Other rights and permissions, and how to obtain them
- Forum and free speech issues

Releases

A *release* is a grant of permission that is commonly provided in written form with signatures of all the people involved. While legal release documents have a variety of applications, property releases and talent releases are commonly used in the broadcasting industry.

Property Release

A *property release* grants the video team permission to shoot on private property. It may be difficult to determine whether a video team has a right to be present at a location when the difference between public property and private property is not clearly defined.

Public property is property owned by local, state, or national government organizations, and generally includes parks, streets, and public sidewalks. It is *usually* legal to have a video crew shooting on public property. If the production involves many people, vehicles, and pieces of production equipment, there may be a negative impact on the property or to other people on the property. Most localities require that a permit be obtained before location shooting begins. The cost and process involved in getting a permit varies by location and may be based on the production's overall impact on the public property. A permit typically holds the production company responsible for the cost of handling any resulting traffic problems, property clean-up, security issues, etc., instead of leaving those expenses to the public property operators.

A public sidewalk is usually the sidewalk that runs parallel to a public street. However, walkways or driveways that lead from the sidewalk to a house are not public property—these are private property (**Figure 12-1**). Additionally, not all streets are public streets. A public street is maintained

Figure 12-1. It is *usually* legal to have a video crew setup and shoot on a public sidewalk (green area). The private walkways and driveways (red area) connected to a public sidewalk, however, are off limits unless a property release is signed by the property owner.

by the city, county, or state transportation department. During a snowstorm, for example, large, publicly-funded Department of Transportation dump trucks fitted with plows and salt spreaders clear and maintain public streets. The parking lot of a local shopping center, however, is most often cleared by a smaller, private company using pickup trucks with plow attachments. The private snow removal company is hired by the owner or manager of the private property.

PRODUCTION NOTE

Ever since the death of Princess Diana, the term *paparazzi* has come into the public focus. "Paparazzi" refers to the photographers and reporters who generally do exposé-type stories on celebrities for tabloid media organizations. In fictional television and film entertainment programs, the paparazzi are often depicted doing things that are not legal in the real world, but that propel the plot of the film. Doing something you saw paparazzi do in a movie is not a valid defense against a trespassing charge. If paparazzi set up on a public sidewalk, it is legal. If they take one step into someone's yard, they can be arrested for trespassing. However, if the owner of the house steps out onto the front porch and motions to the paparazzi to come up to the house, the owner has given *property release by conduct*. His motion to the paparazzi was also likely recorded by their cameras, which documents permission.

Private property is property that is owned by an individual or private organization. Before a video production takes place inside a building that is open to the public, the building owner has the right to require that permission to be on the premises be obtained before shooting begins. However, unless there is a sign near the doorway or public entrance stating that cameras and recording devices are prohibited within the building, a video crew can shoot without permission until told otherwise by an authority figure from the building. At that point, security personnel may require the crew to leave the building. Practically speaking, it is *not* a good idea to sneak into a building in hopes of completing a shoot before the crew is discovered. Once the crew is told to leave the premises, they cannot continue shooting. The crew, most likely, will not finish shooting everything needed and, therefore, the effort is a waste of everyone's time. Asking permission to shoot on premises beforehand may help preserve positive public relations and elicit cooperation instead of confrontation from the property owner or manager.

Even when the crew thinks they are "not hurting anything" by shooting on private property, property owners take a dim view of video crews in their buildings. A video crew can be very disruptive to the normal environment—people stop working to watch the camera crew and some individuals do everything they can to be in front of the camera so they can be seen. Another reason crews are not often welcome involves liability. If a crew member falls and is injured while shooting in the building, the property owner's liability insurance is involved in the medical expenses and other compensation that results.

private property: Property that is owned by an individual or private organization. Permission is required to be on the premises.

PRODUCTION NOTE

Choose your shoot location wisely! Let's say you're shooting an anti-shoplifting PSA and you want to shoot some footage inside a drug store. Carefully consider exactly what you need in the shot. If the purpose is to show someone getting caught lifting merchandise off a store shelf, do you really need to shoot in the national chain drug store in town? The manager there may need to get permission from the corporate level, which might take days to obtain. Instead, you may be able to shoot the same video in a small pharmacy owned and operated by someone who lives in your town. Go to the locally-owned store and ask permission. You are much more likely to be given permission from someone local.

Getting a Property Release

Obtaining a property release is quite simple—go to the property owner or manager and ask. Prepare a simple letter stating that the named person grants permission for the video crew to shoot on the property specified, **Figure 12-2**. Ask the property owner or manager to sign the letter, print their name and title (to ensure legibility), and provide a contact phone number and address.

It is important to get written permission, not just a verbal agreement. Having a signed piece of paper is immediate proof of permission. If you are shooting and a security guard approaches the crew demanding that you leave the property, showing the guard a signed property release letter can resolve a difficult situation instantly.

Talent Release

talent release: A document that gives video producers permission to photograph the talent and/or to use audio of the talent's voice.

A *talent release* is a document that gives video producers permission to photograph the talent and/or to use audio of the talent's voice, **Figure 12-3**. A talent release form should be obtained for all the talent in every production. Talent releases protect the producers from litigation, should the talent later state "I never gave you permission to photograph (or record) me." If the producer can provide a signed talent release form, the proof of permission is on paper and invalidates the talent's challenge.

In practice, talent releases are always obtained before a production begins, whether entertainment or fictitious in nature (dramas, comedies, advertisements, etc.). If performers portray someone other than themselves, a talent release is absolutely required. The talent release must be signed before the cameras begin recording. If a producer records without a signed release and the talent later refuses to sign the release, all the time and money spent in shooting anything that includes the talent is wasted. A re-shoot without the talent is then required.

In the news world, talent releases are not required, but can still relieve many headaches. In a normal interview, the interviewee is quite aware of the presence of a microphone and camera. The interviewee should be asked to do a mic check or state his name and spell it while looking at the camera. This activity is recorded on camera and constitutes "consent by conduct" to be photographed. Of course, it is also necessary that the interviewee be of ordinary intelligence and without developmental disabilities to ensure they are capable of recognizing that recording is taking place.

ABC Production Company
1234 Industrial Parkway
Any Town, USA
(555) 555-9999

I, _Vernon Janson_ , owner/manager of

Janson Hardware , give permission to _Anne St. Claire_

from ABC Production Company to shoot _PSA footage_

on the property at the following address

150 Main Street Any Town, USA

on _September 29, 2011_ .

Signed,

Vernon Janson Date _9/8/2011_

Print name and title: _Vernon Janson, store owner_

Contact address and phone number: _150 Main street_

Anytown, USA

_(555) 555-7799_

Figure 12-2. A signed property release is tangible proof of the agreement between the property owner and the production company.

PRODUCTION NOTE

You cannot get a signed talent release from people recorded on a hidden camera before shooting begins. There is no place for hidden microphones or cameras in an education environment. All laws regarding the use of hidden cameras are state-specific. Every state makes its own laws on this issue; there are no federal laws. Woe to the person who tries to use a hidden camera in a news environment without first checking the applicable state regulations.

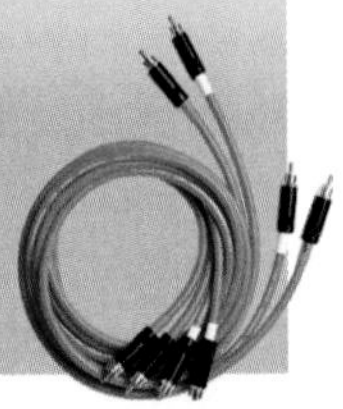

People who appear in the background of a shot (on the street, in the stands of a football stadium, in school hallways) are not required to sign a talent release form, **Figure 12-4**. If people are in a location that a reasonable person would deem to be a public place, a talent release is not necessary to record the activity of people doing what they would normally be doing in a public place. On the other hand, a release is required if a person is recorded

Figure 12-3. A signed talent release form should be obtained for all the talent in every production.

ABC Production Company
1234 Industrial Parkway
Any Town, USA
(555) 555-9999

Date _August 17, 2011_

Production title: _Parkside Center Anniversary Events_

I, the undersigned, grant the ABC Production Company of Any Town, USA, permission to use my name and likeness (with or without my voice) in whole or in part from the photographed, taped, videotaped, and/or digitally recorded material obtained on this date. I understand that the material may be edited, reproduced, exhibited, copyrighted, or otherwise published and circulated for any lawful purpose. I agree to waive compensation for providing this consent and agree that no other compensation is required.
By signing this consent, I waive any and all claims in connection with the above.

(Please Print)

Full Name _Jennifer Wilson_

Address _171 Second Street_

Anytown, USA

_(555) 555-0101_

Signature _Jennifer Wilson_

> *Note: If the person photographed is a minor, the signature of a parent or legal guardian is required on the signature line.*

Guardian's name (Printed) _____

Guardian's Signature _____

in a location where a degree of privacy is reasonably expected. A story about the interesting little things people put in their offices and on their desks to personalize the space, for example, requires signed talent releases. The people featured rightfully expect a small degree of privacy. The same applies to a story on graffiti that is shot in a school restroom—anyone in the restroom undoubtedly expects privacy. No one in the restroom can be photographed without their written permission.

Many of the guidelines regarding consent and releases involve what a "reasonable person" perceives and understands. When dealing with minors and members of special populations, carefully consider how they perceive and understand events and experiences. Parents and guardians are not legally required to sign a talent release form as long as the individual is able to recognize the function of the video camera and understand the intended use of the video being taken. Minors must sign their own talent release form if they are old enough to understand the nature and probable consequences of the program. However, the release forms should also

Figure 12-4. A talent release is not required for people in the background of a shot in a public place.

be signed by a parent or legal guardian as a safeguard for the producers. While there is no legal requirement to obtain permission from the parents of special population individuals, it is reasonable to assume that permission should be obtained. The parents or guardians of these children will likely believe they should be asked before video is made of their child. Video producers can head off many unnecessary problems by simply asking for talent release forms to be signed by the parents or guardians.

Passive Talent Release

Many school systems use a *passive talent release* document, which is a general notice given to all parents indicating that, from time to time, organizations outside the school system may request permission to video record inside the school building. This request may be made for a variety of reasons, including news stories, yearbook pictures, or documentaries on some aspect of education. If a parent does *not* want their child to appear in the footage shot by these third-party organizations, the parent must sign the document and return it to the school principal. Any organization that enters the building must abide by the requests and ensure certain children are not seen on camera. If the passive release form is not returned to the principal, it is assumed that the student is "cleared" and permission to photograph the child is granted.

Passive talent releases apply only to third-party organizations, not to in-school organizations. Broadcast journalism students can video record other students doing various activities, such as participating in football practice, for the purpose of doing a story for a student news program. A school is a public place and the broadcast journalism students are legally on the premises of the school building and property. In the interest of being polite, however, the students on-camera should be asked for permission

passive talent release: A document that serves as a general notice indicating that, from time to time, organizations outside the school system may request permission to video record inside the school building. Parents acknowledge the release by *not* responding to the notice.

before shooting them as primary figures in a video. Shooting video of people who do not want to be recorded likely results in video that is unusable. Recording students in this situation is legal, but is a decision of conscience. Just because you *can* do a shoot, doesn't mean you *should* do the shoot.

VISUALIZE THIS

You are recording video in a hallway between classes. The reporter and quarterback of the football team are in the foreground. In the background are students walking and at their lockers. This is perfectly legal. The fact that the student at locker #489 in the background is in the special education program is not an issue. That student is just like every other student in the hallway—in the background of a shot in a public place.

Now imagine the same shot with a special education teacher instead of the quarterback. As the teacher is talking, the camera operator zooms in on locker #489 and the special education student. This is picking an individual out in a crowd, and is neither acceptable nor legal.

In the first example, the student was as anonymous as any of the other students in the background and may only be known to some of the viewers. In the second scenario, the student was literally pointed out by the camera. This clearly states to the viewership that this student is part of the special education population. In doing this, the story makes public disclosure of private facts, which can quickly lead to liabilities and litigation.

Due to the legal consequences involving permissions and releases, it is wise to have a blanket policy requiring talent release forms be signed by a parent or guardian of all minors and minor members of special populations who are seen prominently in any video program. Signed releases are not legally required where passive talent releases apply, but can prevent misunderstandings and challenges by leaving no question that permission was granted. A safe policy is to require that everyone have a release—period.

Copyright

Copyright Law:
Set of laws that protect the creators of original materials from having their materials and creative work used without proper permission and compensation.

Copyright law can be confusing and is updated without warning. Fundamentally, *copyright laws* protect the creators of original materials from having their materials and creative work used without proper permission and compensation. A predominant example of copyright violations is downloading music and videos from the Internet without permission from the copyright holders. This activity is both legally and morally wrong. Copyright infringement is a federal crime.

VISUALIZE THIS

John bought an older car and spent hours of his free time and thousands of dollars fixing up his car with body work, replacement parts, and various accessories and personal touches. After all his work to make the car both amazing and unique, someone stole his car. Would John be justified in complaining loudly about his loss? Would he be entitled to press charges against the thief? Theft is theft, whether it is theft of an idea, a creative work, or a car.

Fair Use is a section of the Copyright Law that provides guidelines for the limited use of copyrighted materials without obtaining permission from the copyright holder(s). Among other provisions, Fair Use makes certain allowances for educational use of copyrighted materials. There is a common belief that Fair Use allows *any* copyrighted material to be legally used in or duplicated for *any* school or classroom purpose. That is a myth. In truth, some materials may be used in a <u>certain</u> way with a <u>particular</u> audience. If a student is doing a presentation in a speech class about rock-and-roll music, for example, some snippets of relevant music may be included as examples. A teacher showing a movie clip in a television production classroom to illustrate the subjective camera technique is also an acceptable educational Fair Use of copyrighted material.

Educational Fair Use stipulates that copyrighted material may be used without permission, only for direct teacher-to-student contact <u>within</u> a classroom (**Figure 12-5**). However, if the copyrighted material is "aired" <u>outside</u> the classroom, the material is no longer an *educational* tool between a teacher and the students of an individual classroom. "Airing" the program, whether as a newscast or for entertainment, is the *result* of the education the students received in the classroom. This is true even if the material is aired over a closed circuit system within the school building. Additionally, showing student-created programs is not educational Fair Use, in its own right.

Fair Use: A section of the Copyright Law that provides guidelines for the limited use of copyrighted materials without obtaining permission from the copyright holder(s).

PRODUCTION NOTE

The purpose of the television production or broadcast journalism class is to train students to enter the profession. Educational Fair Use does not apply in the television production or broadcast journalism industries. Students need to learn about and operate under the rules of the real world when learning about a profession.

Figure 12-5. Educational Fair Use only applies within the walls of a classroom.

In some cases, copyrighted material *may* be used without permission outside the classroom if the material is used in a transformative manner. *Transformative use* means a work (image or other material) is used for an entirely different purpose than it was originally created and intended to be used. The work is, therefore, "transformed." A diagram of the interior of a clothes washer, for example, was originally created to illustrate the parts and construction inside the machine. Using the diagram for the same purpose in a program about washing machines is *not* transformative of the original purpose, and requires permission for use. Alternately, using the diagram as an illustration of a style of graphics in a program about the career field of graphic arts *may* be acceptable as transformative use of the diagram. However, unless the original artist is contacted to determine the original purpose of the material and how it was intended to be used, there is no way to positively make a case for a transformative use.

Copyright issues emerge in television production and broadcast journalism in several areas, including logos, pictures obtained from other media (including the Internet), and music. The copyright information in the sections that follow assumes that a program containing copyrighted material will be "aired," at minimum, on a closed circuit system within a building.

Logos

Logos are developed by companies to create recognition for their products and services. With product recognition comes a reputation the company has built and marketed. Because companies rely on the public's opinion of their brands for sales and profit, unauthorized use of a logo can have significant consequences. Company and product logos are typically both trademarked and copyrighted. *Trademark law* protects a company's brand identification, which represents their products and services. Because of trademark law, transformative use may not apply to logos, particularly those that are considered "famous." Logos that are recognized worldwide are typically considered "famous," and are protected against third party use except to refer to an item the trademark is used to identify. For example, the ACME Soap Company logo could be used in a story about favorite soap products, if ACME products are discussed in the story.

Clothing with Product Logos

Shooting a person wearing a shirt with a product logo may seem acceptable, especially if the action in the program is not related to the product and does not make statements, true or false, about the product or the company. Now consider the production of a PSA that addresses shoplifting, with the featured shoplifter wearing a shirt emblazoned with the logo for a major brand of athletic apparel. Even though the producers consider the program neutral or unrelated to the product logo, the company may not agree. In the example of the shoplifting PSA, the company could decide that having the shoplifter seen in a shirt bearing their product's logo may be perceived by the viewing public as an endorsement of shoplifting. If the company believes this negative publicity will or has negatively affected their product sales, legal action will likely be initiated by the company's legal team. It is best to avoid shooting someone wearing logo apparel. If the logo is completely unrelated to the video, then it doesn't need to be seen to help get the message across anyway.

Product Placement

Brand name products are often used as props and clearly seen on television programs and in movies. Manufacturing companies are usually very particular in how their product is used in a production, so that only positive publicity is generated for the product and the company. For example, you will not see a psychotic ax murderer drinking a Coke® on a television program or a person drink a Pepsi® and immediately keel over dead in a movie. Both of these situations are horribly negative advertisements for the products. Additionally, manufacturers pay a large sum to production companies for product placement in chosen movies and programs, as a type of advertisement. These companies are not likely to pay for product placement in small, independent, or student productions.

Instead of using a brand name product in the program, take a more generic approach. For example, pour a major brand soft drink into a plain glass or cup with ice and a straw instead of having the character drink from a can of Pepsi® or Coke® (**Figure 12-6**). The time necessary to make this change in the program far outweighs the possible consequences of inappropriately using a brand name product. However, this does not mean that every product name must be removed from a shot. If, for example, there is a Ford vehicle in the program, you do not need to cover up the Ford logo on the trunk. But, the program should not include a prominent close-up of the word "Ford."

Figure 12-6. Using a glass of soda instead of a branded can of soda in a production avoids complicated product placement issues.

Signage

When shooting outside the studio, there may be buildings and business signs of national stores in the background of shots. These signs can be seen by anyone driving down the street in an ordinary public environment, and are placed specifically to be seen publicly. Including business signs in the background of shots is not a copyright or trademark violation. For example, the shoot for a video on how to change a car tire takes place in someone's driveway. Across the street is a shopping center with signs for some of the stores clearly visible in the background of the shot. It is okay to include the signs in the shot because they are intended to be seen by the public.

Pictures from Other Media

Except for fair use situations, it is not legal to use pictures from *any* published work without obtaining permission from the copyright holder. This includes printed works, such as books, magazines, and newspapers, as well as digital media, such as the Internet, movies, and video games.

Print Media

Before planning to use a photo from a printed work, contact the publisher of the work. The publisher may own the rights to the image, or may have received permission from the copyright owner. For example, many photographers hold the copyright to their pictures and grant permission to publishers for their use. Request permission to use the work directly from the copyright owner. Be prepared to answer questions about how the work will be used, how many times it will be "aired," and the size and nature of the viewing audience. If you cannot find the copyright holder, *do not* assume it is acceptable to use the material—find other material.

The Internet

Images found on the Internet are *not* free for anyone to use. Some Web sites require a subscription to download images from their collection, while other sites make images available with stipulations on how they are used. But, most images found on the Internet are subject to standard copyright requirements and protection.

Movies

Programs that provide critics' review of movies typically have movie posters displayed in the background and present clips of the movies reviewed. Both of these are copyrighted materials, but may be used in a program with certain stipulations.

The movie posters on the set of a critic's review program must be only those that are discussed in the episode being recorded. If the movie review segment includes just one movie, then only the poster for that single movie may be displayed on the set. However, if the program includes reviews on multiple movies, posters for all the movies discussed during the episode may be displayed. In either case, the posters must be obtained legally (purchased or received from the production company). Using movie posters in this manner is transformative—the original purpose was to advertise the film, but during a movie review program, the poster provides a visual representation of the film while the critic provides thoughts and opinions about the film.

PRODUCTION NOTE

Be certain to change movie posters on the set with each episode, so only posters for the films discussed are displayed.

Movie clips may be used during a movie review program or a news segment, but *only* if the clips were originally obtained from the official source of the film. The official source may be the official Web site for the film, which may provide a download link for the film's trailer. Using a trailer clip from the film's official site is transformative.

PRODUCTION NOTE

Before downloading video from the Internet, consider the quality and size of the image once it is placed into the program to be aired. The quality of the image may be substantially degraded by the process.

Video Games

Transformative use also applies to using screen shots from video games in a program. For a video game review segment of a newscast, for example, brief screen shots of a game may be used to illustrate the review (making certain that the language, content, and violence levels of the game are appropriate for the audience). On the other hand, a scene in an entertainment program depicting a character playing a video game with the screen visible or the music track of the game audible to viewers is not transformative use of the video game. The game is used exactly as it was intended to be used—as a game to be played—and a license must be obtained to use the video game in the production.

Talk the Talk

When you request permission, the permission you receive is a license. *Permission* is the consumer term for what you ask for; *license* is the professional term for what you get.

Music

The majority of entertainment programs utilize music to enhance the message, mood, excitement, drama, pace, and emotion of the program. Without it, the audience feels that the program is missing something. To use music in a production, the appropriate permissions must be secured. Copyright laws are established by the federal government, as provided in the United States Constitution. Violations are dealt with harshly in the Federal court systems.

Placing full music in the background of news segments to enhance the story (whether reporting on a local event, special interest piece, montage, or sporting event) is using the music exactly as the original artist intended it to be used—to create a mood. In fact, using music to enhance a news story is not good journalism; a news story should be able to stand alone. Additionally, this does not constitute transformative use of the music. To use music in this way, the television station or network must obtain permission to broadcast the music and pay large fees to the copyright holders. Placing full music credits at the end of a program is not an acceptable alternative to getting the proper permission. In fact, including credits for music used in a program without permission indicates knowledge that someone else created the music that was used without permission. The car thief who publicly admits to stealing cars, but hasn't been caught, is still a thief.

Educational Use of Music

Even though some music may be free when used for certain educational purposes, this is never a safe assumption. Some pieces of music, especially those found on the Internet, may have different rules for usage than other pieces of music. Always verify permission with the rights holder before using music in any production. Educational Fair Use does not apply to every type of use within a school building; only in direct student-teacher contact within a classroom as part of a lesson. Using music in student productions is likely not transformative use of the music either. The music was composed and originally recorded to be listened to and evoke some kind of feeling in the viewer. Music used in a student production is usually intended to excite the audience into paying attention to the program, which is not a transformative use of the music.

Music used in productions that are not broadcast school-wide is subject to copyright regulations, as well. Copyright regulations apply to material that is distributed to the public in any form. To use music on a senior class memory video, for example, copyright permission must be secured from the record labels for each piece of music planned to be used. This is true whether DVDs of the program are sold or are given away.

Public Domain

public domain: A status designation applied to material that is no longer copyrighted due to the passage of time (relative to the date of creation) or when rights are relinquished by the copyright holder.

Public domain is the designation applied to material that is no longer copyrighted due to the passage of time (relative to the date of creation) or relinquished rights by the copyright holder. When material enters the public domain, anyone may use it at any time without obtaining copyright permission. The rules determining whether something is in public domain are multifaceted.

- Every work created prior to 1923 is in the public domain.
- Every work created from 1923–1963 that has not had a copyright renewal is in the public domain.
- Every work created from 1923–1977 that does not appear with a copyright notice is in the public domain.
- Every work created from 1978–March 1, 1989 that does not appear with a copyright notice *and* has not had a subsequent copyright registration within 5 years of creation is in public domain.
- For works created after 1977, the copyright lasts for the lifetime of the author/creator, plus 70 years. In the case of joint authors, the

copyright lasts for the lifetime of the longest surviving author/ creator, plus 70 years. For corporate works, the copyright lasts 95 years from the date of first publication.

PRODUCTION NOTE

The following are some resources that offer access to various media online. Many of the works available are in the public domain. Always check the Terms of Use statement and copyright restrictions from any online media source.

- Smithsonian Institution: www.photography.si.edu
- Project Gutenberg: www.gutenberg.org
- LibriVox: www.librivox.org
- Prelinger Archives: www.archive.org/details/prelinger

Once a work is in the public domain, it can be used by anyone. Even though the work is in the public domain, the use or performance of that work may be copyrighted (**Figure 12-7**). For example, Ludwig van

Figure 12-7. An orchestra's performance of a piece of music in the public domain is very likely copyrighted.

Beethoven died over 175 years ago and his music compositions are in public domain. However, to use his Ninth Symphony from a CD recorded by the London Philharmonic Orchestra, permission must be received from the London Philharmonic Orchestra. Their performance of Beethoven's work is copyrighted.

Public Forums and Broadcast Journalism Courses

Much of what television production personnel and broadcast journalists can and cannot do is determined by laws covering issues, such as releases and copyrights. The First Amendment to the Constitution of the United States of America also has a major effect on the media. Freedom of speech and freedom of the press are among the five freedoms guaranteed in this amendment. The First Amendment guarantees that journalists can report on topics they choose, including the government, its people, and its actions, without fear of retribution by the government. This freedom has permitted journalists to probe, analyze, criticize, and report on topics they deem worthy.

The First Amendment was intended to prevent the government from controlling what the press reports. Since the press operates without governmental oversight, it must be self-policing and enforce its own high ethical standards. Historically, the United States has been tremendously affected by the work of journalists and their ability to seek out information and ask difficult questions on behalf of the public.

Another aspect of the law that affects the press is forum status. A forum in ancient times was merely a place to talk, as in the Roman Forum. Today, *forum* refers to the delivery and format of mass communication. Student newscasts are generally considered to be a public forum. There are three types of public forums:
- Public forum
- Limited public forum
- Non-public forum

Forum issues are present in both print journalism classes and broadcast journalism classes. There are significant differences between how print media and broadcast media deal with free speech and newscast content. Since this text is concerned with television media, only issues from the television media environment will be discussed.

Public Forum

public forum: An environment or location, typically public property or media, where an individual can stand and publicly speak their mind. The content discussed in a public forum is not restricted, but the speech cannot incite a riot, violence, or similar activity.

A *public forum* is an environment or location, typically public property or media, where an individual can stand and publicly speak their mind— pure free speech. The content discussed in a public forum is not restricted, but the speech cannot incite a riot, violence, or destruction of property.

A public forum, in its truest sense, does not apply to broadcast journalism because television programs are finite in length and scope and cannot include everything and everyone who wishes to speak. There is no place in broadcast journalism to allow absolutely anyone say anything they want for as long as they would like to say it. A television newscast, for example, cannot run longer than its allotted time. Editing is a key component of journalism.

Limited Public Forum

A *limited public forum* is public property or media that is made available for a specified use. In a limited public forum, the topic or content of speech is restricted to the business at hand or objectives of the particular group. An example of this may be a financial planning seminar held in the conference room of a local park district building. The park district officials can prohibit the discussion of any topics other than financial planning.

In a broadcast journalism course, a limited public forum can allow free expression by the student producers. Student producers may make final decisions on topic choice and content. Students may also fill the role of news director. A different model of limited public forum in broadcast journalism courses may be standards-based. For example, student newscast stories are deemed acceptable as long as the story is similar to the style and technical quality found on the local or national programming of the "big 5 networks." In this model, the standards are enforced by the <u>student</u>.

limited public forum: Public property or media that is made available for a specified use; the topic or content of speech is restricted to the business at hand or objectives of the particular group.

Non-Public Forum

A *non-public forum* is either public or private property or media that is not typically used or made available for public expression. Regulation on speech is allowable in a non-public forum, but must be reasonable and not intentionally exclude any particular or opposing viewpoint. Non-public forums typically include military bases, public schools, and courtrooms.

A broadcast journalism class that operates as a non-public forum may be managed by the teacher, who assumes the role of news director and decides which stories are included in the program (unless the program is long enough to include every story). For example, a teacher reviews all the stories and narrows them down to the top seven. From those seven stories, student producers pick their top five. With this method, students have a say in the decision, but the final decision rests with the teacher. In this example, the teacher is the news director of the program and the process follows the industry model. This classroom management example may also apply in reverse order—student producers make the initial story decisions (as a newscast producer in industry would), and the teacher evaluates the stories after the first cut.

Another model of non-public forum in the broadcast journalism classroom is to apply standards-based criteria. For example, stories are acceptable as long as they match the style and technical quality of local or national programming found on the "big 5 networks." In this model, the standards are enforced by the <u>teacher</u>.

non-public forum: Either public or private property or media that is not typically used or made available for public expression. Regulation on speech is allowable in a non-public forum, but must be reasonable and not intentionally exclude any particular or opposing viewpoint.

In the Classroom

While the First Amendment grants that the government cannot control the press or free speech, it says nothing about private citizens exerting control. No one in the real world of broadcast is given free reign to say whatever they like. Producers, news directors, station managers, station owners, network executives, and the FCC all have the ability to exert some control over what is transmitted over the airwaves in this country. The First Amendment applies to broadcast journalism courses with respect to the teacher's authority in the classroom.

Broadcast journalism courses are usually operated to emulate the broadcast journalism industry. Courses are designed to train students to enter the broadcast journalism industry, or at least acquaint them with the possibilities and demands of broadcast journalism careers. Students learn to make news judgment calls, practice technique, and develop skills. The teacher helps students perfect their skills and technique as they are practicing.

VISUALIZE THIS

On the first day of class, your broadcast journalism teacher gives you access to the school's closed-circuit broadcast system and all the equipment needed to air a program. The only instruction provided is, "Do anything you want and air it to the school. I'll let you know if you've done anything wrong after your program airs." While you may be elated with the amount of freedom this opportunity allows, as an inexperienced broadcast journalism student, you have not acquired the necessary knowledge and skills to be successful. What if you become the subject of ridicule after your program airs? What if you cause offense to others with what you say or how you say it? What are the consequences of unintentionally (or intentionally) saying or doing something illegal on the air?

Just as a student learning to be a trapeze artist uses safety cables and a safety net while practicing stunts, broadcast journalism students learn and practice the "moves" of the trade with the guidance and safety net provided by a teacher in the classroom. The time to make mistakes and learn from them is when all the safety precautions are in place. No one would expect a student who had never been on a trapeze to perform 100 feet up in the air in front of an arena audience without a safety net. Learning about your mistakes after the fact can have serious consequences.

The content and quality of the newscast produced by a broadcast journalism class is a by-product of teaching broadcast journalism standards. The newscast itself is not the objective of the course. Because the First Amendment guarantees free speech and press rights, a teacher cannot *control* the content or quality of journalism. The newscast is a demonstration of the skills students have learned, which means that the teacher can require a story be reworked until it reaches acceptable journalistic standards. Even though a story is not illegal and will not cause a riot or other disruption, a teacher may decide the story is journalistically indefensible and assign a low grade accordingly. The First Amendment does not grant the right to a good grade in broadcast journalism class for bad journalism. In a classroom, indecency and "prime-time standards" are legitimate journalism concerns that may be enforced through grades.

A teacher can, however, control the technical video and audio quality of stories appearing in the newscast. Technical quality disqualifiers may include out of focus images, under- or over-recorded audio, glitches, jump cuts, shaky camerawork, and inadequate lighting. The minimum standards for technical quality are part of the skills students acquire in a broadcast journalism course. Teachers do have the right to set a standard minimum level of technical quality to qualify a video project for broadcasting. To differentiate between the content of a story and the technical quality, ask yourself "Can the product be improved to meet the technical standards without altering the intended message of the program?"

In general, a broadcast journalism teacher's role mirrors that of a production's news director. Both provide training and guidance to help the production staff make good news judgments, check facts before broadcasting a story, credit people in stories correctly, write stories that are balanced and fair, spell titles accurately, and discuss the newsworthiness of a story idea. As with any training, initial assignments are rather simple. As students' skill set grows, so does the complexity of stories and responsibilities given. Increased responsibilities are earned with successful completion of tasks, as determined by the teacher/news director.

As students take more responsibility in class productions and decision-making abilities are proven over the course of the academic term, the forum status may gradually evolve from a non-public forum to a limited public forum. A student newscast will not evolve to an open public forum because the newscast team is limited to the students in the broadcast production class. It is important to note that forum status is granted in a broadcast journalism course, and can be taken away if abused.

Wrapping Up

There are many legal issues concerning television production and broadcast journalism, and this chapter has addressed many of the most common issues in both the classroom and the broadcast journalism industry. One of the most important rules to take from this chapter is that *you must ask for and receive permission.*

While the First Amendment guarantees the freedom of speech and press, these rights come with awesome responsibilities. Broadcast journalism classes are ground zero for learning these rights and responsibilities and how to appropriately apply them.

Review Questions

Please answer the following questions on a separate sheet of paper. Do not write in this book.

1. What is the difference between *public property* and *private property*?
2. What permission does a talent release provide and when should it be obtained?
3. How does passive talent release apply in a school setting?
4. How does Fair Use apply to the use of copyrighted material in education?
5. Describe problems that may arise from including product logos and brand name products in a program.
6. What is *transformative use*? Give an example.
7. When does material become part of the public domain?
8. Identify and explain each type of public forum.
9. Explain how a broadcast journalism teacher's role in the classroom mirrors the responsibilities of a news director in the broadcast journalism industry.

Activities

1. Investigate what a copyright protects and what it does not. Create an outline summarizing the information you discover.
2. Learning to apply good judgment in journalism and following established journalism standards are key skills in being successful in the broadcast journalism industry. Research the career-changing judgment calls made by the following two broadcasting figures while at their "former" jobs. Be prepared to discuss your findings in class.
   - Don Imus, former host of "Imus in the Morning" radio program
   - Dan Rather, former anchor of the CBS Evening News

STEM and Academic Activities

1. Find several websites that offer royalty-free images, but require a subscription to download the images. Investigate the guidelines provided by each company for use of the images and note the specific permissions granted to subscribers.

2. Research the total number of copyrights issued per year in the last 15 years. Create a graph that depicts the yearly totals and illustrates the change in numbers over time.

3. Choose a brand name product that could be used in a program you are producing or have produced. Contact the corporation's public relations department to find out what is required to get a release to use the product in your program. Be prepared to explain the production and specifically how their product will be used or featured.

4. Create a display of some graphic-only corporate logos. Present your display in class and ask your classmates to identify the company represented by the logo

www.splc.org The Student Press Law Center provides free legal advice to high school and college journalists, and low-cost educational materials for student journalists on a wide variety of legal topics.

Chapter 13

Music

Objectives

After completing this chapter, you will be able to:

- Summarize the difference between background and foreground music.
- Identify the guidelines for using background music in a production.
- Explain how copyright licenses apply to student-produced programs.
- Recall the types of contracts available when using a music library service.
- Recognize the unique characteristics of the different music rights available.

Professional Terms

background music
broadcasting rights
cablecasting rights
cover music
foreground music
recording rights

re-recording rights
streaming rights
synchronization rights
transitory digital
   transmission rights

Introduction

The vast majority of programs utilize music to some extent because without it, the audience feels that the program is missing something. This chapter presents general items to consider when selecting music for a production, but does not offer legal advice. None of the information contained in this chapter should be construed as legal advice.

Using Music in a Production

foreground music:
Music in a program that is the subject of the production.

background music:
Music in a program that helps to relay or emphasize the program's message by increasing its emotional impact.

The music used in any type of program can be categorized as either foreground music or background music. *Foreground music* in a program is when the music itself is the subject of the production. The song in a music video, **Figure 13-1**, is an example of foreground music—the music is the primary element and the video is secondary. *Background music* helps relay or emphasize the message of a program by increasing its emotional impact. A common example is the exciting or intense music heard during a chase scene.

Foreground Music

Foreground music is the focus of a production, such as in music videos and a critics' music review. In making a music video, typically the artist and/or recording label hold the copyright. If the person making the music video holds the copyright to the music, acquiring permissions is not necessary. In the case of a music review program or segment, the question of copyright permission can be complex. There are no laws that define a specific percentage of music, number of measures of music, or number of seconds of music that may be used without copyright permission. However, only a small portion of the entire piece should be played—typically, about ten seconds. In addition to possible copyright violations, it is bad journalism to play the entire piece of music during a program or segment.

Choosing which portion of a music piece to play is an important decision, as well. The "heart" of a song may not be used without permission. The "heart" of a song is the most recognizable portion, which may or may not contain lyrics. For example, the rhythm "stomp, stomp, clap; stomp, stomp, clap" makes most people think immediately of Queen's *We Will Rock You*. Carefully consider which portion of a song to include in a program, as the recording industry won a lawsuit over the unauthorized use of only 15 notes of music!

Figure 13-1. The music is the primary element in a music video and is considered foreground music.

Background Music

The following is a list of general guidelines regarding the use of background music in television productions:

- Avoid music that is widely recognized. When the audience hears a piece of music they recognize, they reminisce about things they associate with the song, such as special events, particular people, or childhood memories. But, while the audience recalls these moments, they are not watching and paying attention to the program.
- Avoid music containing lyrics during a dialog scene. The audience will find it difficult to separate the music lyrics from the talent's dialog.
- Use music only to enhance a mood. Music should not be placed in a program merely because "it is a really good song" or someone "just wanted to put it somewhere in the program." Sometimes, no music is best. The sound of silence has a powerful effect on mood.
- Do not use catchy or busy music during a dialog scene. Once again, if the music distracts attention from the dialog, the audience has to consciously separate the dialog from the music.
- Do not mix styles of music within a program. In other words, do not include heavy metal, rock and roll, country, opera, classical and blues all within the same program. A wide array of styles within a single program produces an amateur-sounding audio track.

Exceptions to the Rules

As with almost any set of guidelines or rules, there are exceptions. Breaking the rules is not something teachers usually encourage, but there are often valid reasons for doing so. The following are examples of some legitimate reasons to break the rules:

- Use music that is widely recognized. Recognizable music is sometimes required in a program. The famous nostalgic music used in the classic movie *American Graffiti* reinforced the program's era setting and increased the impact of action in the program.
- Play music with lyrics during a dialog scene. If a scene takes place in a nightclub, for example, it is very unlikely that the music playing would be exclusively instrumental music—vocals accompany most popular music. In this situation, the audio engineer's challenge is to keep the music levels far below the level of the talent's dialog.
- Use music for reasons other than mood enhancement. Breaking this rule often produces effects that are jarring to the audience, but also demands their attention. For example, playing Johann Strauss' *Blue Danube* waltz in the background of a scene that depicts a bombing run during World War II, makes the effects of the bombs even more horrifying.
- Use catchy or busy music during a dialog scene. A myriad of background sounds may be necessary to create a convincingly real setting, depending on the environment in which a scene is set. If the scene takes place on the midway of a carnival or on a beach boardwalk during the summer, the environment is naturally full of many different sounds, including music. Omitting these sounds is an error that leaves the audience questioning the realism of the setting. The challenge for

the audio engineer becomes keeping the dialog clearly audible over, called "on top of," the natural sound of the setting.

- Mix styles of music within a program. Various styles of music may be used in a program to be consistent with the action and setting of particular scenes. In a documentary about an opera star, it is expected that opera music be in the background of most scenes. In a scene that portrays the opera star having dinner in a country-western themed restaurant, the most appropriate background music is country-western style music. To have opera music playing in this type of restaurant would be laughably wrong.

Music in Student and School Productions

Student videos shown *only* in the classroom as part of a class project may include copyrighted music with instructor approval. However, if the video will be played outside the classroom, copyright licenses must be obtained for the program. The phrase "outside the classroom" literally means any location beyond the walls of the television production or broadcast journalism classroom. This includes showing the video

- on a school-wide distribution system (CCTV), such as a student-produced newscast or "morning announcement" program, **Figure 13-2**.
- as part of a film festival in an auditorium or theatre-type presentation, whether or not admission fees are charged.
- over a cable system to the community.
- on the Internet, such as on YouTube or SchoolTube.
- at an outdoor event on a projection screen.
- at any public location, including in a church.

Broadcasting Student Productions

The issues of copyright and permissions must be addressed for *any* student production that contains music and is presented or broadcast outside the classroom.

Figure 13-2. Student newscasts are considered "outside the classroom." Appropriate permissions are required to use copyrighted music in a student newscast.

In a student-produced program of a school football game that will be cablecast, for example, the music played by the band at the game will be heard in the program, even if the band is not seen on camera. The cable company carrying the football game may have a blanket license for music, which *may* allow the music to be included in the program. If the cable station has a blanket license with ASCAP (American Society of Composers, Authors, and Publishers) or BMI (Broadcast Music, Inc.), for example, that license might provide legal permission to record music from ASCAP or BMI in the program. Examine the cable company's license to verify that it covers music that may be recorded incidentally in the student production. If the cable company does not have a blanket license that extends to the student production, the production team should reconsider recording the music under *any* circumstances. If the band plays only a small portion of a song during a timeout or other short break in the game and the camera continues recording activity on the field, it may be acceptable to include the music in the program. However, shooting the band while they are playing (such as during the half-time show) is considered synchronizing the music with video of the performance. Without permission from the music publisher, synchronizing the music and video is a serious copyright violation. When sheet music is sold to the band, live performance rights are granted to the band at the same time. Live performance rights do not extend to the video production crew and do *not* include video rights.

PRODUCTION NOTE

To get permission to record the band's half-time show performance, ask the band director for the name and contact information for the music leasing company used for the band's performance music. Contact the company's licensing department and explain that you want to video record a half-time show where their music is being played. If the licensing department says you can record the half-time show, get the permission in writing *before* you shoot the show; a faxed letter of permission is sufficient. On the other hand, if the band director tells you the music is ASCAP or BMI, you may be covered by the cable station's blanket license with ASCAP or BMI. Ask to see a copy of the cable station's contract and verify it yourself.

How the footage containing music is used in a program may affect the necessary permissions. Consider the music played by marching bands and other groups during a parade, **Figure 13-3**. Using the parade footage in a news story that is a package in a larger newscast is acceptable, because the music heard is background to the reporter's stand-up or interview with bystanders or parade officials. If the parade moves continuously, only a small piece of larger compositions will be heard in the background, with the interview dialog remaining more prominent in the video. However, if the entire parade is shot for cablecasting, the recording industry states that a license must be obtained for each piece of music played on the video. Again, check with the cable company to see if they have a blanket license that covers the student-produced recording of the parade.

Figure 13-3. The music from a marching band can be heard in the footage of a parade.

PRODUCTION NOTE

When networks cover events like football games and parades, the network pays for blanket rights to use an almost unlimited amount of music. It is usually not financially possible for a broadcasting class to purchase blanket rights to music.

Sources of Music

The music used in productions may come from various sources:

- Professional and commercial recordings.
- Unrecorded sheet music.
- Original music.
- Music in the public domain.
- Music libraries.

Using any of these sources requires that specific permissions be obtained *before* the music is placed in a program.

PRODUCTION NOTE

In the industry, the director is involved in choosing music for a production and works closely with the composer in making decisions about original music. The production company's attorneys are responsible for acquiring the necessary permissions and handling the corresponding contracts.

Recorded and Copyrighted Music

If it is absolutely necessary to use copyrighted music from any analog or digital recording of a live concert, you must contact the company listed on the recording label. Remember, a recording you made with your own equipment at a live concert is *not* a legal recording for duplication and distribution. Some professional rock bands give the audience permission to record their live concert, but that permission applies only to a single, personal recording. You can find the company name on the label of the CD and contact information can likely be found on the Internet, **Figure 13-4**.

PRODUCTION NOTE

The Library of Congress Online Catalog (http://catalog. loc.gov) is a searchable resource that provides quite a bit of information on songs, including the name of the recording company.

Contact the recording company's licensing department and request a "copyright license" for each piece of their music you would like to use. You will need to explain exactly how you want to use the music and that it will be recorded onto video media and synchronized with visual images. Inform the company where the program will be seen and provide an estimate of the size of the potential audience. It is also important to indicate whether anyone will receive payment for their work, or whether you (the production company) are creating the video for a profit. The licensing department will also want to know if duplications will be made and sold, and what the price of the duplications is expected to be. After making initial contact with a recording company, follow-up with a letter that clearly states the intended use of the music. Do not proceed with video recording until the recording company has provided a letter granting the copyright

Figure 13-4. The recording company's name can be found on the CD label.

licenses. The program cannot be aired without confirmation of the copy-right licenses.

It is not recommended that amateurs attempt to get permission from major recording artists for use of their music in student projects. The process is lengthy and can be discouraging. Most often, the result is a resounding "no" or a very high usage fee, **Figure 13-5**.

PRODUCTION NOTE

Small, independent record labels are typically easier to deal with than major recording companies. You are more likely to speak to an actual person when contacting smaller recording companies, instead of an automated answering system or recorded voice message. Smaller labels are also typically faster in responding to rights requests.

Music heard on a radio broadcast is also copyrighted, which means that a recording of an actual radio station broadcast cannot be used as background music. Copyright permission is required to use the music played on a radio station, as well as the voice of the announcer or DJ. Instead, approach a few local bands and get recordings of their music, with written permission to use the music. Create a radio station recording with the local bands' music and a friend pretending to be a radio DJ. Add the "radio station" background to the scene in the editing phase of production.

Sheet Music

If music is published in sheet music form, but not already recorded (such as the music used in a school band concert that will be shown on cable), permission must be obtained from the sheet music publisher. Contact the publisher directly, provide the requested program information, and request that permission be provided in writing.

Figure 13-5. The fees charged for using copyrighted music depend on various factors.

Music Usage Fees

- There is no standard in the amount charged to use a piece of copyright music.

- The fees are often set by the agency or the artists themselves.

- The fee may depend on the size of the potential viewing audience.

- Whether a program is "for profit" (has commercials), "not for profit," or public access programming affects the fee amount.

Original Music

Using original music from local musicians is a less complicated way to acquire music for a program. Most local artists love the free publicity of having their music included in a television program and, in some cases, will not charge you for recording it. Permission from the band is required to use their original music in the production. The letter of permission should include:

- the broadcast medium for the program (CCTV or cable television, for example).
- the date permission was granted.
- the duration of time permission is granted.
- signatures of the composer(s), lyricist(s), and musicians, accompanied by legibly printed names.
- contact information for each person that signed the letter of permission.

Only the band's original music may be used in the production. The band's rendition of a copyrighted song, often called a "cover," cannot be used. *Cover music* may sometimes be used legally, but assistance from an attorney is required to navigate the process. Musical works that are in the public domain, however, may be performed by a local band and legally used in a program.

cover music: A band's rendition of another band's copyrighted song.

ASSISTANT ACTIVITY

Choose a popular song that would be appropriate for use in a student production that is broadcast locally. Research the necessary permissions and fees involved in using that song. Record each step of the process, including names, dates, and all the information provided to you.

Music Libraries

Another effective and less complicated source of music for programs is a music library service. An Internet search for "music library" will return dozens of options. Many large music libraries contain various types and styles of music that can be purchased by a studio facility, **Figure 13-6**. Pre-recorded music libraries vary widely in size and cost. These companies use several types of contracts with their clients:

- With a "buy-out" contract, the production facility purchases music and is free to use the music as often as they like without additional fees.
- "Needle-drop" is a term left over from the days of vinyl records. In a "needle-drop" contract, a fee is paid to the music library company every time a piece of music is recorded for a program (drop the phonograph needle on a record).
- When using a "lease" contract, a flat fee is paid for unlimited use of certain CDs in the music library or a specified number of downloads from the company's website. For online access to music downloads, clients typically must establish a user name and password. A lease term is usually one year. When the lease is up, all the CDs must be

Figure 13-6. There are dozens of music libraries available for purchase. Once the music is purchased, depending on the contract, the user is free to use the music at any time.

returned, access to online music downloads is terminated, and music from the library can no longer be used in any new videos.

PRODUCTION NOTE

After the first year of a lease contract, it may be possible to negotiate a better rate for a longer contract, such as a 3 or 5 year lease contract. At the very least, you can usually "freeze" the yearly lease fee for a multi-year contract, making it inflation-proof.

Make Your Own Music

Using your own original music in a program requires no permissions, rights, or licenses. Music composition and creation computer programs allow you to create original pieces of music. Many software companies offer trial downloads to "test drive" the program before purchasing it.

Necessary Rights

There are several different types of rights available to use copyrighted music. There are also different types of rights necessary to provide television programs to the public by any means. The combination of television and music rights is called a "rights package." The applicable package depends on how copyrighted music will be used in the program.

recording rights: Permission to record music from a live performance.

- *Recording rights.* Permission to record music from a live performance.

- *Re-recording rights*. Permission to copy music from its current format to a video medium. Re-recording rights are included in a majority of rights packages because nearly everything done in the video industry involves copying material from one medium to another.
- *Synchronization rights*. Permission to synchronize video with the music. This means that video of something other than the creation of the music itself is added to the music. This is the difference between radio and television—pictures!
- *Broadcasting rights*. Permission to broadcast the music to the public.
- *Cablecasting rights*. Permission to cablecast the music to the public.
- *Streaming rights*. Permission to stream material on the Internet with settings that do *not* allow the material to be downloaded or recorded—it can only be streamed.
- *Transitory digital transmission rights*. Permission to place material on the Internet in a format that permits downloading and recording from the Internet.

PRODUCTION NOTE

When you contact a record label for permission to use music and provide all the information on how you will use the music, the company will tell you which rights package applies.

re-recording rights: Permission to copy copyrighted material from its current format to a video medium.

synchronization rights: Permission to synchronize video with the music.

broadcasting rights: Permission to broadcast copyrighted material to the public.

cablecasting rights: Permission to cablecast copyrighted material to the public.

streaming rights: Permission to stream material on the Internet with settings that do not allow the material to be downloaded or recorded.

transitory digital transmission rights: Permission to place material on the Internet in a format that permits downloading and recording from the Internet.

Wrapping Up

Many novice producers choose to use music for the wrong reasons in their television programs. Because a piece of music is a favorite or is popular for the moment are not justifiable reasons to use it in a video program. It is important to remember that background music has one purpose—to increase the impact of the scene. "Favorite" music likely did not become the producer's favorite because it blends into the background. The guidelines for using music in a program should be followed at all times.

Copyright issues related to music typically do not apply to teacher-assigned programs that are seen only within a single classroom. Copyright permission *must* be obtained, however, for use of music in programs shown in any manner to viewers outside of that single classroom. If you use music without obtaining the appropriate copyright license, you are liable for legal action initiated by the copyright holder. "Your honor, I couldn't locate the copyright holder," is not an excuse that will stand up in court. If the CD in your possession doesn't have a label, it is likely a bootleg CD and you're on the wrong side of the law to begin with.

Review Questions

Please answer the following questions on a separate sheet of paper. Do not write in this book.

1. What is the purpose of background music?
2. What is the "heart" of a song?
3. List three of the guidelines for using background music in a program.
4. Explain the challenge created in using catchy or busy music in a dialog scene.
5. How do copyright laws apply to student and school video programs?
6. Identify the elements that should be contained in a letter of permission from a local musician.
7. What types of contracts are available to use the services of a music library?
8. Which music rights apply to use on the Internet?

Activities

1. Go to the Public Domain Information Project website and review the various songs considered to be "in the public domain." Make a list of song titles that you were surprised to see included on the website.
2. Research Title 17 of the U.S. copyright code. Locate Chapter 1, Subject Matter and Scope of Copyright and read Section 115. Write a composition that explains Section 115 and how this portion of copyright law applies to using music in student productions.

STEM and Academic Activities

1. Choose an Internet-based music file sharing or download site. Review the site's Terms of Use. To what extent may the downloaded files be shared? What conditions apply to the use of downloaded files?

2. Research the contracts available through three different music library companies. Compare the cost of the same type of contract from each of the three companies. What is the difference in contract prices? Determine the per song price for each contract. Which company offers the lowest cost per song?

3. Write a formal letter to a band requesting permission to use a piece of their music in one of your productions. Provide details about your program, explain how their music will be used, and follow standard guidelines and format for writing a formal letter.

4. Identify a movie where the use of background music breaks the guidelines discussed in this chapter. Explain the effect the use of music has on the scene or message of the movie.

www.school-video-news.com School Video News is an online magazine that provides television and video production information and resources for schools, students, and teachers.

Chapter 14

Image Display

Objectives

After completing this chapter, you will be able to:

- Explain the appropriate use of still photos in a video production.
- Understand how fps affects the television image.
- Recall the guidelines for creating text to display on a television screen.
- Summarize the application of aspect ratio in creating the television image.
- Explain how contrast ratio affects television graphics.

Professional Terms

aspect ratio	film scanner
character generator (CG)	fps
contrast ratio	graphics
crawl	hot
credits	luminance
digital intermediate	pop the contrast ratio
essential area	roll
film chain	telecine
film island	titles

Introduction

Graphics include any artwork required for a production, from paintings that hang on the walls of a set to the opening and closing program titles. Charts, graphs, sports scores and statistics, election results, weather statistics, and any other electronic text that is part of a visual presentation are also considered graphics in a program. Most graphics are computer-generated, but some are still created on paper or canvas with ink, paint, or other medium used by the artist.

graphics: All of the "artwork" seen in a program, including the paintings that hang on the walls of a set, the opening and closing program titles, computer graphics, charts, graphs, and any other electronic representation that may be part of a visual presentation.

Talk the Talk

In some facilities, the terms "visuals" and "graphics" are used interchangeably.

Copyright

Any picture taken from a magazine or book, Web site, or a motion picture still frame is almost always copyrighted. This means that these images may not be used in a video program without the copyright owner's permission. The simplest way to obtain images for a production is to create your own—take original photographs or make unique pieces of art. If existing copyrighted works must be used, find the copyright holder and get permission. Refer to the copyright information in Chapter 12, *Legalities: Releases, Copyright, and Forums*.

Still Photos

If used sparingly, still photography can work well in a video program. Excessive use of still photography, however, makes a television program look like a slide show. To make interesting use of still photos, move the camera around on the picture to create a sense of motion.

Assistant Activity

Watch a few documentaries to see this "roaming the camera on a still photo" technique. Also notice that sound effects and music are added while the camera roams across a still image. The net result is quite effective.

Photos may be used in a video production, if certain precautions are taken. Take the picture holding the camera horizontally. A horizontally-oriented picture more closely matches the shape of the television screen than a vertically-oriented picture, **Figure 14-1**. A horizontal picture is a rectangle with the long side on the top and bottom, just like a television screen. If using a photo print, use a satin finish instead of glossy. Glossy photo paper reflects the glare of lights into the lens of the video camera.

Photographic slides may be used instead of printed photos, which eliminate the issue of lighting glare. A photographic slide must be oriented in the slide projector horizontally, rather than vertically.

With the prominence of digital video, many non-linear editors (discussed in Chapter 24, *Video Editing*) accept image files in various formats, such as .jpg, .tif, and .gif. There are also computer programs available that allow the user to crop and change a photo in many ways before sending it to a non-linear editor, **Figure 14-2**.

Figure 14-1. The horizontally oriented picture is more closely shaped to a television screen than the vertical image.

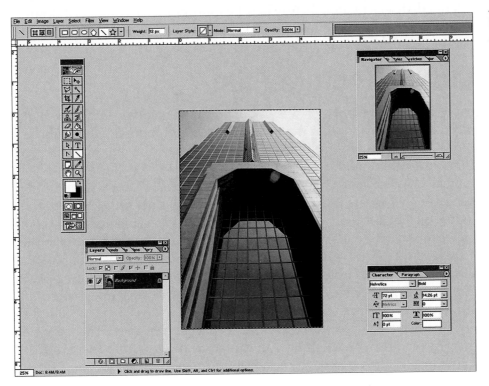

Figure 14-2. Some computer programs allow photographs to be cropped and otherwise manipulated before placing them in a program.

Talk the Talk

Adobe® Photoshop® is one of the most well-known image manipulation software programs. This product is so well-known, in fact, that the verb "photoshopped" is now used in common language. For example, "He photoshopped his girlfriend into the picture."

Still images that are not manipulated by a computer must be shot with a video camera. Printed photographs should be placed on an easel to be captured by the video camera. With photographic slides, the camera captures images that are projected onto a screen. The video camera must be positioned with the lens pointed directly at the center of the photo. If it is not "dead on" the center, the image captured will be distorted. Neither of these low-cost techniques produces the high quality video image output of a telecine (pronounced "tell-eh-seen").

In concept, a *telecine* is a slide projector or movie projector pointed directly at the television camera lens. A telecine is also referred to as a *film chain* or *film island*. This machine is quite expensive and large; about the size of a full-size refrigerator. Many optical maneuvers must be performed to actually get an image onto the recording medium. The most advanced telecines take theatrical motion pictures and convert them to video media for broadcast, purchase, or rental. A telecine can work in two directions—it can create a video of a film and it can create a film of a video.

A *film scanner* works like the scanner connected to a desktop computer and is used for motion picture film. Each frame of the film is digitally scanned and typically creates a high-quality, digital version of a motion picture called a *digital intermediate*. Many films are transferred to video for editing purposes, particularly if digital effects will be added. The digital intermediate of a motion picture may be transferred back to film, if desired, for further processing, duplication, and distribution.

telecine: A device that facilitates the transfer of film images onto videotape. Telecines are used, for example, to transfer theatrical motion pictures to DVDs for purchase or rental. Also called a *film chain* or *film island*.

film scanner: A digital device designed to copy/scan motion picture film.

digital intermediate: A high quality digital version of a motion picture created by digitally scanning motion picture film.

Motion Picture Film

While the majority of television production today involves digital media, a great deal of film is also used. A piece of motion picture film is a series of still pictures. Each picture is only slightly different from the next. When projected rapidly, one picture after another, the illusion of motion is created. When you watch a movie, 24 individual pictures are projected per second. This rate is expressed as 24 *fps* (frames per second). Each individual picture is actually flashed twice, for a total of 48 pictures per second. However, there are only 24 different frames. Each frame is projected for a flash, is followed by black, and projected again for a flash followed by black, while the projector advances one frame and flashes a new picture. The pictures flash in such rapid succession, the blackness is not perceptible to the viewer's conscious mind.

fps: The rate at which individual pictures are displayed in a motion picture and on television, expressed as frames per second.

The television picture also uses the concept of frames. The television image frame is not something you can actually see, like a frame of film. Individual pictures are not visible on a piece of videotape. Videotape simply looks like a ribbon of brown or black. The television signal is either a magnetic signal on a videotape or an optical digital signal on a DVD, and is recorded one picture at a time. The frame rate for television is 30 fps.

The difference between motion picture frame rate (24 fps) and television frame rate (30 fps) causes the image on a CRT television or computer monitor to flicker or roll when seen in a movie. A CRT television or computer monitor uses a picture tube to create the image. Newer, flat-panel televisions and computer monitors are digital and do not use a picture tube. Instances of the "black" between motion picture frames and the "black" between frames of the television picture can coincide so the image on screen is black long enough to be detected by the human eye. The viewers' minds register this blackness as a roll, seen in different places on the picture.

High-definition television (HDTV) technology can use the motion picture frame rate of 24 fps, as well as the television frame rate of 30 fps. The 24 fps capability of an HDTV is the reason that image flicker or roll is now rarely seen in a movie. Image flicker and roll may still be seen in home videos, unless a digital camera and monitor are used—digital equipment may eliminate the effect entirely.

Text

Most of the text seen on the television screen is created on a computer and electronically fed into the video switcher to be recorded with the program. Relatively small text is easy to read when viewed on a computer screen, because the viewer is less than three feet away from the monitor. When sitting further away from the monitor, however, the letters of small text are too small to read and the words seem to run together. Most television viewers sit between 8' and 15' feet away from the television screen. Any words that appear on screen must be large enough to be clearly read at that distance, and should remain on the screen long enough to be read out loud twice.

Letters on a television screen need to be relatively large, with only a few words on the screen at a time. No more than 5 lines of writing and no more than 5 words per line should appear on the screen, **Figure 14-3**. However, there are exceptions to this rule. The most notable example of an exception is the opening text narrative of the *Star Wars* films. In these movies, there are more words per line and more lines per page in the opening sequences than the rule states. However, if the words or lines are revealed to the audience one at a time, more lines and words can be placed on a page. When the producer forces viewers to look only at a certain place on the screen, more words can be displayed than the rule states.

PRODUCTION NOTE

Misspellings on graphics are an announcement to the viewing public that you cannot spell. Viewers will likely make immediate assumptions about the program or product, as well as your credibility and intelligence.

Figure 14-3. A basic rule for using text on the television screen.

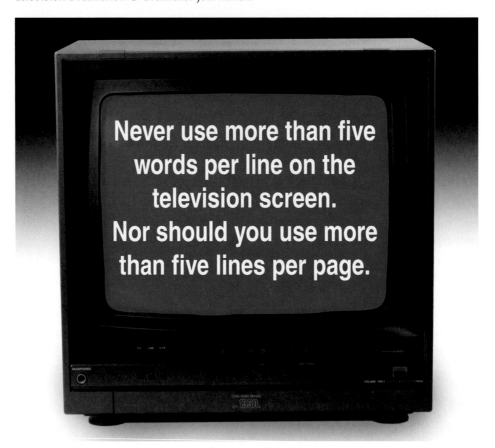

If many lines of text are presented on screen at once, the audience will find it very difficult or impossible to read. For example, the "fine print" displayed at the end of a car sale or lease promotional spot on television explains the details of the advertised deal. By law, the advertiser must display this information in the spot. But, the text is created in such small type and is displayed so briefly on the screen, it is unlikely that many viewers have ever been able to read the contents. The law does not specify that the information must be presented in an easily legible format.

Just as in word processing programs, numerous font styles are available for creating television titles. Many of the fonts, however, should not be used. Only bold, simple letters can be clearly read on the television screen. Fancy and elaborate fonts simply do not display well on television, **Figure 14-4**. Even though these fonts appear clear and crisp on a computer monitor, they may not translate well to the poorer quality of some television receivers. While many consumers purchased new, digital television receivers, several million television viewers opted to use converter boxes for their analog televisions. Therefore, graphics must be created for viewing on lower-quality, analog television screens. With titles created on a computer, always view the titles on a television monitor to verify that they are readable.

Aspect Ratio

aspect ratio: The relationship of the width of the television screen to the height of the television screen, as in 4:3 (four by three) or 16:9 (sixteen by nine).

Aspect ratio refers to the relationship of the width of the television screen to the height of the television screen. From the early 1950s to the

Figure 14-4. Fancy or thin fonts are not clearly visible on the television screen, even though they appear clear on a computer monitor.

present day, television sets common in most homes had an aspect ratio of 4 units wide by 3 units high, or a 4:3 aspect ratio, **Figure 14-5**. This ratio applies to any unit of measure, such as centimeters, inches, feet, or yards.

Talk the Talk

When referring to aspect ratios while speaking, it is proper to say "four by three" or "sixteen by nine." For example, "Is that program in sixteen by nine or four by three?"

The recorded video images of a motion picture are approximately 16 units wide by 9 units high, or in 16:9 aspect ratio. See **Figure 14-6**. Images this size cannot be viewed in their entirety on a 4:3 aspect ratio television screen. The original video image is too wide to completely fit on the television screen. Only about three-fourths of the video image actually appears on the television screen, **Figure 14-7**. Currently, the only way to see the entire image on a 4:3 screen is to "letterbox" the image. Letterboxing displays the entire image, from left to right, on the television screen. The top of the image is pushed down and the bottom of the image is pushed up. The result is a narrow horizontal strip on the screen, with a black bar on the top and bottom of the screen, **Figure 14-8**.

Figure 14-5. A television screen with an aspect ratio of 4:3 (4 units wide by 3 units high).

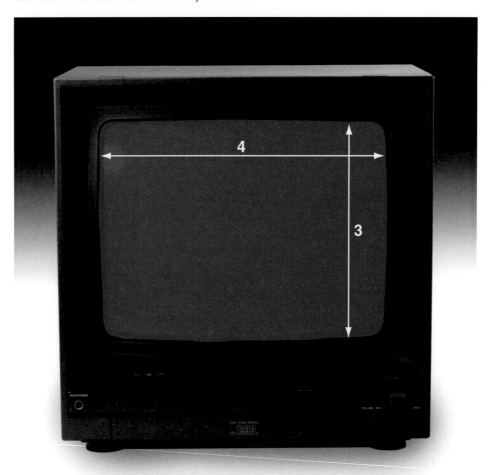

Figure 14-6. The standard format for broadcast television is 16:9.

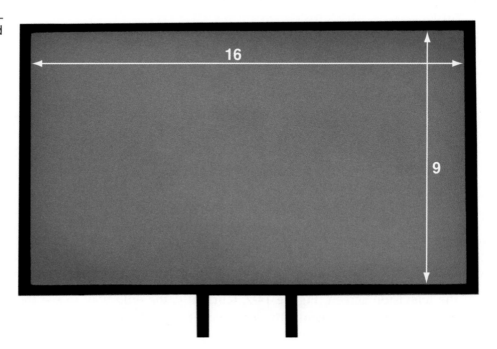

Figure 14-7. About one-quarter of a 16:9 picture does not fit on a 4:3 aspect ratio screen.

Figure 14-8. An entire movie image can be scaled to fit on a 4:3 screen by "letterboxing" the picture.

VISUALIZE THIS

You are sitting in a classroom and your teacher is setting up a film to show on a film projector. The screen at the front of the room is square, and the film is wide screen format. The movie starts. The picture fills the screen from top to bottom, but the left and right edges of the picture (almost two thirds of the total image) are projected off the screen, onto the blackboard behind the screen and are difficult to see. To solve the problem, the teacher moves the projector closer to the screen, so the left and right edges of the picture are completely on the screen. However, the entire image is now smaller and there is now a dark gray bar at the top and bottom of the square screen (where nothing is projected at all). This is an example of the result of letterboxing.

PRODUCTION NOTE

The recorded images on 35mm and 70mm theatrical film are much larger than the older 4:3 aspect ratio. The 16:9 aspect ratio format incorporates as much of the original film image as current technology permits.

The standard format for broadcast television is 16:9. The 16:9 aspect ratio is more closely shaped to a 35mm motion picture or still photo image. On 16:9 format televisions, viewers see much more of the original film images when watching movies at home. On a digital television, viewers also see digital images, which are strikingly clear. The clarity of images viewed on a computer monitor is noticeably sharper than the average picture on an analog television set. Consider that images on a digital television are greatly improved from the display on a high-resolution computer monitor.

PRODUCTION NOTE

The 16:9 aspect ratio is not synonymous with large, "home theater wide-screen" television sets. Many 16:9 screens are large, but they can be rather small as well. The 16:9 ratio refers to the shape of the screen, not the amount of real estate it occupies.

The federal government organized a digital conversion project that provided a schedule and guidelines for all broadcast stations in the United States to convert from broadcasting an analog signal to broadcasting in digital format. To view television after the conversion on June 12, 2009, viewers needed a digital television or a converter box connected to an analog television. Many cable and satellite systems offer to convert the digital signal to analog before sending it on to subscribers. The converter equipment takes the superior digital image, changes it into an analog image, and sends that signal to an analog television set.

Essential Area

essential area: The area of an image or shot that must be seen on any television set, regardless of aspect ratio or age, and must include all the words in a graphic.

The concept of aspect ratio is very important when generating graphics. While all graphics need to be in 16:9 format, the *essential area* of the graphic must be in 4:3 aspect ratio. Essential area is the area that *must* be seen on any television set, regardless of aspect ratio or age, and includes all the words in a graphic. Words cannot be cut off by the left and right margins of a television set. For example, a graphic created to advertise a local business does not communicate the correct information if the essential area is not considered, **Figure 14-9**. The background of a graphic is not included in the essential area. Therefore, the background may be created in 16:9 aspect ratio.

Talk the Talk

The 4:3 aspect ratio essential area that is centered horizontally on a 16:9 screen is referred to as the "4 by 3 hot spot."

Essential area

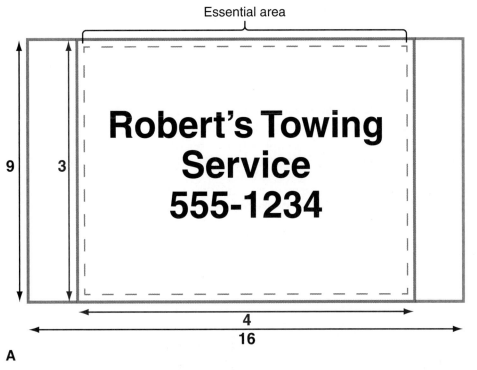

9   3

4

16

A

9   3

4

16

B

Figure 14-9. The essential area is the region that must be seen by the viewer. A—Even when created in 16:9 aspect ratio, a graphic must stay within the essential of a 4:3 television screen. B—Graphics formatted for 16:9 display fall outside of the 4:3 aspect ratio essential area.

Character Generator

The *character generator*, or *CG*, essentially creates letters (generates characters). Think of a CG as a video word processor. The letters and characters generated by the CG are used to create *titles*. Titles may be very simple pages that appear on the screen, or they can move across the screen. For example, the titles may move around on the screen in seemingly three-dimensional motions or the letters used in the titles can be animated. The degree and style of movement should contribute to the overall effect of the

character generator (CG): A device that creates (generates) letters (characters), primarily for titles.

titles: The letters and characters generated by a CG that are displayed on-screen.

video program, rather than being a display of the CG operator's capabilities, **Figure 14-10**.

Talk the Talk

The term "CG" also means "computer generated." When used in this context, CG is usually mentioned in conjunction with special effects. When referring to computer generated special effects, one might say "those effects are CG." The context of the sentence indicates how "CG" is used in conversation or direction.

credits: The written material presented before and after programs, listing the names and job titles of the people involved in the program's production.

roll: Titles in a program that move up the screen.

crawl: Words that appear either at the top or the bottom of the screen and move from the right edge of the screen to the left, without interrupting the program in progress.

The simplest and most common types of titles are a credit roll and a crawl. *Credits* are the written material presented before and after programs. Credits list the names and job titles of people involved in the program's production. In a *roll*, titles move up the screen, as if they were printed on a long roll of paper. A credit roll, for example, usually occurs at the end of a program. In order to be easily read by the audience, the titles must move up the screen. If titles move down the screen, the viewers must constantly jerk their eyes up and down to read the titles. This creates a feeling of discontent with the audience. A *crawl* appears either at the top or bottom of the screen, without interrupting the programming or footage. Words move from the right edge of the screen to the left edge of the screen, presenting the words the way we naturally read—from left to right. Running a crawl from the left to the right also creates a feeling of discontent with the audience. Local news programs typically use a crawl to display current traffic conditions and weather updates at the bottom or top of the screen.

There are many different CG computer programs available. Some are independent hardware and software units, others are programs that can be loaded onto a desktop computer, and many non-linear editing programs

Figure 14-10. The CG creates titles for television programs.

Studio Rules Of Operation

have built-in modules with CG capabilities. The titles are created on a computer and edited into the video program.

Many CGs and CG programs have customizable templates that can be used to create graphics for weather forecasts and sports scores. A variety of premade backgrounds are also available; both still backgrounds and dynamic, moving backgrounds. Premade graphic backgrounds are, generally, extremely affordable and offer nearly limitless possibilities.

Contrast Ratio

Contrast ratio is the relationship between the brightest object and the darkest object in the television picture. The human eye can see 100% contrast between black and white objects, expressed as a contrast ratio of 100:1. This means that we have no difficulty seeing black objects on a white background or white objects on a black background. The text of this book is black on a white page, for example. Some digital cameras actually have a contrast ratio of 100:1. Some extremely high-end video cameras produce a contrast ratio of 55:1. Consumer and most professional cameras produce a contrast ratio of 40:1. Some 35mm motion picture film stock can produce a contrast ratio of up to 90:1. Differences in the ability to capture and display varying contrast ratios partially accounts for the difference in the "look" of film and video. For the foreseeable future, the television industry must operate within the constraints of the 40:1 contrast ratio. It is vitally important to understand what 40:1 contrast ratio actually means.

On a scale of 1 to 100, "1" represents no-luminance (black) and "100" represents total luminance (white). *Luminance* is the brightness or lightness of the video image. Imagine a long board with equal sections marked and numbered 1 to 100, and a shorter board with sections of the same size numbered 1 to 40. The board that is 40 units long represents the television scale of 40:1. The 100-unit board represents a realistic scale of 100:1. The first unit on each board is black. Each subsequent unit is a lighter shade of gray, with the last unit on each board being white, **Figure 14-11**. Even though each board begins with black and ends with white, the 40:1 television scale is missing many shades between.

If number 1 (black) on the television contrast ratio scale is lined up evenly with number 1 (black) on the realistic contrast ratio scale, the difference in the color range is apparent. See **Figure 14-12**. The right end of the

contrast ratio: The relationship between the brightest object and the darkest object in the television picture.

luminance: A measure of the brightness or lightness of a video image.

Figure 14-11. With the 40-unit board beside the 100-unit board, the ends of the 40-unit board represent the upper limit of "lightness" and the lower limit of "darkness" possible on the television set.

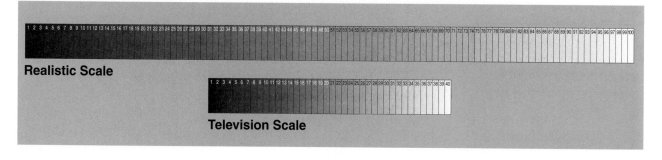

Realistic Scale

Television Scale

Figure 14-12. The lightest end of the television scale falls in the medium-gray range on the realistic scale. Anything lighter than unit #40 on the television scale is seen as white.

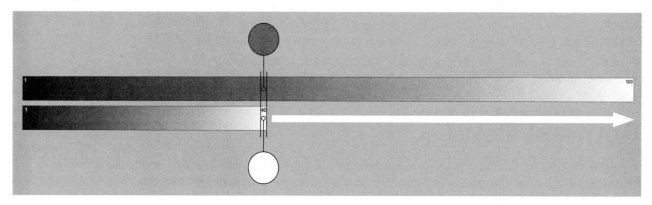

television scale is white, but the corresponding realistic color is medium gray. If there is a substantial amount of black in the picture, that medium gray color is what the television can display for the lightest object in the picture. In this scenario, any object that truly is a medium gray color displays as white on the television screen, because white is the lightest color a television set can produce. Any object lighter than medium gray on the realistic scale in the picture will glow with an otherworldly light.

If number 40 (white) on the television contrast ratio scale is lined up evenly with number 100 (white) on the realistic contrast ratio scale, the black end of the television scale aligns with a lighter medium gray on the realistic scale, **Figure 14-13**. The lighter medium gray is the darkest object a camera can see when the majority of the picture is white. Objects of this color appear black on the television screen. Anything darker than lighter medium gray on the realistic scale appears solid black and without detail; like a flat, black silhouette of an object.

In most cases, the middle of the television contrast ratio scale (20) should be aligned with the center unit of the realistic contrast ratio scale (50). To produce a quality, realistic image, the picture should not contain anything lighter or darker than the shades of gray on the realistic scale that correspond to the units at either end of the television scale (unit 31 through unit 70 in **Figure 14-11**). If an object is darker, it appears black. If an object

Figure 14-13. The darkest end of the television scale falls in the light-gray range on the realistic scale. Anything darker than unit #1 on the television scale is seen as black.

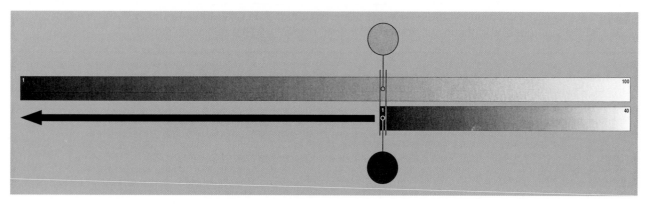

is lighter, it appears white. The further away an object falls from either end of the 40-unit stick, the more negative attributes (black or white) the object acquires.

VISUALIZE THIS

A student decided to produce a program about the degrees of wear and tear on automobile tires as her class project. In preparing for the shoot, she painted the backdrop of the set a relatively bright yellow. The tires were arranged on the set against the yellow background and the studio lights were turned on. But, the tires were solid black shapes on the monitor. The student realized that the camera could not "see" any detail of the tire treads because she exceeded the contrast ratio in the image. There was too much bright yellow in the image. The iris closed so much to prevent over-exposure of the yellow that the details of the black tires were lost.

There are two ways to control movement of the television contrast ratio scale.

- Place the camera in auto-iris mode. The camera will position the scale automatically, considering the amount of lightness or darkness in the total picture. With more of one or the other of these extremes in the picture, the camera adjusts the television contrast ratio scale further down or up the realistic contrast ratio scale. To control movement of the scale, alter the brightness of objects in the picture.
- Put the camera in manual iris mode to manually override the automatic setting. Even though you can force the camera to accept an image higher than 40:1, there will be negative side effects. Blacks may be flattened and have no detail; whites may glow and have no detail.

PRODUCTION NOTE

For scenes that require a greater amount of darkness, like exploring a cave, remember that the television contrast ratio scale has moved down the realistic contrast ratio scale. The talent should not be dressed in bright white costumes. If shooting in an extremely light environment, such as Antarctica, the television contrast ratio scale has moved to the upper end of the realistic contrast ratio scale. The actors should not wear dark navy blue parkas.

Popping the Contrast Ratio

Exceeding the contrast ratio limitations, or *popping the contrast ratio*, of video results in an image that contains glowing light-colored objects (**Figure 14-14**) or flat, dark-colored objects without detail, like silhouettes. When the contrast ratio is popped, cameras equipped with a zebra stripe circuit display diagonal zebra stripes on the viewfinder to indicate that an object in the scene is too brightly lit, **Figure 14-15**. When zebra stripes appear, it is not always necessary to make adjustments. The image is not

pop the contrast ratio: When the brightness or darkness of objects in a shot exceeds the contrast ratio limitations of video.

Figure 14-14. Popping the contrast ratio of a predominantly dark picture causes light-colored items to glow.

Figure 14-15. The camera operator sees zebra stripes through the camera's viewfinder when something in the shot is too brightly lit.

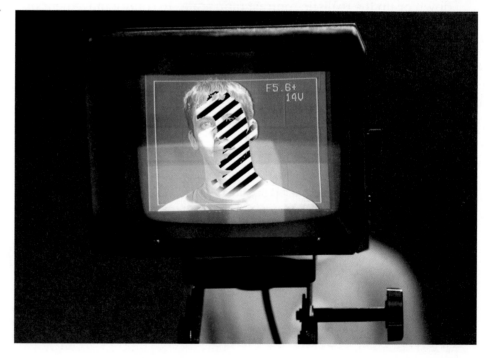

considered "bad" unless the negative effects intrude on the main focus of the picture. The zebra stripes only indicate that there may be a problem.

The zebra stripe circuit is typically found on higher-quality cameras, and is not available on low-end cameras. In mid-grade cameras, the circuit is pre-set at the factory and operates as described in the previous paragraph. Adjustable zebra stripe circuits can be found in higher quality (and cost) cameras. An adjustable circuit allows the user to manually set the level at which zebra stripes appear. For example, the factory preset may prompt zebra stripes to appear when any object in the picture reaches the 40:1 contrast ratio. The factory preset of 40:1 indicates that the image is <u>now</u> distorting due to contrast ratio problems. There is no warning that the image is getting close to distortion. An adjustable zebra stripe circuit allows the operator to set the zebra stripe display threshold to a different level, such as 35:1. At 35:1, the operator has a safety margin. In this case, the zebra stripes are a warning that the image is close to distortion.

PRODUCTION NOTE
Zebra stripes appear only on the camera's viewfinder. They will not be seen on any other monitor in the recording system or in the recorded program.

The camera operator must alert the director when zebra stripes appear; the program's director decides if the problem is worth correcting. For example, the director may choose to dismiss a "glow" that appears on the visible part of white shirts on a group of men wearing black suits. If these men are relatively minor characters in the scene, the zebra stripes on the small portion of their shirts can be ignored to continue shooting.

When shooting outside a ski lodge, however, the entire image is quite *hot* (very bright) due to all the white snow in the environment. The main character skis into the frame of the camera wearing a dark green outfit. Both the character's face and outfit appear completely black because the outfit has popped the contrast ratio. This is not an acceptable shot. One solution is to paint the snow a gray color, but this is not a sensible or realistic option. Practical solutions to this problem include the following:

hot: A term used to describe an image or shot that is very bright.

- Shoot the scene at night. This, however, might not work for the scene.
- Close the aperture of the camera or place filters on the camera lens to reduce the amount of light coming in. As the amount of light coming in is reduced, the television scale moves lower on the realistic scale. The costume and talent's face will eventually be visible.
- Change the character's costume.
- Zoom in on the subject of the shot. Zooming in reduces the amount of bright snow in the picture and magnifies the center of the picture—there is less white snow and more dark green clothing. This lowers the contrast ratio.
- Increase the shutter speed. "Light eating" is a negative consequence of increasing shutter speed. In this situation, however, the consequence becomes an advantage. Excess light from the bright snow can be "eaten" by the increased shutter speed until the contrast ratio and exposure on screen is acceptable.

The concept of contrast ratio also applies to production graphics. High contrast ratios should be avoided for television images, such as white letters on a black background. Additionally, colors with similar luminance values should not be used together in a picture. Luminance refers to the degree of lightness in a picture related to the degree of darkness. For example, red and green are two colors most associated with the winter holiday season. Therefore, these colors are very likely to be included in a written graphic—red letters on a green background or green letters on a red background. On a black and white television, this graphic appears as an almost entirely gray screen, **Figure 14-16**. The shades of red and green used in the graphic may have identical levels of luminance, which makes them appear to be the same color on a black and white television screen.

Colors of contrasting luminance should be used in graphics, but the contrasting colors cannot exceed the 40:1 contrast ratio in luminance. Many professional cameras are equipped with a black and white viewfinder, which allows the camera operator to easily evaluate the quality of any graphic shot. However, if graphics are generated on a computer, they should always be fed into a black and white monitor for review before being recorded.

PRODUCTION NOTE

If a black and white monitor is not available to review a computer-generated graphic, a color monitor may be used. Simply adjust the image to black and white by turning the "color" or "chroma" down using the accessible controls on the monitor.

Television systems reproduce most colors very well. However, in analog television production, the colors red, pink, and orange are difficult to accurately reproduce for technical engineering reasons. Avoid large amounts of these colors as much as possible in analog television production. These colors do not pose a problem in digital television production. However, consumers with analog television sets will still be affected by the red, pink, and orange colors.

Figure 14-16. Different colors with the same luminance value appear to be almost the same shade of gray on a black and white screen.

Wrapping Up

Graphics for television are an important aspect of production because they include anything the viewer needs to read. In order for viewers to read the information, the graphic must be simple enough and sufficiently large to be seen from a couch or chair that is 8′ to 15′ away from the screen. Graphics should also be displayed on the screen long enough to be read out loud twice. The contrast ratio of the graphic must remain within the limits of the television system. Due to the emergence of digital television, graphics must be generated to fit on the screen of every style and size television set. Graphics must appear satisfactorily on both 4:3 and 16:9 aspect ratio television screens.

Review Questions

Please answer the following questions on a separate sheet of paper. Do not write in this book.

1. What are the conditions for using still photos in a video program?
2. How is a digital intermediate produced?
3. Why does the image displayed on a CRT television or computer monitor flicker or roll when shown in a movie?
4. What is the maximum number of lines of text that should be displayed on screen at one time? What is the maximum number of words per line?
5. What is *aspect ratio*?
6. What are some benefits of 16:9 aspect ratio displays compared to 4:3 aspect ratio displays?
7. What is the difference between a roll and a crawl?
8. What is *contrast ratio*? What is the greatest contrast ratio possible with analog television systems?
9. What is *luminance*?
10. What happens when an image pops the contrast ratio?
11. Why should color graphics be evaluated on a black and white television monitor?

Activities

1. Watch 15 minutes of two versions of the same movie: one in full screen format (4:3 aspect ratio) and one in wide screen, or letterbox, format (16:9 aspect ratio). Write down the noticeable differences in various scenes. Be prepared to share this information in class.
2. Visit an electronics retail store. Compare the picture on several digital televisions that are tuned to the same image feed, in both standard definition and high definition. Make note of the various specifications on several digital and HDTV models and indicate the prices of each. Be prepared to share your findings in class.

STEM and Academic Activities

Technology

1. Research how home movie recording technology has changed. Describe important technological advancements and products in the evolution of home movies.

2. Investigate the technology behind image manipulation software products. How does image manipulation software edit or alter still photos?

Mathematics

3. When you watch a movie, 24 individual pictures are projected per second, which is expressed as 24 fps (frames per second). Each individual picture is actually flashed twice. How many pictures are flashed per second? Between each frame, there is an instance of black. Including the 24 individual pictures that are flashed twice and the instances of black between each frame, how many total images are flashed per second?

Social Science

4. List some television programs that display a crawl on screen. For each program, identify the information presented in the crawl. How is this information useful to the viewing audience?

Chapter 15

Lighting

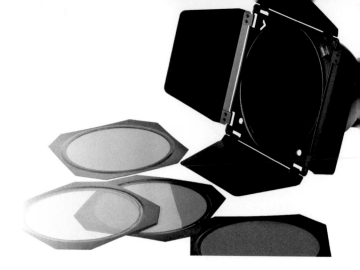

- Explain how the color temperature of light affects the video image.
- Recall methods to control lighting intensity.
- Identify the steps in the procedure to light a set.
- Describe the television lighting techniques presented and identify the instruments used with each technique.

Introduction

There are two main functions of lighting for television production:

- To meet the technical requirements of the camera. There should be enough light to produce an acceptable picture on the screen.
- To meet the aesthetic requirements of the director. Sufficient lighting is necessary to create the desired mood, from an artistic standpoint. A romantic dinner, for example, should have a different lighting design than a football game.

Ultimately, the television screen is a flat piece of glass. Industry professionals try to create the "illusion" of a three-dimensional image by manipulating many objects and aspects of a program. To create three-dimensions:

- Shoot a person in a three-quarter angle, rather than straight on or in profile.
- Apply makeup to the talent to create lines of light and shadow.
- Paint a production set to create the illusion of three dimensions.
- Arrange set elements to create a foreground, middle ground, and background.
- Make certain areas more prominent on the screen through the creative use of light and shadow.
- Creatively use a shallow depth of field when shooting.

Special lighting is necessary in television production because, typically, the lens aperture is closed significantly to accomplish great depth of field. Closing the lens aperture requires that the light level be increased, or a wonderfully focused picture will be a wonderfully focused *dark* picture.

Professional Terms

3200° Kelvin (3200K)
back light
background light
barndoors
basic hang
bounce lighting
C-clamp
cross-key lighting
diffusion
dimmer
fill light
flag
flood light
floor stand
fluorescent lamp
four-point lighting
Fresnel
gel
grid

hard light
honeycomb
incandescent lamp
instrument
Kelvin color temperature
   scale
key light
lamp
light hit
light plot
limbo lighting
raceway
rough hang
scoop
scrim
soft light
spotlight
three-point lighting
triangle lighting

Objectives

After completing this chapter, you will be able to:

- Identify the various types of lighting instruments and cite unique characteristics of each.
- Compare the characteristics of incandescent lamps with the characteristics of fluorescent lamps.

Using Professional Terms

It is imperative that consumer terms *not* be used in a professional studio environment. Using the correct terminology is considered an entrance exam for broadcasting employees. You risk losing respect and credibility among industry peers and superiors if you use consumer terms or misuse professional terms in the workplace. The importance of correctly using industry terms cannot be overstated.

Average consumers call the lighting fixture on the side table in their living room a "lamp" and the part inside that glows a "lightbulb." Television production industry professionals do not use these terms. The lighting fixture is called an *instrument* and the part that glows is a *lamp*. In the industry, a lamp illuminates a set when installed in a lighting instrument.

instrument: The device into which a lamp is installed to provide illumination on a set.

lamp: Part of a lighting instrument that glows when electricity is supplied. The consumer term for this item is "lightbulb."

Talk the Talk

When referring to a lighting instrument, just the word "instrument" is typically used. "This instrument needs a new lamp." The word "lighting" is understood and is not actually spoken when industry professionals use this term. Other terms used to refer to a lighting instrument are "fixture" or "head." Sometimes, the name used refers to the wattage of the lamp, such as "1K," "2K," or "baby." It is important to learn the naming conventions used in your workplace when you are a new employee.

In the television production industry, the word "light" has two definitions:

- "Lights" refer to the collection of all the instruments used in the studio or on location. "Let's turn the lights on now." Most professionals say "instrument" only when referring to a specific lighting instrument.
- "Light" also refers to the illumination created by turning on a lamp. For example, the lighting engineer may use a light meter to measure the amount of light hitting or reflecting off an object on the set.

Types of Light

The two types of illumination used on a studio set are defined by the type of shadows they produce—hard and soft. See **Figure 15-1**.

Hard light creates a sharp, distinct, and very dark shadow. Hard light is the type necessary to create shadow puppets against a wall. If a hard light instrument is hung from the ceiling of a TV studio and pointed straight down onto an object, a perfectly-shaped shadow is created on the floor below the object. The line on the floor between areas of light and shadow is thin and distinct.

Soft light creates indistinct shadows. Pointing a soft light instrument straight down onto an object creates an indistinct shadow pattern on the floor. There is no definitive line between areas of light and shadow. The lighted area gradually fades into shadowed area.

hard light: Type of illumination used in a studio that creates sharp, distinct, and very dark shadows.

soft light: Type of illumination used in a studio that creates indistinct shadows.

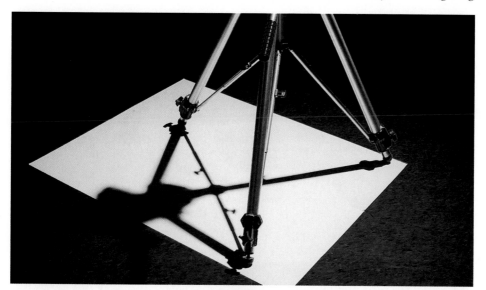

Hard light

Soft light

Figure 15-1. Hard light creates sharp, distinct shadows. Soft light creates indistinct shadows.

Types of Lighting Instruments

Nearly all studio hard lights are a type of *spotlight*, **Figure 15-2**. Spotlights create a circle of light in varying diameters. These instruments can be fixed to a pipe on the ceiling or wall, placed on a stand, or be very moveable. Moveable spotlights can be moved by hand or be motorized and operated by remote control. Moveable spotlights, "spots" for short, are often used in theatrical presentations when the spotlight follows a person walking around the stage.

Convertible spotlights have a sliding lever on the body of the instrument, **Figure 15-3**. These instruments may be referred to as "focusing fixtures" or "focusable fixtures." The instrument is a hard light in one setting and converts to a somewhat softer lighting instrument by sliding the lever.

The *Fresnel*, pronounced "fruh-NEL," is a hard lighting instrument, **Figure 15-4**. It is a lightweight instrument that is easily focused and can

spotlight: Type of hard light instrument that creates a circle of light in varying diameters.

Fresnel: A hard light instrument that is lightweight and easily focused.

Figure 15-2. Spotlights are instruments that create hard light.

Figure 15-3. A convertible spotlight is very versatile because it can create hard or soft light.

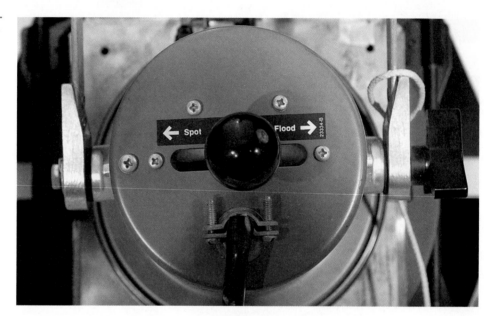

Figure 15-4. A Fresnel is a lightweight, focusable, hard lighting instrument.

produce a great deal of light. Fresnel instruments are named for the inventor of the Fresnel lens, Augustin-Jean Fresnel. Because the instrument is named after a person, "Fresnel" should always be capitalized.

A *flood light* is a soft lighting instrument that provides general lighting in a large area. One of the most common flood lights is a *scoop*, **Figure 15-5**. It is a half-spheroid shaped instrument that creates a great deal of light.

Accessories

Some situations may require light to be projected in a specific shape or be blocked from hitting a particular object on the set. *Barndoors* are the most commonly used items to shape and block light, **Figure 15-6**. Barndoors are fully moveable metal flaps that attach to the front of an instrument. The operator moves the barndoors into the beam of light to block or reshape the light.

flood light: A soft light instrument that provides general lighting in a large area.

scoop: A common type of flood light with a half-spheroid shape that produces a great deal of light.

barndoors: Fully moveable black metal flaps attached to the front of a lighting instrument; used to block or reshape the light.

Figure 15-5. The scoop lighting instrument is named for its domed, or scoop, shape. *(Mole-Richardson Co., Hollywood, CA)*

Barndoors

Figure 15-6. Barndoors allow the light to be shaped, rather than merely projecting light in a large circle.

Safety Note

The entire lighting fixture gets extremely hot when turned on, including the barndoors. If the instrument is on or has recently been turned off, wear gloves when handling the barndoors.

Cinefoil™ and Blackwrap™ are two professional products used by most lighting designers to reshape light. These products are flexible, rolled sheets of aluminum that can be wrapped around the front of an instrument. Cinefoil and Blackwrap are flat black in color, so they do not reflect light. Budget-conscious production environments often use aluminum foil to produce the same effects as Cinefoil, Blackwrap, and barndoors, **Figure 15-7**. Use heavy-duty foil instead of regular aluminum foil—heavier foil resists accidental tears better and can withstand the extreme heat of the instruments for a longer period of time. While aluminum foil works to reshape light, aluminum foil is shiny and reflects light—sometimes uncontrollably. Be aware of wayward reflections and correct any lighting problems and hot spots on the set.

Some very interesting and creative shadow patterns can be created using inexpensive aluminum foil on instruments. To use aluminum foil to shape or block light:

1. Tear off a sheet several feet long.
2. Shape the aluminum foil into a cylinder.
3. Attach the cylinder to the front of the instrument with metal paper clips (not plastic or vinyl coated clips).
4. Turn the lighting instrument on.
5. Shape the foil by hand.

Safety Note

Do not use transparent tape, masking tape, or duct tape to attach foil to an instrument. The tape may ignite! This poses a serious safety risk to every person on the set and in the building. Also, if a fire ignites, water from the sprinkler system will damage every piece of video equipment, the set, costumes, props, and all other valuable items in the studio.

light hit: A white spot or star shaped reflection of a lighting instrument or sunlight off of a highly reflective surface on the set.

flag: A flexible metal rod with a flat piece of metal attached to the end; used to block light from hitting certain objects on the set.

Brightly polished objects, like a silver serving tray or brass lamp, may be among the elements included in a shot. However, highly reflective surfaces create a white spot or star-shaped reflection of a lighting instrument or sunlight, which can be reflected into the camera lens. This reflection is called a *light hit* and is generally considered an undesirable effect. The easiest solution is to remove the reflective object from the set. If this is not an option, a *flag* needs to be placed between the lighting instrument and the reflective object, **Figure 15-8**. A flag is a metal rod with a flat piece of metal attached to the end. The metal rod is flexible, about 2′–3′ long, and has a clip on the end. This rod is attached to the side of a lighting instrument with the excess length extending in front of the instrument. A small, flat piece of metal cut into a shape is attached to the clip. The lighting designer bends the rod until the flag is positioned between the light source and the reflective surface on the set. The flag blocks light from the reflective object, but the rest of the set remains illuminated by the lighting instrument.

Figure 15-7. Heavy aluminum foil can be used to shape the light projected from an instrument.

Figure 15-8. A flag prevents light hits by blocking light from hitting a particular item on a set. A—Shiny surfaces on a set reflect light from the lighting instruments used. B—A flag is placed between the lighting instrument and the reflective surface on the set. *(Courtesy of Matthews Studio Equipment)* C—With the light blocked from shiny surfaces, light hits are avoided.

A

Lighting instrument

Flag

B

C

Other ways to remove a light hit include:
- Spray the item with dulling spray available from photo supply stores. The spray can be removed with a damp cloth after the shoot.
- Spray the item with inexpensive hair spray. Hair spray is also water soluble for easy removal.

Fluorescent Lamps

incandescent lamp: Type of lamp that functions when electricity is applied and makes a filament inside the lamp glow brightly.

fluorescent lamp: Type of lamp that functions when electricity excites a gas in the lamp, which causes the material coating the inside of the lamp to glow (fluoresce) with a soft, even light.

The types of instruments discussed to this point in the chapter use incandescent lamps. *Incandescent lamps* contain a filament inside the lamp that glows brightly when electricity is applied. Incandescent lamps used in television production are usually tungsten, tungsten halogen, or quartz halogen.

A *fluorescent lamp* functions when electricity excites a gas in the lamp, which causes the material coating the inside of the lamp to glow (fluoresce) with a soft, even light. Older fluorescent lamps were unsuitable for use in television production environments due to the bluish or greenish color temperature of the lamps. Professional television lighting fluorescent lamps are available in various shapes, sizes, and color temperatures, **Figure 15-9**. The most important color temperature in the television industry is 3200° Kelvin. Color temperatures are discussed in detail later in this chapter.

Figure 15-9. Fluorescent instruments can hold multiple lamps and can be hung from a grid or placed on lighting stands. *(Photo courtesy of Lowel-Light Mfg., Inc.)*

Fluorescent instruments and lamps have several advantages over incandescent varieties.

- Fluorescent lamps cost less to purchase, cost less to replace, and provide tremendous energy savings.
- Fluorescent lamps typically produce three to four times more light per watt, compared to tungsten halogen sources.
- Fluorescent lamps produce only a fraction of the heat generated by incandescent lights. With less heat produced by lights, the air conditioning system does not have to work as hard to keep the studio at a cool, comfortable temperature.
- Fluorescent lamps are longer-lasting than incandescent lamps. While lamp life is usually rated by the manufacturer between 8,000 and 10,000 hours, lamps gradually darken with age, reducing their efficiency, and should be inspected periodically.
- Fluorescent lamps can be touched with a bare hand while turned on. The lamp is warm, but not warm enough to cause any discomfort.

To the human eye, a fluorescent lamp appears considerably less bright than an incandescent lamp. While the illumination is less bright, the images created under fluorescent instruments appear beautifully on video. This is because professional video fluorescent lamps provide the exact frequency of light required by the camera. An incandescent lamp spreads a wide frequency of light, most of which the camera does not need. Also, since fluorescent lamps are not as bright as incandescent lamps, talent is less likely to squint at the camera due to bright lights on the set. Incandescent instruments are still needed on a set, but using fluorescent instruments can reduce the number of incandescent instruments needed and, therefore, reduce the cost of operation.

Supports for Lighting Instruments

Lighting instruments may be attached to floor stands in the studio. A *floor stand* has three or four legs and a long vertical pole to which a lighting instrument is clamped. Studio floor stands often have wheels on the legs for ease of movement. Even though floor stands are convenient, they have several disadvantages (**Figure 15-10**):

- They are top-heavy when an instrument is attached to the top of the pole and may be tipped over easily.
- The power cord for the instrument lies on the floor of the shooting area and is a tripping hazard. To be safe, the power cords should be taped to the floor.
- They occupy valuable floor space.
- Studio personnel can accidentally cast a shadow on the entire set by walking in front of a floor stand and instrument.

On a remote shoot, however, floor stands are a necessity. The crew should be aware of where the floor stands are positioned and be cautious around them. Some smaller, portable lighting systems come with large spring-loaded clamps. The clamps allow instruments to be attached to a flat, steady object, like the edge of a bookcase, door, or table, **Figure 15-11**.

In the studio, the best mounting option for lighting is to use a *grid*, **Figure 15-12**. Most studios have a grid hanging about twelve inches below

floor stand: A lighting support with three or four legs and a long vertical pole to which a lighting instrument is attached.

grid: A pipe system that hangs from the studio ceiling and supports the lighting instruments.

Figure 15-10. A floor stand can support small lighting instruments, but there are several precautions the crew must observe.

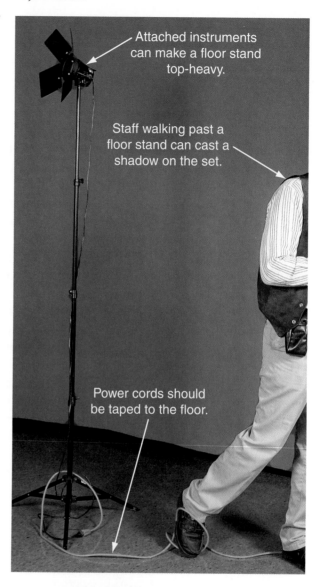

Attached instruments can make a floor stand top-heavy.

Staff walking past a floor stand can cast a shadow on the set.

Power cords should be taped to the floor.

Figure 15-11. This large clamp allows an instrument to be attached to a sturdy shelf, door, or table.

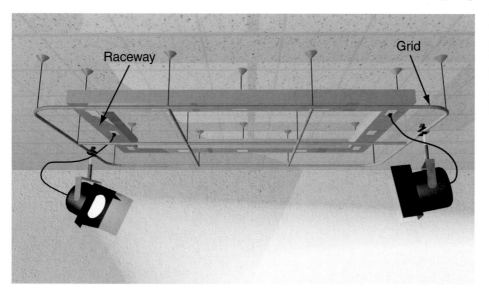

Figure 15-12. A grid is the pipe system that hangs from the ceiling and supports the instruments. The raceway supplies electricity for the instruments.

the ceiling. The grid is made of pipe that is at least two inches in diameter. Lighting instruments attach to the grid using a *C-clamp*, **Figure 15-13**, which is built into the instrument. The bottom of the "C" attaches to the instrument and the top hooks onto the pipe. A large threaded screw is tightened to firmly press against the pipe and secure the instrument safely to the grid.

C-clamp: A clamp in the shape of a "C" that is used to attach lighting instruments to the grid.

Safety Note

Any instrument hanging on the grid should have an additional safety chain attached. The chain should loop around part of the instrument and around the grid pipe. If the C-clamp comes loose, the safety chain prevents the instrument from falling to the ground and possibly injuring someone standing beneath it. Since barndoors can be removed if desired, some barndoors also accommodate a safety chain to prevent them from accidentally falling off an instrument.

A   **B**

Figure 15-13. A—C-clamps are used to attach lighting instruments to the grid. B—This fluorescent instrument is hung from a grid by a C-clamp and can be swiveled in any direction. *(Photo courtesy of Lowel-Light Mfg., Inc.)*

raceway: The system of electrical cables and outlets used to power lighting instruments on the grid. The raceway either hangs beside the grid pipes or is mounted to the ceiling above the grid.

The *raceway* either hangs beside the grid pipes or attaches to the ceiling above the grid, **Figure 15-12**. The electrical cables and outlets that power the instruments on the grid are part of the raceway. Each of the many outlets on the raceway is numbered and corresponds to a dimmer or switch on the lighting board. The lighting instruments are plugged into the outlets and are powered selectively from the lighting board. This eliminates the need to climb ladders to turn the instruments on and off. The electrical wiring is not placed within the grid piping due to the possibility of puncturing or crushing the pipe while tightening a C-clamp. If a C-clamp is overtightened and punctures a grid pipe, the electrical current that might be running inside the wiring would pose a great danger to the lighting director or gaffer on the ladder.

Safety Note

Placement of instruments on a grid must be carefully considered in studios with a ceiling-mounted sprinkler system. Instruments that get very hot cannot be positioned near sprinkler heads. Heat from the instrument can melt the sensor in a sprinkler head, which will activate the entire sprinkler system. This would be a real danger for studio personnel and a disaster for all the electronic gear.

Colors of Light

Colors reflect different frequencies of light. A frequency is measurable and can, therefore, be graphed. In 1848, the scientist Lord Kelvin devised a system to quantify and measure color. At that time, a black carbon rod was considered the blackest item available. Using the concept that black is the absence of color, Lord Kelvin applied heat to the black rod. Each time his eye could discern a color change, he noted the color and measured the amount of heat applied to produce that color, **Figure 15-14**. Based on the

Figure 15-14. Approximate values of the Kelvin Color Temperature Scale.

Temperature	Color
2000K	Red
2500K	Yellow
3000K	Pale Yellow
3200K	White
4000K	Green
4500K	Greenish Blue
5000K	Blue
6500K	Cobalt Blue
7000K	Violet
10,000K	Black

data collected, Lord Kelvin created a scale for measuring colors known as the *Kelvin Color Temperature Scale*. The Kelvin Color Temperature Scale measures color temperatures in degrees Kelvin.

Modern technology allows us to use combinations of materials, such as tungsten, quartz, and halogen gas, to produce light of the same color temperature without applying the extreme levels of heat Lord Kelvin used. However, some instruments that use lamps made of these materials still get incredibly hot and will immediately burn the skin if touched after being on for as little as 15 seconds. As mentioned earlier, the fluorescent lamps used for television lighting do not produce a great amount of heat while operating.

Kelvin Color Temperature Scale: A scale developed by the scientist Lord Kelvin for measuring color temperatures of light in degrees Kelvin.

PRODUCTION NOTE

The colors of light are not similar to the colors of paint. The principles that apply to each are different. With light, the color white is created when all the colors of light are combined. Black is the absence of all colors. A television screen is black until it is turned on. The screen becomes bright white, even though the lights creating the image are red, green, and blue.

White Light

The most important result of Lord Kelvin's research to television production is that *3200° Kelvin* (*3200K*) equals white light. In order to reproduce colors and flesh tones properly on television, the light hitting an object must be white.

3200° Kelvin: The temperature of white light in degrees Kelvin. Also noted as *3200K* or "32K" when spoken.

Talk the Talk

When temperature is written in degrees Kelvin, the word "Kelvin" is replaced with an upper case "K," such as 4500K. When this same temperature is spoken aloud, the last two zeroes of the temperature reading and the word "degrees" are omitted, as in "45K."

Most home videos taken indoors have a yellow hue to them, **Figure 15-15A**. This is because the lamps inside the instruments in most homes are considerably cooler than 3200K and produce light that is less than white. The Kelvin temperature of most incandescent lightbulbs for home use is about 2000K. On the other hand, video taken under regular fluorescent ceiling lights has a greenish hue, **Figure 15-15B**. The fluorescent lights used in classrooms and professional buildings are considerably warmer than 3200K. These lights are between 4000K and 4500K. Light in that temperature range is blue greenish and produces a gray, unhealthy look on natural flesh tones. Video shot outside under sunlight appears to be tinted with a shade of blue, **Figure 15-15C**. Sunlight is 5000K and up, which produces various shades of blue.

Figure 15-15. A—Color temperatures below 3200K cause the yellowish tone of pictures taken in consumer house lighting. B—A standard fluorescent light creates a blue-greenish tint due to a color temperature in the 4000K–4500K range. C—The camera sees sunlight in a bluish tint.

A

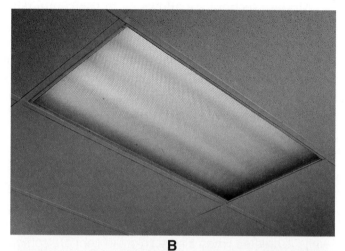

B

C

VISUALIZE THIS

Think of the Bunsen burner used in science classes. The hottest part of the flame is blue in color and the coolest part of the flame is yellow in color. This corresponds to the colors of light. Light sources warmer than 3200K create images with a blue tint. Light sources cooler than 3200K create images with a yellow tint.

PRODUCTION NOTE

In art classes, reds and yellows are referred to as "warm" colors and blues and greens are referred to as "cool" colors. The terms "warm" and "cool" in art class refer to the emotional feeling attributed to a particular color. Using the Kelvin Color Temperature Scale, however, warm and cool colors are opposite from those in art class. To create an "emotionally" warm color with lighting, we have to reduce the temperature (make it cooler) of a 32K light. To create an "emotionally" cool color, like blue, we have to increase the temperature (make it warmer) of a 32K light.

Objects on television appear as their actual color only when pure white light, 3200K, is used. Most television lamps are rated at 3200K. This temperature rating refers to the color of light emitted by the lamp, not the brightness of the light emitted. Brightness is indicated by the wattage of the lamp. A 1000-watt, 3200K lamp is much brighter than a 300-watt 3200K lamp. However, both produce the same white light.

To get the necessary 3200K white light at home or when shooting outside, two options are available:

- Bring enough lighting instruments to flood the shooting area with white light and overcome the natural light of the area.
- Trick the camera into thinking it is getting white light, even though it is not.

To "trick" the camera, activate the white balance circuit while the camera is pointed at a white object on the set that is under the lighting you are balancing for, such as sunlight, incandescent light, or studio lights. The camera is forced to see the object as white, without regard to the type of light hitting it. The camera then sees all other colors correctly because it has been balanced to one color—white. Other colors then fall into place on the scale. For example, assume that the color orange is two shades up (warmer) from red and two shades down (cooler) from yellow on the Kelvin scale. Each color is identified by its relationship to adjoining colors. Accurately identifying one of the colors places all the other colors into their proper position because the one color specified is used as a reference point.

Interesting effects can be produced by intentionally throwing off a camera's white balance circuit. For example, point a camera at a red object and white balance on it. The camera tries to turn anything red in color into white. In the process, every other color of the spectrum shifts, as well. This effect can only be achieved when the camera is in "manual" white balance mode. If the camera is in "automatic" white balance mode, the camera operator cannot control the circuit in any way. The automatic white balance circuit can be disabled on some high-end consumer cameras and all professional cameras.

Colored Light

Some programs or specific scenes require colored lights to be used on the set. For example, a rock concert typically has many different colored lights on the set. For a nightclub scene in a dramatic production, the audience expects to see various mood-enhancing colored lighting instruments. To turn the white light from a lamp into a colored light, a heat-resistant plastic sheet called a *gel* is used. Gels can be purchased from theatrical lighting stores and are available in hundreds of shades and colors. The plastic material is cut into a small rectangle that fits into a special gel holder on the front of a lighting instrument, **Figure 15-16**. The white light passes through and becomes the color of the gel.

Because colored lighting instruments are often used to create a specific mood or effect, it is important that the audience is able to see the colored lighting on the screen. In order for the audience to properly perceive the colored lights, the camera should be white balanced with the colored lights turned off and only white lighting instruments turned on. When the colored lights come on, the camera sees each color. White balancing on an

gel: A heat resistant, thick sheet of plastic placed in front of a lighting instrument to turn white light from a lamp into a colored light.

Figure 15-16. A gel can change the color of the white light emitted from an instrument.

object that is under a colored light throws off the camera's color reproduction circuits.

A production location may have a window or glass door to the outside. If the program is shot during daylight hours, blue sunlight will stream into the room while shooting. One solution is to place a CTO (color temperature orange) gel on the inside of the window. The CTO gel converts sunlight coming in the window from 56K to 32K. Another solution is to convert all the lighting instruments at the location to sunlight using CTB (color temperature blue) gels. CTB gels convert the 32K of the lighting instruments to 56K. When all the light at the location is the same color temperature, perform a white balance and the recorded image will appear correctly lit.

Lighting Intensity

Once lighting instruments have been set up, either on a set or on location, controlling the intensity of the light becomes an important task. For example, if a person or object is too brightly lit in a shot, they may appear to glow. Depending on the time of day, natural sunlight may cast dark shadows on the talent's eyes, nose, and chin. If the lighting instruments are the convertible type, the easiest solution is to move the lever from "spot" to "flood" or "flood" to "spot." If the lighting intensity issue is not resolved, several other techniques are effective in reducing the amount of light that hits an object on the set.

Move the Instrument

Moving lighting instruments farther away from or closer to the set is a simple solution to control the intensity of light. Move a lighting instrument farther away from the set to reduce the amount of light hitting objects on the set and decrease lighting intensity. To increase the lighting intensity, move the instrument closer to the set.

Replace the Lamp

Replacing the lamp with one of lower wattage decreases the intensity of the light. For example, remove a 1000-watt lamp from an instrument and replace it with a 400-watt lamp. Make certain the new lamp is rated at 3200K. A lower-wattage light rated at 3200K will provide the same color of light, but less light overall.

Use Diffusion or a Scrim

Diffusion material is placed on the front of a lighting instrument to soften the light and reduce lighting intensity, without reducing the color temperature. When placed in front of a lighting instrument, diffusion material appears translucent, **Figure 15-17**. These devices are most commonly attached using a gel holder, but metal paper clips may also be used.

diffusion: A translucent material that is placed in front of a lighting instrument to soften and reduce the intensity of light, without altering the color temperature.

PRODUCTION NOTE

A piece of spun fiberglass cloth may be used as diffusion material. The fiberglass material may be found at an auto parts store, as it is commonly used to repair fiberglass car bodies. Fiberglass cloth cannot be placed in a gel holder, however. Heat from the lamp is too intense for the fiberglass and can darken or melt the material. Fiberglass cloth should only be used with lamps that are less than 1000 watts. Clip the fiberglass cloth to the front of the barndoors with metal paper clips.

A *scrim* is a device used to reduce the intensity of light, **Figure 15-18**. Scrims are made of wire mesh or black woven, heat resistant-material and may be purchased from a theatrical supply house. Metal window screening (not nylon) may be used as an economical scrim and is available at

scrim: A wire mesh or woven material placed in front of an instrument to reduce the intensity of light.

Figure 15-17. Diffusion material reduces and softens the light coming from an instrument.

Figure 15-18. A wire mesh scrim is mounted in a frame that snaps into the front of an instrument. *(Photo courtesy of Lowel-Light Mfg., Inc.)*

hardware stores. Metal window screening may be cut to size and inserted into a gel holder.

Use Bounce Lighting

bounce lighting: A lighting technique where a lighting instrument is not pointed directly at the subject of the shot, but the light is bounced off of another object, such as a ceiling, wall, or the ground.

Bounce lighting is produced when an instrument is pointed at a photographic reflector, the ceiling, a wall, or the ground instead of directly at the subject of a shot, **Figure 15-19**. Two things occur when light is bounced off another object:

- The light takes on the color of the object it was bounced off. Always bounce light off a white or light gray object.
- The light's intensity is reduced.

One technique is to bounce light off a highly reflective surface, such as a mirror. This increases the distance between the light source and the object, which reduces the light level.

A white tablecloth or bed sheet placed on the ground at the talent's feet reflects light back at the face from below. The reflected light fills in dark shadows created by the sunlight. In addition to their use for bounce lighting, white tablecloths and bed sheets are handy sources for white balancing a camera.

Other common tools used for bounce lighting are vehicle sunshields and aluminum foil. Many people place a folding vehicle sunshield inside the windshield to reduce the effects of direct sunlight and heat when a car is parked. The reflective side of a sunshield is an effective bounce lighting tool. A reflector may also be made using regular kitchen aluminum foil.

Figure 15-19. A white surface can bounce light to help fill in a dark shadowed area of the subject.

Key light only   **Bounce lighting**   **Final image**

Crumple up the foil, flatten it out into a rough rectangle, and tape the foil to a piece of cardboard. If it is not crumpled first, the foil will be too reflective, like a mirror. The wrinkled foil creates a more subtle reflective surface.

Bouncing light off a mirror or other highly reflective surface is a technique that may compensate for insufficient lighting in certain areas on a set. For example, if it is not possible to place a lighting instrument in a particular spot, hang a mirror in that spot and position it to add illumination to the area needed.

PRODUCTION NOTE

Mirrors can be used with cameras, as well. If there is not enough room to position a camera in a tight space, a mirror can be placed where the camera needs to be. The mirror reflects the image you want to record and the camera shoots into the mirror. When using a mirror in this way, do not include any items in the shot that contain writing; letters are reversed when reflected in a mirror.

Use a Dimmer

A *dimmer* is attached to the power control of a lighting instrument and can reduce or increase the amount of electricity flowing through the instrument. Dimming an incandescent instrument is the least desirable solution. A lamp glows because the electricity flowing through it heats the filament. In television production, that filament glows white. Using a dimmer to reduce the amount of electricity flowing through the lamp causes the light to dim and cool. When a lamp cools, it progressively takes on a reddish tint. The reddish tint becomes perceptible to the camera if an incandescent instrument is dimmed more than 10%. This is why many television studios do not use the dimmer function on the lighting board. Instead, the instruments are simply turned on or off. Fluorescent lamps may be dimmed with no negative effects.

dimmer: A device attached to the power control of a lighting instrument that regulates the amount of electricity that flows to the lamp.

Preserving the Life of Incandescent Lamps

Exercise extreme care to ensure that incandescent lamps last as long as possible to avoid costly replacement. Incandescent lamps operate at very high temperatures because their glow is created by a white hot metal filament. The high operating temperatures of these lamps require that some basic precautions be taken for use and handling.

Never turn incandescent lamps on and off in rapid succession. Regular studio lamps burn out in a very short time if they are flashed on and off. To achieve a strobe light effect, use a strobe light.

Do not move incandescent instruments while the lamp is hot, whether turned on or recently turned off. The filament in an incandescent lamp is very fragile when hot—any jarring movements can cause it to break. If the filament breaks, the lamp must be replaced.

The barndoors should never be completely closed with the lamp turned on. The lack of ventilation and build-up of heat can cause the lamp to burn out prematurely.

Some instruments may be swiveled in any direction, including upside down. With instruments of this type, ensure that the heat of the lamp is not directed at the base of the instrument, so the heat does not go into the base. Consult the manufacturer's material for information and cautions about this issue.

An incandescent lamp should never be handled with bare hands or fingers. Handle the lamp using the foam it is packed in, a paper towel, or tissue paper, **Figure 15-20**. No matter how clean, there is always a certain amount of oil on the skin that is transferred to the surface of the lamp. When the lamp is turned on and reaches high operating temperatures, the oil boils to the point of evaporation. This creates a cooler spot on the hot glass surface of the lamp, which may cause the hot glass to shatter. Many instruments have a special groove in front of the lamp to accommodate a wire safety mesh that will trap glass fragments in case the lamp shatters. If a lamp is accidentally touched, it should be wiped clean with alcohol and a soft cloth before being turned on.

Figure 15-20. Oils on the skin will ruin a lamp. Never touch a good lamp with your bare fingers.

Planning the Set Lighting

Before the set is built, the lighting designer meets with the set designer and program director. In this meeting, the director describes what the set looks like, provides a set diagram, and explains the movement of the talent on the set. The director may also express specific lighting preferences.

During set construction, the lighting designer (LD) studies the set diagram and determines the placement and aiming of lighting instruments. When the lighting decisions are final, the LD develops a diagram for instrument placement, or *light plot* (**Figure 15-21**).

After the set is built and dressed, the LD lights the set. Instruments are hung over the set according to the light plot and plugged into the raceway.

light plot: A diagram developed by the lighting designer that indicates the placement of lighting instruments on the set of a program.

Figure 15-21. A light plot indicates the placement, color, and aiming of the lighting instruments.

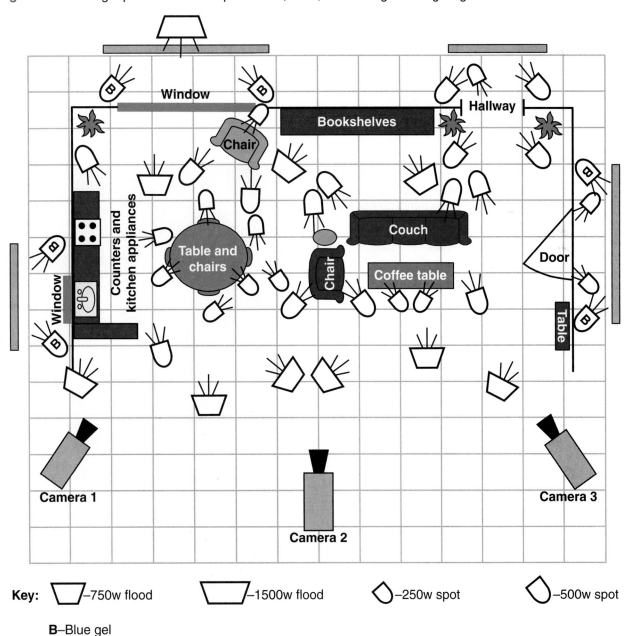

Key: ⬜–750w flood   ⬛–1500w flood   ◇–250w spot   ◇–500w spot

B–Blue gel

basic hang: The initial process of hanging instruments over the set according to the light plot and plugging them into the raceway. Also called a *rough hang*.

This is called the *rough hang*, or *basic hang*. The goal of a rough hang is to place instruments in the general location they need to be for the production. After the rough hang is complete, the LD notifies the crew and turns out all the lights in the studio. The set must be completely dark so the LD can evaluate the placement and see the effect of the moving each instrument. To light the set, the LD and crew follow a general procedure:

1. The LD turns on one instrument.

2. The LD or a gaffer puts on heavy work gloves, climbs a ladder, and manually aims and focuses the instrument on a specific area of the set. Move the instruments very gently when hot. If jarred sharply, the incandescent lamp may burn out.

Safety Note

Always wear protective gloves when adjusting a lighting instrument that is turned on to prevent fingers from being burned, **Figure 15-22**. Do not use gloves with rubberized palms or finger grips. The rubber will soften or melt when in contact with a hot instrument. Inexpensive work gloves with leather palms and finger tips are a good option.

Figure 15-22. Always wear protective gloves when handling hot instruments.

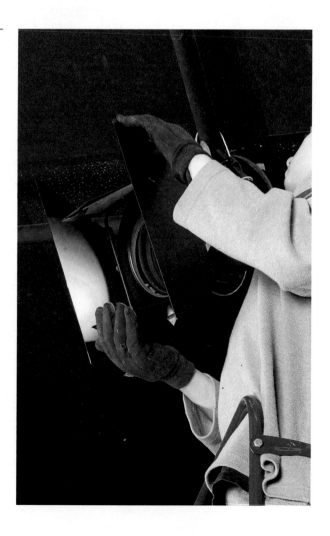

3. Once aimed, the instrument is turned off.

4. The next instrument is turned on and adjusted. The steps are repeated for each instrument on the set.

5. When the entire light plot has been aimed and precisely focused, all the instruments are turned on at once and any final adjustments are made.

PRODUCTION NOTE

Some lighting designers prefer to light the set with general illumination using mostly floodlights, turn off all the general lighting, and then position spot lighting in the specific areas that talent will be moving or standing. General lighting is then turned back on and final adjustments can be made. Other lighting designers prefer to start with areas specific to the talent and then set up the general lighting. Both techniques work well.

Techniques of Television Lighting

Each of the following basic lighting techniques may be used in production studios of any size. Properly using each technique is crucial to creating quality shots for a program.

Three-Point Lighting

Three-point lighting, sometimes called *triangle lighting*, is the most commonly used photographic lighting technique in television production. Three-point lighting is designed to both make a person look attractive and create the appearance of three-dimensionality on the flat, two-dimensional television screen. Three-point lighting uses three instruments for each person/object being photographed:

- key light
- fill light
- back light

Each of the three instruments performs a specific function, **Figure 15-23**.

three-point lighting: A common lighting technique that uses three lighting instruments for each person or object photographed: a key light, a fill light, and a back light. Also called *triangle lighting*.

Key Light

The *key light* provides the main source of illumination on an object, **Figure 15-24**. It is usually in front of and above the object, and on an angle to the left or right. A key light is never placed directly above or directly in front of an object. Imagine the face of a clock. The subject of the shot stands in the center of the clock and the key light is located at either four-thirty or five o'clock on the right side of the stage or at seven o'clock or seven-thirty on the left side.

key light: The lighting instrument that provides the main source of illumination on the person or object in a shot.

When placing the key light, location of the primary light source must be considered. If lighting a living room for a scene that is supposed to take place at noon, the primary source of light would probably be the windows of the room. If the window is on the talent's right side, the key light is placed on the talent's right side to augment the sunlight coming in from the window.

Figure 15-23. This illustration presents the general placement of instruments in a three-point lighting setup.

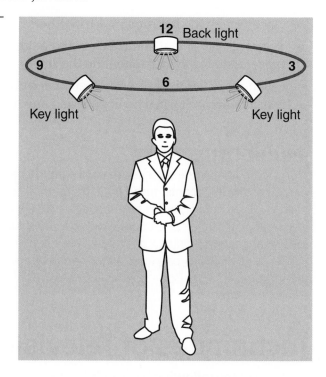

Figure 15-24. The key light is a hard light that supplies the primary source of illumination on an object. The fill light somewhat reduces the harsh shadows created by the key light. The back light illuminates the top of the head and shoulders. The proper use and placement of each of the three instruments produces a well-lit, three-dimensional subject.

Key light

Fill light

Back light

Three-point lighting

A key light is a hard light—it produces sharp shadows. If a person is lit only with a key light, the side of their face opposite the key light will have heavy nose and chin shadows and their eye will be completely shadowed.

Fill Light

The *fill light* is placed opposite the key light and above the talent to light the other side of the talent's face, **Figure 15-24**. If the key light is positioned at five o'clock, for example, the fill light is placed at seven o'clock. The fill light is a softer light that is lower in intensity and reduces the dark shadows created by the key light to a certain extent. If the fill light completely eliminated the shadows created by the key light, it would be a second key light and would create a flat image on the television screen. The fill light must be lower intensity than the key light to leave some shadows that create a three-dimensional appearance. Notice the difference in brightness between the key light and fill light in **Figure 15-24**.

fill light: A lighting instrument that is placed opposite the key light and above the talent to provide illumination on the other side of the talent's face or object in the shot.

Back Light

The *back light* is placed above and behind the talent at the twelve o'clock position, **Figure 15-24**. It must be positioned fairly high so that none of the cameras on the set shoot directly into the instrument. The purpose of the back light is to provide some illumination on top of the talent's head and shoulders. It also serves to separate the talent from the background.

Inexperienced lighting personnel often confuse a back light with a background light. However, these instruments pointed in opposite directions on a set. A *background light* is pointed at the background of the set, but a back light is pointed at the back of the talent.

When using three-point lighting, there is a three-point lighting setup for every member of the cast, at every spot on the set they move or remain stationary. This is why there are so many lighting instruments on the lighting grid of a television studio.

back light: A lighting instrument that is placed above and behind the talent or object in a shot, at the twelve o'clock position, to separate the talent or object from the background.

background light: A lighting instrument that is pointed at the background of a set.

Assistant Activity

The areas on a set where talent or important objects will be positioned for a period of time are lit using three-point lighting. Look back to **Figure 15-21**. Locate the instruments configured for three-point lighting and determine where talent or important objects will be most consistently positioned on the set.

Four-Point Lighting

Four-point lighting includes four instruments (two key lights and two fill lights) that are placed in a square around the talent, **Figure 15-25**. The two key lights are positioned diagonally opposite each other, and the two fill lights are placed in the remaining two corners. In four-point lighting, the camera can arc all the way around an object and the lighting levels remain sufficient.

An advantage of four-point lighting is that it is easy to set up, somewhat easier than three-point lighting. However, this technique requires more instruments, electricity, and lamps, which results in increased cost.

four-point lighting: A lighting technique that uses four lighting instruments for each person or object photographed: two key lights and two fill lights. The two key lights are positioned diagonally opposite each other, and the two fill lights are placed in the remaining two corners.

Figure 15-25. Four-point lighting is easier to set up than three-point lighting, but requires an additional instrument.

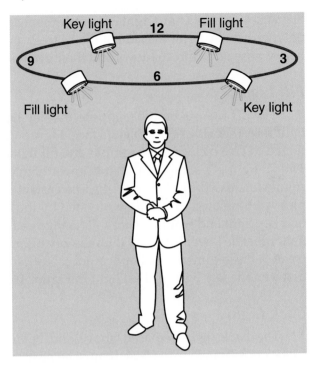

Cross-Key Lighting

cross-key lighting: A lighting technique that covers more than one person or object in the lighting spread using only two key lights and one back light.

Cross-key lighting can cover more than one person or object in the lighting spread using only two key lights and one back light, **Figure 15-26**. As an example, picture two people sitting in chairs on the set. Both chairs are angled slightly toward each other, but both are generally facing six o'clock. In cross-key lighting, a key light is placed at four o'clock and another is placed at eight o'clock. A back light is positioned at twelve o'clock. Two back lights may be

Figure 15-26. In cross-key lighting, *Key light 1* provides key lighting for person A and the fill light for person B. *Key light 2* provides key lighting for person B and the fill light for person A.

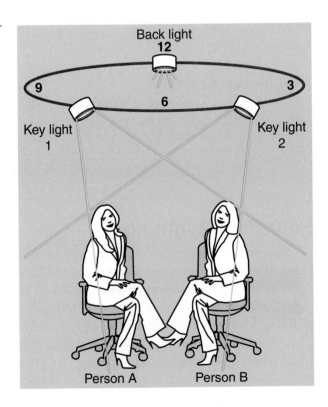

used in some cases. The cross-key technique works because the key lights hit both characters, but are not at the same distance from each character. The amount of light that hits the nearest talent is brighter than light from the same instrument that hits the talent farther away. The key light nearest the talent on the right is the fill light for the talent on the left. The key light for the talent on the left becomes the fill for the talent on the right.

A few more instruments may be necessary when using cross-key lighting, but the final number of instruments is significantly lower than using multiple three-point lighting setups. Cross-key lighting reduces energy costs, as well as the heat produced by the instruments.

Lighting with Fluorescents

The principles of studio lighting with fluorescent fixtures are the same as lighting with conventional fixtures—revealing shape, form, and texture to create an interesting three-dimensional image. See **Figure 15-27**. Studio fluorescents are soft lights. When used properly, soft light is pleasing on

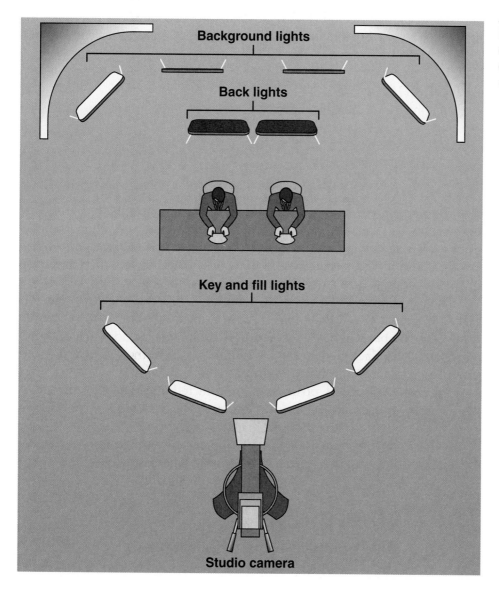

Figure 15-27. Placement of fluorescent instruments on a news set. *(Courtesy of Lowel-Light Mfg., Inc.)*

Figure 15-28. A honeycomb is placed in front of a fluorescent instrument. The hundreds of small holes act as miniature barndoors and make the soft fluorescent light much more directional. *(Photo courtesy of Lowel-Light Mfg., Inc.)*

honeycomb: A device that attaches to fluorescent instruments to reduce the shape and size of the light beam, making the light more directional and easier to control.

people's faces and easier on the eyes than hard edge spotlights. If used incorrectly, however, soft light can produce a flat, uninteresting image.

The placement of fluorescent instruments is identical to the three-point lighting technique previously discussed for incandescent instruments. Lighting starts with a key light. The key is the primary source of illumination that reveals the shape of the subject. When using a large soft source as a key, the light scatters all over the studio. To control the "scattered" light, most fluorescent instruments are equipped with a *honeycomb* that reduces the shape and size of the light beam (**Figure 15-28**). A honeycomb does not make the light much harder, but it makes the light more directional and easier to control.

Talk the Talk

A honeycomb may also be called a *grid.* The word "grid" also refers to the support system used to hang lighting instruments. Use the context of the sentence to determine the intended meaning of the word.

The second fixture is the fill light, which illuminates the opposite side of the subject's face and fills in any objectionable shadows. Often in multi-camera setups, a ring of lights is placed in front of the subject to provide a key light and fill light combination that covers the subject when they turn in any direction to address various cameras on the set.

The next fixture is a back light, sometimes called a *hair light*, which is placed directly behind the subject. A very narrow honeycomb is usually used with the back light to prevent light from spilling forward and creating lens flare. Back light provides separation from the background and produces a "glamorous" halo of light.

The final lights to be positioned are directed at the back of the set. A very simple set uses several fixtures aimed at the back wall to produce an even wash of light for an indistinct background.

A studio can be adequately lit with fluorescent fixtures alone, however focusing spotlights may be added for both the talent and areas on the set.

The Camera Light

Many consumer and news cameras are equipped with a built-in light. When a reporter is doing a location shot, it may seem logical to use the on-camera light for additional lighting, as needed. If other lighting options are available, do not use the on-camera light. Because of the light's proximity to the camera lens, the talent will have a "deer in the headlights" look—the talent's eyes will appear overly large and their face will appear very flat. In general, using the camera light by itself produces an unattractive image of a person.

An acceptable image may be created using the on-camera light if an additional light can be added on one side of the camera. For example, a hand-held sun-gun light can be pointed at the talent from a distance of about four feet on the left or right side of the camera. Car headlights can also be used as a side light, if necessary.

PRODUCTION NOTE

You have, undoubtedly, seen images on evening news programs in which one of the cameras on location pans a large crowd of reporters and news cameras. All the other news cameras in the background have the camera lights on. In this instance, the on-camera lights from all the news cameras function as additional side lighting for every other camera at that location!

Contrast Ratio

The two extremes of light are black (the absence of all light) and white (the presence of all light). Contrast ratio (Chapter 14, *Image Display*) is the relationship of the amount of darkness to the amount of lightness in a picture. A television camera and a television set have difficulty reproducing black and white simultaneously. Therefore, large quantities of these colors are rarely seen in the same picture.

Some television studios have solid black curtains that surround the sides of the studio. A completely solid black or white background can be very useful. If, for example, the talent is staged with a black, white, or any solid color background, viewers consider the talent to be removed from reality. A solid-colored background gives no visual clues to where or when the action is taking place. Lighting with this kind of background is called *limbo lighting*. Limbo lighting creates a background that is a solid, indistinct color. For example, many car commercials are shot on a white floor with an indistinct white background. Also, singers often perform in front of a solid black background, **Figure 15-29**. A solid background concentrates the viewer's attention on the performance or main object in the shot.

limbo lighting: A lighting technique in which the background of the set is lit to create the illusion of a solid-color, indistinct background.

Lighting Check

Always check the lighting setup on a monitor with the contrast, brightness, color, and tint controls set correctly. If the shot does not look good on the monitor, stop shooting immediately. The cause of the poor image displayed on the monitor must be determined before shooting continues. A likely culprit of this problem is the monitor itself.

Figure 15-29. Limbo lighting is used when the talent or subject is in front of a completely indistinct background.

To ensure that the contrast, brightness, color, and tint controls are set correctly on the monitor, perform the following:

1. Turn on the color bar generator in the video camera.
2. Verify that the color bars display in the proper order on the monitor. Color bars are discussed more fully in Chapter 25, *Getting Technical*.
3. If the color bars are incorrect, adjust the contrast, brightness, color, and tint on the monitor until the bars display correctly.
4. Turn the color bar generator off and the camera will operate as usual.
5. If the images on the monitor still do not appear properly, the camera may need adjustment before shooting can continue.

PRODUCTION NOTE

To check a monitor's contrast, brightness, color, and tint settings when using a camera that is not equipped with a color bar generator, perform the following:

1. Find a tape that you have watched and know the images have good color.
2. Place the tape in the camera and play it.
3. Watch the monitor and check the colors.
4. If colors on the monitor are incorrect, adjust the monitor settings until the colors are displayed correctly.

Do not make any adjustments to the camera's color controls until you verify that the monitor has not caused the display problem. If the monitor is operating correctly, either the video engineer or a camera repair shop will adjust the color controls on the video camera. This adjustment requires the use of test instruments and procedures that are not familiar to most production staff.

Adjusting any of the color controls on a camera to improve the picture on a monitor that needs adjustment will affect the actual recorded image. As a result, viewers will need to adjust the contrast, brightness, color, and tint controls on their own televisions to improve the picture. You probably have never needed to adjust these controls on your television at home while watching a network television program. This is because all programs adhere to industry-established standards regarding contrast, brightness, color, and tint levels. The goal is to shoot a program in the correct lighting situation, so the viewing public can sit back and watch your award-winning television program.

Wrapping Up

Unfortunately, many inexperienced television production personnel consider lighting to be an afterthought. Proper lighting is extremely important. As part of an assignment, a student produced a commercial for a burglar alarm system. In the commercial, the burglar climbed into a house at night and triggered the burglar alarm. When viewing the commercial, it was difficult to miss the lack of a picture. There was only a black screen with sound. The student said that he intended the picture to be dark because it took place at night. He had successfully produced a radio program, but not a television program. Audiences accept a slightly darker scene as "night," especially when shot with a dark blue filter on the camera. But, a completely black screen is not an acceptable depiction of "night." If viewers are unable to clearly see what the camera is shooting, they change the channel and the program's effort to communicate fails.

Review Questions

Please answer the following questions on a separate sheet of paper. Do not write in this book.

1. What is the difference between *hard light* and *soft light*?
2. Name the types of lighting instruments commonly used on a production studio set.
3. What items can be used to redirect or change the shape of light?
4. What are the advantages of using fluorescent lamps instead of incandescent lamps?
5. How is power supplied to the lighting instruments that hang from the studio ceiling?
6. How do different frequencies (colors) of light affect a recorded video image?
7. Describe three methods used to reduce the intensity of production lighting.
8. What are two effects of bounce lighting?
9. What precautions should be taken to preserve the life of incandescent lamps?
10. List the steps involved in lighting a set.
11. Identify the instruments used in three-point lighting and explain the function of each.
12. What is *cross-key lighting*?
13. How is a honeycomb used with fluorescent instruments?
14. What is *limbo lighting*?

Activities

1. Look around your home and identify the light created by the following instruments as either hard light or soft light:
   - The tabletop lighting instruments in your living room.
   - The instrument that illuminates your desk.

- The lighting fixture in your bathroom.
- The lighting instrument over your kitchen table.
- The lighting instrument over the stove in your kitchen.
- The instrument that generally lights your bedroom.

2. Research the experiments and discoveries of Lord Kelvin. Choose one of his accomplishments (other than the Kelvin Color Temperature Scale) and write a report on it. Be prepared to present this information in class

STEM and Academic Activities

 1. Identify the gas (or gases) used in fluorescent lights. How does the gas inside a light create illumination?

2. What is the *frequency* of a color? What is the unit of measurement used to indicate color frequency?

 3. To convert Kelvin values to degrees Celsius, subtract 273.15 from the Kelvin value or °C = K − 273.15. Using this formula, convert the following Kelvin values to Celsius:

   A. 2000K

   B. 3200K

   C. 4300K

4. To convert degrees Celsius to degrees Fahrenheit, multiply the Celsius value by 1.8 and add 32 to the product or °F = (°C × 1.8) + 32. Using this formula, convert the Celsius values calculated in the previous activity to Fahrenheit.

 5. Create a set diagram for one of your productions or a production you have been involved with. Include all the fixed elements on the set. Write an explanation of the program, the program's message, and the movement of talent on the set to be used by a lighting designer.

www.rtnda.org Website of the Radio Television Digital News Association and Foundation. The RTDNA (association) sets standards for and provides programs that encourage excellence in electronic journalism. The RTDNF (foundation) offers development opportunities and educational resources for journalism professionals and educators.

Chapter 16

Studio and Remote Shooting

Objectives

After completing this chapter, you will be able to:

- Recall the specific characteristics of both studio and remote shooting.
- Identify the types of monitors set up in the control room and the function of each.
- Explain the differences between ENG and EFP.
- Identify the items to be evaluated during a location survey.
- Summarize the advantages and challenges of both studio and remote shooting.

Introduction

Studio shooting is the method that students most commonly associate with television production. Shooting in a studio environment may not, however, be the most effective location for every type of program. There are advantages and disadvantages to both studio shooting and shooting at a remote location. This chapter examines some of these issues in order to help you determine which method is more appropriate for particular production types.

Professional Terms

audio booth
audio console
camera monitor
confidence monitor
control room
editing suite
EFP (electronic field production)
ENG (electronic news gathering)

flat
jack
location
location survey
master control room
preview monitor
production meeting
program monitor
remote shoot

The Production Meeting

production meeting:
A meeting with the
entire crew in which
the director lays
out the program's
main message and
either the director or
producer assigns each
task involved in the
production to members
of the crew.

Whether the shoot takes place in a studio or at a remote location, a *production meeting* must take place before shooting can begin. During the production meeting, the director lays out the program's main message for the entire crew. The director/producer assigns each task involved in the production to members of the crew. The success of the entire project depends on the completion of each task. Responsibility and dependability are requirements of every member of the production team.

At this meeting, the director distributes a tentative production schedule. In broadcast journalism, the time frame for the entire production may be less than 8 hours from beginning to end. On the other hand, the production time frame for feature stories or more creative/entertainment-oriented programs may be weeks or months. The crew members review their commitment calendars and arrange schedules to meet the production schedule, **Figure 16-1**. Flexibility is very beneficial when putting a production schedule together. During the course of the production meeting, a rough production schedule is completed. The director takes all the notes and compiles them into one master calendar. This calendar indicates who is needed, on which day, at what time, at which location, for which scene, and what they need to bring, wear, and do when they are there. The schedule is printed and distributed to the crew. Last-minute corrections are made and the entire schedule is finalized.

Figure 16-1. The production meeting results in a production calendar that serves as the road map for completing the entire program.

Studio Shooting

A studio shoot occurs in a controlled environment where outside conditions and sounds do not affect the shooting schedule. All the necessary video equipment and supplies are readily accessible and, usually, do not require any major setup. The set, lighting, and cameras are planned and positioned according to the program and production requirements.

The Studio Environment

At first glance, a studio may seem to be an intimidating space. Each feature of the studio and each piece of equipment is purposely constructed and placed to ensure a quality production, **Figure 16-2**.

The Ceiling

The television production studio is usually a large room with a high ceiling. The minimum ceiling height requirement is typically 10′, but 14′–18′ is average. With lower ceilings, the lighting grid hangs too low and instruments may

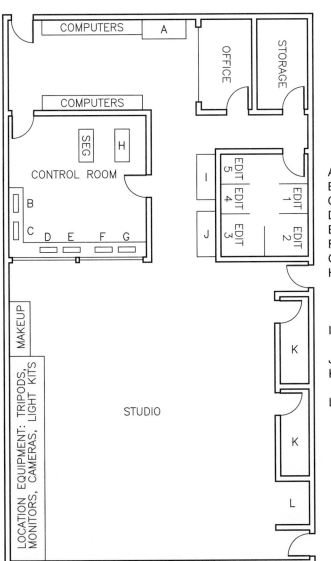

Figure 16-2. Each area of the studio is designed and organized to make the production process as efficient and effective as possible.

A. Duplication system
B. CG
C. Audio mixers
D. CD/cassette
E. Video recorder
F. Video player
G. Light board
H. Camera monitors, program monitors, preview monitor, CCUs
I. Camera batteries and chargers
J. Cable storage
K. Audio/narration booths
L. Flat storage

appear in shots. Additionally, if the talent walks near an instrument that is hung too low, their head glows. The heat generated by incandescent instruments also becomes significant if the ceiling is not high enough to help the heat dissipate.

The Walls

The studio walls should be treated with any number of materials that are designed to do two things:
- Keep sound within the studio from bouncing around the room, **Figure 16-3**.
- Keep unwanted outside sounds from entering the studio.

A studio is a cavernous space. Without sound treatment on the walls, sound in the studio will echo. Sound should not echo in the studio under any circumstances. Hearing an echo in the studio is an indication that the production team has lost control over the audio.

Safety Note

Local fire laws are an important consideration when choosing sound treatment materials. The decision of the local fire marshall always prevails in a dispute over suitable materials.

Along the walls of the studio are various places to plug microphone cables into jacks, **Figure 16-4**. These jacks are hardwired to the audio console in the control room.

Curtains

The studio walls should, ideally, be covered on all four sides with a curtain. In truth, most studios curtain only two or three walls. Having all four walls curtained provides even more flexibility, **Figure 16-5**. The curtain color is entirely up to studio management. Black curtains are integral to limbo lighting, but bring down the contrast ratio in all other situations. White curtains

Figure 16-3. The sound deadening material on the walls of the controlled studio environment prevents unwanted outside sound from entering the studio. This material also prevents sounds from echoing inside the large open space of a studio.

Figure 16-4. Studio microphones plug into these studio wall-mounted jacks, which are connected to the audio mixer in the control room.

Figure 16-5. The studio curtain can be pulled tight or left loose to hang in folds.

also create a limbo effect, but invariably bring up the contrast ratio. Curtains in varying shades of gray are most often used. A gray curtain should be selected to match a shade in the middle of the gray scale. (The gray scale is explained in Chapter 14, *Image Display*.) Cobalt blue or emerald green curtains are also sometimes used. Some studios have curtains in various colors hanging from the ceiling on multiple tracks. This arrangement provides maximum flexibility.

PRODUCTION NOTE

Never touch the front of a studio curtain with bare hands. Oil from the skin transfers to the curtain and attracts dirt. The inconvenience and cost of dry cleaning such a huge piece of fabric is significant! If a curtain must be grabbed to pull along a track, it should only be grasped from the back side. Studio curtains are quite expensive. Avoid puncturing, snagging, or ripping the cloth. The repair bill may be as staggering as the cost of a complete replacement.

The Floor

The studio floor should be as level and smooth as possible to allow studio cameras to easily move across in dolly, truck, and arc movements. The floors should be maintained without using a high-gloss polish or wax. Maintaining a non-reflective floor surface is critical in controlling the lighting in the studio. A highly reflective floor bounces light all over the set.

Scenery Units

The scenery unit most commonly used on a television studio is called a *flat*. A flat is usually a 4′ × 8′ frame constructed with 2×4 or 1×3 boards and braced in the center. 4′ × 8′ sheets of plywood are attached to the frame. Flats are found in almost any theater department. These flats may be placed beside each other and painted to create the appearance of a wall. Two triangle-shaped braces, called *jacks*, are fastened to the back of a flat. The jacks are set up and provide stable support so the flat can stand on its own, **Figure 16-6**. The joint space between flats can be concealed before painting with masking tape or drywall tape. Storage for scenery units and other props should be located near the studio floor, **Figure 16-7**.

flat: A scenery unit that is usually a simple wood frame with a painted plywood shell.

jack: A triangle-shaped brace that is fastened to the back of a flat to provide stable, upright support.

Talk the Talk

The term "jack" also refers to a female connector. Use the context of the sentence to determine the intended meaning of the word.

Figure 16-6. Jacks allow a flat to stand upright and stable.

Figure 16-7. A flat rack allows scenery flats to be stored like books in a bookcase.

Cameras

A studio has several cameras permanently located within the room. Any camera that is part of a studio system should be kept in the studio environment at all times. Ideally, studio cameras should be identical in brand, model, and age. When cutting between camera shots, there is little to no variation in color balance or video signal quality when extremely similar cameras are used within the studio.

The Control Room

The *control room* contains several monitors and the special effects generator, **Figure 16-8**. In smaller facilities, the control room also houses the audio mixer, all of the sound equipment, video recorders, the CG, CCUs, and even the light board. The director is stationed in the control room during a shoot and communicates with the control room and studio production teams via headsets. A very noticeable aspect of a control room is the number of television monitors in the dimly lit room. The director is positioned at or near the special effects generator and faces the wall of monitors.

The control room contains at least one monitor for each camera in the studio, **Figure 16-9**. Each monitor displays the image that its corresponding camera is shooting. Therefore, a three-camera studio has one *camera monitor* for Camera 1, another for Camera 2, and a third for Camera 3. A *program monitor* displays the image going to the recorder. There may also be a *preview monitor*, which allows the director to set up an effect before

control room: A room in the studio containing several monitors and the special effects generator. In smaller facilities, the control room also houses the audio mixer, all of the sound equipment, video recorders, the CG, CCUs, and even the light board.

camera monitor: A monitor that displays the image shot by the corresponding camera.

program monitor: A monitor that displays the image going to the recorder.

preview monitor: A monitor that allows the director to set up an effect on the SEG before the audience sees it.

Figure 16-8. A control room is the location of the major equipment used to process a studio shoot.

Figure 16-9. The director calls the shots for the program by watching the individual camera monitors. The camera monitors are on the bottom row, with program and preview monitors above them. The final output going to videotape is displayed on the large top monitor, called a "confidence monitor."

Confidence monitor

Program monitor

Preview monitor

Camera monitors

confidence monitor:
A monitor connected to the output of the video recorder. Seeing the image on this monitor ensures that the video recorder received the signal.

the audience sees it. A *confidence monitor* is connected to the output of the video recorder. Seeing the image on the confidence monitor ensures that the video recorder received the signal. However, the confidence monitor is not an indicator that the image was actually recorded. Playback is the only sure way to know that the signal was recorded. Since the CG is located in the control room, there may be a corresponding CG monitor in the grouping, as well. The CG should also be connected to a black and white monitor, which displays the CG output and allows the contrast ratio of graphics to be reviewed and verified.

PRODUCTION NOTE

Even though the director is stationed in the control room, headsets are still used to communicate with the control room staff. There are several sources of background noise in the control room—audio of the program, chatter of control room staff and observers, and commands being relayed from the studio floor to the control room. The headphones used have multiple channels. Everyone wears headphones to filter the noise and hear only the information that applies to them.

The Audio Booth

In smaller production facilities, the control room and audio booth are combined into one large control room. In larger facilities, the *audio booth* is a separate room that contains the audio console. The *audio console* includes many different pieces of audio gear, including the microphone mixer, mp3 players, audio cassette players, and CD players. Any equipment capable of adding sound to the program is located in the audio booth, **Figure 16-10**. Any music and sound effects library CDs the facility owns are also stored in the audio booth.

Master Control Room

The *master control room* is where all the hardware is located. The video recorders and other equipment needed to improve and process the video and audio signals are housed in the master control room. In larger facilities, the master control room is not located near the control room. This

audio booth: A room in the studio that contains all of the equipment capable of adding sound to the program.

audio console: A unit that includes many different pieces of audio gear, including the microphone mixer, audio cassette players, CD players, and turntables.

master control room: A room in a production facility where all the hardware is located, including video recorders and other equipment needed to improve and process the video and audio signals.

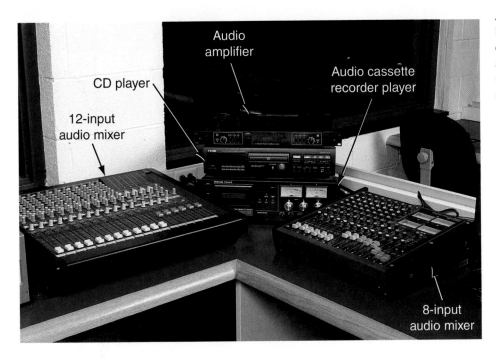

CD player

Audio amplifier

Audio cassette recorder player

12-input audio mixer

8-input audio mixer

Figure 16-10. The audio console includes the audio mixer(s), amplifier, CD players, and audio cassette decks.

equipment is usually placed in a separate room because it creates a fair amount of noise and heat. In smaller studio facilities, the control room, audio booth, and master control are combined into one large control room.

Specialized Areas

Studio facilities usually have a carpentry shop—an area where the sets can be built and painted. Dressing rooms and makeup areas are located near the studio. The requisite number of offices and conference rooms are also part of the facility. *Editing suites* are cubicles or small rooms where the program is put through post-production processing, **Figure 16-11**. Post-production processes performed in editing suites include: video and audio editing, voiceover, music and sound effects recording, and graphics recording. Individual studios may have many other types of specialized rooms or areas on the premises.

Remote Shooting

A *remote shoot* is any shoot that takes place outside of the studio. A remote shoot has its own unique features and details, just like the studio shoot.

Types of Remote Shoots

Remote shoots are divided into two categories:
- ENG (electronic news gathering)
- EFP (electronic field production)

editing suite: A cubicle or small room where the program is put through post-production processing, such as video and audio editing, voice-over, music and sound effects recording, and graphics recording.

remote shoot: Any production shooting that takes place outside of the studio.

Figure 16-11. An editing suite is a cubicle or small room where the program material is put through post-production processing.

ENG

ENG *(electronic news gathering)* is the process of shooting information, events, or activity that would have happened whether a reporting/production team was present with a camera or not. For example, a major car crash on the local freeway covered by a news crew is ENG. The crew arrived fifteen minutes after the crash and, obviously, did not shoot the crash itself. The crash cannot be staged again to be captured on tape. If a news crew arrives 30 minutes late for a parade, the parade officials will not back everyone up and start again. In the ENG environment, the camera is an observer of an event that would happen whether the camera is there or not. There is no possibility for a second take. Another unique characteristic of ENG is that only one or two crew members are necessary for a shoot—a reporter and a camera operator. It is possible for a photojournalist (one person) to perform functions of both the reporter and camera operator at the same time. The camera operator is usually in charge of lighting, audio, and the satellite truck, in addition to the camera. The reporter is in charge of makeup and wardrobe, writing the news story, and editing the video.

ENG (electronic news gathering): The process of shooting information, events, or activity that would have happened whether a reporting/production team was there with a camera or not.

EFP

EFP *(electronic field production)* is the opposite of ENG. In EFP, the video crew and production staff are in total control of the event. If the director is not satisfied with the way the cars crashed in a particular scene, additional cars can be acquired and stunt drivers can redo the scene. Depending on the scope and details of the scene, restaging it may not be financially feasible. Purchasing an additional building to rig with explosives and demolish, for example, is not likely to be within budget constraints. In cases where a second take of a scene is prohibitively expensive, the director will arrange for multiple cameras to shoot the scene from several angles.

EFP (electronic field production): A shoot in which the video crew and production staff are in total control of the events and action.

EFP is much more expensive than ENG, but it is also dramatically more effective. Where ENG offers only a single viewpoint of an event, scenes in EFP may be reshot until the director is satisfied and several camera angles can be acquired. All the footage is edited together to create a scene that draws the attention and interest of the audience. In EFP, extra compensation is allotted for performers, technicians, producers, directors, staff members, and all other production members. If the director calls a "Take 2," everyone must be paid for their additional time and services.

The Location Survey

A *location* is any place, other than the studio, where production shooting is planned. Anytime an EFP shoot takes place on location, a *location survey* is required before the shoot date, **Figure 16-12**. The day of the shoot is too late to complete a location survey. ENG shooting, by its very nature, rarely allows time for a location survey. An ENG news crew typically operates using battery power and only short clips of the footage they record is usually used in the news program. A location survey is only performed when the location of a shoot is known in advance of the shooting date.

To avoid a trespassing charge, the first order of business for a location survey must be to get the property owner's permission to shoot at

location: Any place, other than the studio, where production shooting is planned.

location survey: An assessment of a proposed shoot location that includes placement of cameras and lights, available power supply, equipment necessary, and accommodations needed for the talent and crew.

Figure 16-12. Important items to consider when performing a location survey.

Location Survey Checklist

- ☑ Permission from property owner.
- ☑ Availability of electrical power supply.
- ☑ Placement of cameras.
- ☑ Placement of lighting instruments.
- ☑ Note the natural sunlight present at the location.
- ☑ Note noises that are part of the environment during various times of the day.
- ☑ Availability of necessary facilities.
- ☑ Equipment necessary for the shoot.
- ☑ The number of crew members necessary on location.

the selected location. (Review release information in Chapter 12, *Legalities: Releases, Copyright, and Forums.*) If ownership is unknown, refer to the appropriate real estate records; these are usually public record.

After securing permission to be on the property, visit the location and determine the placement and positioning of cameras, lights, and other equipment. Always be mindful of electrical power supply issues. If using electricity, the property owner must also grant permission to plug equipment into the property's power outlets. If using batteries for all electrical equipment, the additional permission is not necessary.

PRODUCTION NOTE

Portable video production gear uses very little power over all. Excluding the lighting instruments, the gear for most small productions uses about the same amount of electricity as four television sets. Portable lighting equipment does draw more power. However, if the instruments are turned on only when actually shooting, energy costs can be significantly reduced.

Consider the lighting situations at the location. If shooting exterior shots, keep in mind the position of the sun during each part of the day. Assess how the sunlight will affect the shooting schedule. The production equipment cannot be set up to shoot into the sun, but if the sun is directly at the camera operator's back, talent will be squinting at the camera.

Make a point to perform the location survey on the same day of the week and time of day that the shoot is scheduled. Note the various noises that are part of the environment, such as traffic and low-flying aircraft.

Other considerations for a location include:

- Are necessary facilities, like restrooms, water, and food, readily available?
- Where will the electrical, camera, and audio cables be run?

- What kinds of microphones are needed?
- How much cable is necessary?
- How many crew members are necessary for the production on location?
- Where will the crew and cast park?

Comparing Studio and Remote Shooting

Both studio and remote shooting are effective in gathering footage for a program. The type of program being produced is a large factor in deciding whether studio or remote shooting should be used. In addition, the specific advantages and disadvantages associated with both studio and remote shooting must be considered.

Advantages of the Studio Shoot

- Arrangements for on-site transportation, food, and lodging are not necessary.
- The forces of nature rarely affect the inside of a studio; it never rains in a TV studio.
- All equipment and supplies are easily accessible. For example, if a cable goes bad, it is readily replaced with back stock.
- Major set up of equipment is usually not required. Equipment is already wired into a full system, including a switcher/special effects generator.
- Control of the people within the studio environment; a person mistakenly walking through the background of a shot is not a major concern.
- Precise control of the lighting situation; the sun does not cast shadows that move as time passes inside a studio.
- Extraneous sounds do not enter the studio to interrupt the shooting schedule.
- Proper use of the video switcher greatly reduces editing time—the program is essentially edited while being shot.

Disadvantages of the Studio Shoot

- Building a set can be expensive and time-consuming.
- The ambient sound of "the great outdoors" is difficult to recreate in a sound-treated studio.
- Recreating an outdoor feeling in the studio presents serious lighting concerns. As the sun moves during the course of the day, the shadows it casts also move.
- The amount of equipment necessary for a full-scale studio production is extensive and expensive. This is primarily because the video equipment must be linked together with the signal from each matched to the others.
- A shoot in the studio usually requires more personnel than a remote shoot. In addition to the talent and the director, the minimum staff required for a studio shoot includes: up to three camera operators, a floor manager, an audio engineer, video engineer, technical director, and lighting director, **Figure 16-13**.

Advantages of a Remote Shoot

- In many cases, it is not necessary to build a set. A location is usually chosen because the set already exists.
- Natural light can often be used.
- Everything about an existing set is realistic. This supports the illusion of reality created in the program.
- A remote shoot usually requires less equipment overall than a studio shoot.
- A remote shoot usually involves fewer crew members.

Disadvantages of a Remote Shoot

- Murphy's Law, "If anything can go wrong, it will," is a constant threat in location shooting.
- Inevitably, something goes wrong with the equipment. The crew must plan for that eventuality and bring spares of everything (extra cables, adapters, etc.).
- Inclement weather may completely halt production, **Figure 16-14.** Pay attention to weather forecasts and always have an alternate plan.
- In a remote interior location, power supply for equipment and lights may be insufficient. Always check for circuit limitations.
- Permission from the property owner must be granted before setting foot on the location selected for the shoot.
- If something breaks or is forgotten at the studio, there is significant downtime while someone travels back to get it.
- The terrain is often not suitable for simple dollying, trucking, or arcing. Special track must be installed to perform these camera moves.

Figure 16-14. An unexpected change in the weather can make a selected location unusable and cause a remote shoot to be cancelled.

- All equipment must be transported to and from the location and must be repeatedly set up and torn down. This increases wear and tear on the equipment and, in turn, contributes to equipment failure.
- In order to get varying camera angles, a single camera must reshoot each scene several times from different angles. This also requires numerous hours in the editing room, cutting together a cohesive and interesting program.

Wrapping Up

When deciding whether to shoot in the studio or on a remote location, carefully consider the program type and the benefits and risks that each environment presents. A remote location offers the excitement of the real world. The studio offers a stable and controlled production environment. Programs in a studio can usually be produced in much less time than those shot remotely. A studio shoot utilizes the camera switcher, which can save hours in the editing room. On location, a single camera must shoot something multiple times to get various angles. Each option must be weighed carefully before deciding to shoot in the studio or on a remote location.

Review Questions

Please answer the following questions on a separate sheet of paper. Do not write in this book.

1. What topics are addressed in the production meeting?
2. Explain the special features of studio walls.
3. Name each of the television monitors found in the control room and state the function of each.
4. What do the letters "ENG" stand for? What does the term mean?
5. What are the unique characteristics of EFP?
6. List the items that should be evaluated during a location survey.
7. What are the disadvantages of a studio shoot?
8. What are the advantages of a remote shoot?

Activities

1. Research the fire laws and regulations in your area. Make a list of all the fire laws and regulations that apply to the construction and use of a television production set.
2. Perform a location survey for shooting an interview. For this activity, assume that the interview is with someone you know and will be shot in your living room.
3. The power requirement (number of amperes) for electronic equipment is typically stated on either a label on the equipment itself or the specifications page in the instruction manual. Make a list of the power requirements for some of the equipment in your home, such as the largest television set, smallest television set, electric dishwasher, electric clothes washer and dryer, desktop computer, and laptop computer. Bring the list to class and compare the power consumption of the items in your home to the video gear used on a remote shoot.

STEM and Academic Activities

1. Evaluate the environmental impact of a typical remote shoot. What environmental aspects should be considered when performing the location survey (pollution, waste disposal, etc.)?

2. Identify some advances in technology that have affected remote shoots in terms of efficiency, convenience, and cost-effectiveness.

3. Design the floor plan for a television news studio. Indicate the placement of all production equipment and specialized areas.

4. Compare the costs involved in shooting a program in the studio to the costs of shooting the same program at a remote location.

5. Watch several evening news programs and list instances of ENG footage and EFP footage used in news stories. Explain why each piece of footage is ENG or EFP.

www.nab.org The National Association of Broadcasters supports local broadcasting professionals through education, advocacy, and readily-available resources.

Chapter 17

Remote Shooting

Professional Terms

crab dolly
dolly grip
film-style shooting
multi-camera shooting

nickel cadmium (NiCad)
single-camera shooting
track dolly
windscreen

Objectives

After completing this chapter, you will be able to:

- Explain the options available to solve lighting problems when shooting on location.
- Identify general safety precautions related to the handling of cameras and batteries.
- Compare the features and procedures of both remote shooting techniques.

Introduction

Shooting outside of the studio was almost unheard of in the early days of television—the size and weight of equipment and cabling, as well as power requirements, prevented remote shooting. Any remote work was done on film and transferred to videotape once back at the studio. For many years, all news footage was shot on film and transferred to tape. The disadvantage of using film, especially for news, is the considerable time needed to develop and print film before it could be transferred to tape and used on a newscast. This process made it impossible to immediately release a story to the public. Today, digital cameras in the field can uplink directly to stations or networks through hard-line connections, microwave communication, or satellite transmission. This technology makes live video and audio instantly available from virtually anywhere in the world.

Remote shooting is actually the majority of production work done today by many television production facilities. Much of the location footage seen on television is shot on video. A few network television dramas still use film, but the number is steadily declining due to changes in technology.

Television technology is progressing rapidly. Bulky cameras attached to large recorders have given way to relatively small, self-contained camcorders. The most sophisticated camcorders are now being replaced by very small cameras that have built-in transmitters to send a radio signal to the recorder, **Figure 17-1**. Footage from political convention floors, football fields, and war zones is recorded using cameras with transmitters. Cameras on a convention center floor cannot trail cables through the crowds, creating tripping hazards along their path. Even more sophisticated examples of this technology are the cameras built into headgear, such as football helmets, and even smaller units that can be placed into the frame of eyeglasses, **Figure 17-2**.

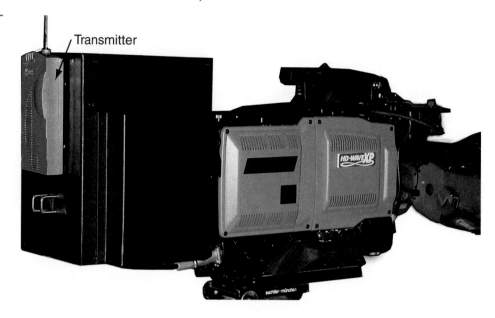

Transmitter

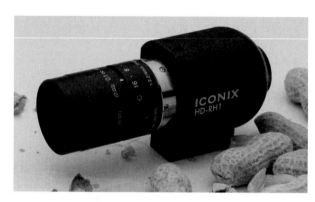

Figure 17-2. Current technology has produced tiny camera devices that were once depicted as "futuristic" equipment in science fiction films. Notice the size of this very small HD camera compared to the peanuts lying around it. *(Photograph courtesy of Iconix Video, Inc.)*

Camera Mounts

Extensive camera movement is not usually seen in many programs shot on location. Dolly shots are more realistic than a zoom, but the camera movement required with a dolly on location is difficult to accomplish. On location, the ground is usually not smooth and level. While using a dolly is possible, both a larger budget and additional equipment are necessary.

Several companies manufacture a type of track, similar in concept to railroad track, that is relatively lightweight and portable. Once the track is assembled, a four-wheeled cart, called a *crab dolly* or *track dolly*, is set in place. The camera, along with the operator in some cases, is placed on the crab dolly to capture the movement shots, **Figure 17-3**. The crab dolly may be pushed or pulled along the track by a crew member known as a *dolly grip*. The advantage of this track system is the ability to accomplish smooth dolly, truck, and arc movements in the field. The disadvantages of the track system include:

- Time consumed by constructing and leveling the track and cart.
- Cost of the track and related equipment.
- Additional personnel needed to assemble and operate the track system.
- Camera movements are restricted to the path of the track. If a shot is changed, production is delayed to reposition the track.

crab dolly: A four-wheeled cart that travels on a lightweight track and enables the camera to smoothly capture movement shots while being pushed or pulled along the track. Also called a *track dolly*.

dolly grip: Member of the crew who pushes or pulls a crab dolly along the track during production.

Since the development of camera stabilization systems, **Figure 17-4**, a decreasing number of location shoots require the track system. (Mounting the camera is discussed in Chapter 3, *The Video Camera and Support Equipment*.) Camera stabilization systems can be rented or purchased, but a system-trained operator must also be hired.

Figure 17-3. A crab dolly on its track makes smooth dolly and truck movements possible on uneven terrain when location shooting.

Figure 17-4. The Steadicam® Pilot® is a body-mounted camera stabilization system that facilitates very smooth camerawork. *(Steadicam Pilot photo courtesy of The Tiffen Company)*

PRODUCTION NOTE

A high-end camera stabilization system is an expensive investment for any production studio, let alone for a classroom. It is not likely that this type of equipment will be readily available while learning the processes of television production, or even when initially entering the industry. Remote shooting must be accomplished using either tripod-mounted cameras or hand-held cameras. With time and experience, you may come to work for a company that uses the track system or a stabilization system. The experience gained by that time will help in tackling the special challenges each system presents.

A creative workaround is to seat a camera operator in a wheelchair and have a grip smoothly push the wheelchair while shooting. However, this technique is critically dependent on a smooth floor surface at the shoot location.

An Internet search for "Glidecam®" or "Steadicam®" will also return results for websites with homemade versions of these camera stabilization systems. Some of the sites include plans and instructions to build your own device. As with many things, you get what you pay for. However, even the diminished results from a homemade camera stabilization system may be of acceptable quality for your production.

Lighting for a Remote Shoot

Lighting is always a serious issue for remote shooting. If something is not sufficiently lit in the studio, additional instruments are easily added and aimed. This is not the case at a remote location. Shooting inside someone's home, for example, does not provide unlimited power supply, ventilation, ceiling height, or a convenient grid and raceway. Portable light kits are available from many manufacturers in different configurations and are specifically designed for field use, **Figure 17-5.** Any lights brought to a location must have mounts, such as a light stand or large spring-loaded clips that attach to a door, bookcase, or table.

Figure 17-5. Portable light kit and instruments. *(Photos courtesy of Lowel-Light Mfg., Inc.)*

Portable light kit

Portable instrument

Portable instrument

As an example, consider a shoot scheduled in the living room of a house from 1:00 p.m. through 4:00 p.m. The lamps on the end tables in the living room cannot be used for illumination because they are not the correct color temperature and will produce an orange tint. However, some compact fluorescent lamps (CFL) can produce 32K light. The normal incandescent bulbs in the end table fixtures can be replaced by a CFL lamp rated at 32K. The crew must bring white (32K) lights to use at any location.

Outside windows in the room create additional challenges. Windows in the living room let in sunlight, which casts blue light in the shot even with the additional lighting instruments. Solutions to these lighting problems include:

- Change the color of the sunlight by attaching a gel to the inside of the window. This lowers the color temperature, so the light coming through the window is essentially white on the inside of the room. The camera can then be white balanced. CTO gel will convert 56K to 32K light.
- Place gels on the additional lighting instruments to raise the color temperature up to match the sunlight. Then white balance the camera. CTB gel will convert 32K to 56K light.
- Cover the window with a light-blocking material, such as a black or blue plastic tarp. Close the curtains in front of the tarp to create a neat appearance. Turn on the television lighting instruments and white balance the camera.
- Shoot at night. At night, however, any exposed window turns into a mirror unless covered. Turn on the television lighting instruments and white balance the camera.

In smaller budget productions, the last two options are most often used.

Audio

On remote indoor shoots, microphones must be placed very close to the talent. Lapel mics, mics on booms, or mics hidden on the set are ideal. If the mics are too far away from talent or are attached to the camera itself, the talent sounds as if they are speaking from the bottom of a well.

When shooting outside, using the wrong type of mic or incorrectly placing a mic may cause the audio of the environment to completely overpower the talent. If the talent is standing in a forest, for example, the birds chirping may be louder on the recording than the talent's voice. When shooting outside, a lapel mic is the best option if a directional mic on a boom is not used.

PRODUCTION NOTE

Always check the audio levels on the recorder to ensure that the mic signal is being picked up at an appropriate level. For analog systems, the reading should fall between −3 and +3. For digital systems, the audio level should be approximately −20. Some camera viewfinders or recorders only have a digital bar meter or LEDs to indicate the audio level. These typically do not have level reading numbers for reference. Before the shoot begins, find out what the bar meter reading or LED display *should* be for this type of equipment.

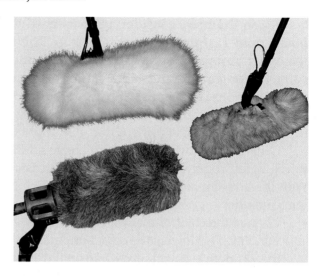

windscreen: A covering, usually foam, placed over a mic to reduce the rumble or flapping sound created when wind blows across the mic.

When shooting outdoors, the wind creates another obstacle. Using a *windscreen* is an effective solution, **Figure 17-6.** A windscreen is a covering, usually foam or furry fabric, placed over the mic. It reduces the rumble or flapping sound created when wind blows across the mic. The size of the windscreen is directly related to the amount of wind in the environment. A news reporter standing outside during a hurricane uses a more substantial windscreen than the reporter in the middle of a field on an average spring day.

PRODUCTION NOTE

On a remote shoot, either the camera operator or audio technician must wear good quality, full-muff headphones to monitor the audio while recording. Pay particular attention to unwanted environmental and other extraneous sounds that may, inadvertently, end up on the recording.

Power

Plan for power sources when shooting on location. The electrical circuits in most homes are either 15 or 20 amps. Plugging too many lighting instruments into one circuit may trip the circuit breaker or blow a fuse in the house. Electrical professionals use a technical, complex formula to determine the amount of amps a lighting instrument draws. A simplified formula is not as accurate, but works very well for television production purposes (**Figure 17-7**). This easy-to-remember formula is safe because it always over-estimates the amount of power drawn by an instrument. To use the formula:

- Check the wattage of the lamp to be used. For this example, assume the lamp is 650 watts.
- Divide the wattage value by 100 (move the decimal point two places to the left). The result is 6.50 amps.

This means that three 6.50 amp instruments may be connected on a 20 amp circuit. According to the formula, all three have a combined draw of 19.5 amps ($6.50 \times 3 = 19.5$). Because the formula provides a safety margin, slightly fewer amps are actually drawn.

Converting Watts to Amps for Location Shooting

Watts = 650

Move decimal two places to left:

Amps = 6.5

Figure 17-7. This simple conversion provides a safe overestimation of the power requirements.

Batteries

Plugging a cord into the wall outlet is the most convenient power option. However, remote shooting does not always include conveniently placed wall outlets. Therefore, videographers must rely on battery power. There are many different types of batteries used for cameras; the most common are *nickel cadmium* batteries, usually called *NiCad*. NiCad batteries are rechargeable, but may take as long to recharge as they take to discharge. Buying a charger with two sets of batteries allows the camera to run almost uninterrupted. The only time lost is when batteries need to be changed. The operator must remember to change batteries when one begins to get low on power. Rechargeable lithium ion (Li-ion) and nickel metal hydride (NiMH) batteries are becoming more popular, **Figure 17-8**.

nickel cadmium (NiCad): A type of rechargeable battery commonly used to power cameras.

General Cautions

- Do not drop batteries. The plastic shell of the battery can split open and the cells inside can break, which makes the battery unusable.
- Batteries do not last very long when they are in a cold environment.
- Do not completely discharge batteries. Always leave a small amount of power in the battery before removing it and placing it on the charger. Pay close attention to battery level indicators.

Figure 17-8. All the various types of rechargeable batteries available must be handled with care.

- Many professional batteries perform better if they are placed on a plugged-in charger whenever they are not in use.
- Do not *ever* leave batteries in the trunk or on the seat of a car parked in the hot sun. *Batteries can rupture when heated.*

PRODUCTION NOTE

Always carefully check the charging instructions provided with any type of battery. While most professional batteries should be placed on a charger when not in use, this does not apply to *all* professional and consumer-grade batteries. Some batteries may actually be damaged if they are continuously left on the charger.

Cold Temperatures and Batteries

Fundamentally, batteries create power through chemical reactions that occur within a battery cell. Electrons move from one molecule to another, creating usable electricity in the process. As the temperature of a battery lowers, the chemical reactions slow down and the amount of power created diminishes. As power output diminishes, video equipment stops operating. A flashlight becomes dimmer and dimmer and a battery-powered toy moves slower and slower as power diminishes. However, video equipment demands a certain amount of power at all times. When a battery gets cold and the chemical reactions slow down, the power demand of the equipment is not met and the equipment fully shuts off—it does not operate slower. Imagine what a wake-up call that would be if you are conducting the interview of a lifetime with the President of the United States on a ski vacation and the video camera shuts off in the middle of the interview.

Because batteries become ineffective so quickly in cold conditions, precautions should be taken to keep the batteries warm. Some effective solutions when on location include:

- Wear a belt with your pants and a tucked-in shirt. Place the batteries inside the shirt, next to your skin. Your body temperature keeps the batteries warm.
- Wrap the camera in a blanket to keep the cold out of it for as long as possible.
- Take frequent breaks inside a warm environment.
- Keep extra batteries in a cooler with activated chemical hand warmers. Place a towel between the hand warmers and the batteries.

Condensation

Circuit boards inside cameras, recorders, and camcorders are so small that even a single drop of moisture can short them out. That short causes the equipment to require servicing before it can be used again. Condensation is a problem because many of the internal components of the equipment are metal and portable equipment is exposed to a variety of environments. Remote shooting environments can include a wide range of humidity and temperature levels. When equipment is stored in a cool, air-conditioned environment and is then taken outside into humid summer weather, condensation forms on the inner components and on the camera lens. If shooting is planned in this kind of condition, the camera should be powered off and placed in the environment an hour or more before it is needed. This allows the equipment to acclimate to the humidity and temperature levels. Condensation still forms inside the camera, but evaporates as the camera warms to the temperature of the environment.

Most cameras and camcorders have a special circuit that senses condensation before it shorts out the equipment. This sensor overrides the power switch and shuts the power off completely. When this happens, many people panic thinking that the camera just stopped working for no apparent reason. The equipment is not operable again until the condensation evaporates. If you attempt to turn the equipment on after this override, the phrase "Dew," "Auto Shut-Off," or "Self-Protect Mode" is momentarily displayed in the camera's viewfinder, **Figure 17-9**.

To speed up evaporation:

1. Turn the camcorder on.
2. Press the "Eject" button. The door opens and the camera, once again, turns itself off.
3. Use a blow dryer set on "cool" to blow air into the open panel, **Figure 17-10**. Never blow hot air into a camcorder—the heat can damage internal camera components.

Condensation can also form on the lens. Like the mirror in a steamy bathroom, wiping the lens with lens paper does not prevent the condensation from reforming. The moisture must evaporate and leave the affected area. As the lens acclimates to the temperature of a room, the moisture will evaporate. Once the moisture has completely evaporated, the lens may be used again.

Remote Shooting Techniques

There are two methods used when shooting on location:
- multi-camera shooting
- single-camera shooting

Figure 17-9. When this message is displayed in the viewfinder, condensation has developed inside the camera body. Once the moisture evaporates, the camera will operate normally.

Figure 17-10. A blow dryer may be used to circulate cool air into the camcorder body to speed up evaporation.

Multi-Camera Shooting

Multi-camera shooting is exactly what its name implies—shooting with multiple cameras. Multi-camera shooting is either recorded using a separate recorder for each camera or by using a mobile control room.

When a separate recorder is used for each camera, a scene is shot from different positions and angles by several cameras at the same time. A multi-camera appearance can then be created in the editing room during post-production. This technique requires less shooting time than other options, but greatly increases the amount of time spent in the editing room. Using this kind of shooting is dependent on the production budget and the amount of time performers are available, compared to the time available for work in the editing room. Extra personnel and equipment are also necessary to have multiple cameras shooting at the same time. When using a separate recorder for each camera, the mixed audio is recorded on only one of the recorders. So, all the cameras must be synced up. This allows the editor to match the recorded audio to the lips of the performers.

PRODUCTION NOTE

An easy way to sync up multiple cameras is to have someone walk onto the set with a flash camera. Start all three cameras recording and have the person snap a picture with the flash. The camera flash provides a visual "mark" the editor can use to synchronize all three cameras during the editing process.

A mobile control room is typically a truck parked as close as possible to the shoot location, **Figure 17-11**. All of the cameras are run to a live video

Figure 17-11. The mobile truck is essentially a control room on wheels.

switcher on the truck that provides a synchronizing signal back to the cameras. The crew runs hundreds of feet of cable from the truck to the cameras shooting on location. When a network covers a professional football game, for example, the truck is parked outside the stadium and the cameras are located inside the stadium. The director and, often, the announcers are outside in the truck. In this case, the game pictured behind the announcers is a simple chromakey background wall. (Chromakey is discussed in Chapter 23, *Electronic Special Effects*.)

Single-Camera Shooting

single-camera shooting: A technique of remote shooting that involves only one camera and is most often used for event recording.

Single-camera shooting involves only one camera and is the least difficult type of shooting. The only setup required is to set the camera on a tripod, turn it on, and cue the performer(s). After the scene is over, the camera can be moved, set up again, and turned on to capture another scene. This type of shooting is not suitable for entertainment television programs because it is not interesting and does not keep the audience's attention. On prime-time television programs, the average camera shot lasts seven seconds. Unless the picture on screen is extremely compelling in action or dialogue, the audience tends to look away when shots last longer than seven seconds.

Single-camera shooting is most often used for general news coverage from reporters in the field and event recording, **Figure 17-12**. A videographer of event recording is usually hired by a client to shoot a concert, play, speech, seminar, presentation, or wedding. This type of client is not interested in what is normally considered entertainment television (with many cuts and effects). They merely want an event recorded as it happens, without all the production aspects of entertainment television.

Film-Style Shooting

film-style shooting: A type of single-camera shooting in which a scene is shot many times with the camera moving to a different position each time to capture the scene from various angles. The finished scene is edited together to look like it was shot with several cameras.

Film-style shooting is a unique subcategory of single-camera shooting. In film-style shooting, only one camera is used and it is moved very frequently. While there are many methods to this style of shooting, the following example is one procedure for using film-style shooting.

1. A master shot is recorded of the entire scene. The master shot is a relatively long shot of all the action in a scene.

2. After the master shot is completed, the entire scene is shot several more times from different positions and camera angles.

3. Certain segments of the scene are restaged to get close-ups of key actions.

4. Cutaways are then shot, but may also be shot before the entire process starts. (More information on cutaways is presented in Chapter 19, *Production Staging and Interacting with Talent*.)

Figure 17-12. Single-camera shooting is easy to set up, but results in uninteresting images.

Each of the different shots recorded are edited together and the finished scene looks as if it was originally shot with many cameras. The finished product is far more interesting than a single-camera, single-shot production because the cut rate is much higher. The disadvantage of film-style shooting is having the talent repeatedly perform each take of the scene with the exact same action and dialog. Without exact repetition, the program cannot be edited together properly. Also, the editing process of film-style shooting is considerably more complex.

Wrapping Up

Remote shooting is very enticing because it eliminates the need to construct a set. On the other hand, remote shoots have their own collection of quirks. The real key to successful television production is extensive and detailed pre-production planning, in conjunction with strategy and planning throughout the process.

Review Questions

Please answer the following questions on a separate sheet of paper. Do not write in this book.

1. What is a crab dolly? When is it most often used?
2. What are some challenges of planning the audio for a location shoot?
3. What is the formula presented to estimate the power requirements of an instrument? Why is this imprecise formula safe to use?
4. List three precautions that apply to the use and care of camera batteries.
5. What is the built-in safety feature that protects a camera from condensation?
6. Explain the characteristics of the two remote shooting methods discussed in this chapter.
7. How are the various camera angles captured when using film-style shooting?

Activities

1. Visit an electronics or other specialty equipment store and locate the rechargeable batteries section of the store. Review the recharging instructions on several types and brands of batteries. Note the cautions on each package and the differences in the directions provided.
2. Research the various plans and instructions for homemade camera stabilization systems available online. Choose one of the plans and summarize the features of the system and the components required to construct it.

STEM and Academic Activities

Technology

1. Research the changes in battery technology. How have these changes impacted battery life, the materials used to manufacture batteries, and how power is delivered from the battery to a device?
2. Investigate the technology used in a mobile control room. Explain how video and audio signals are sent and received from the mobile control room. Is any specialized equipment needed to run a production from a mobile control room?

Mathematics

3. Inspect a main living area in your home and list the wattage of all electric-powered items in the room (lighting, entertainment equipment, charging units, etc.). Using the formula presented in this chapter, calculate the total number of amps required to power the room.

Social Science

4. Programs are often shot on location to create the appearance of authenticity. You may have seen a production in progress in your local area. Describe the impact that a location production (large or small) has on the day-to-day lives of the people and businesses residing in the production area.

Chapter 18

Props, Set Dressing, and Scenery

Professional Terms

cyclorama (cyc)
moiré
props
set decorator

set design
set dresser
set dressing
strike

Objectives

After completing this chapter, you will be able to:

- Identify factors to be considered when selecting furniture for a production.
- Recognize the difference between set dressing and props.
- Explain how the pattern of materials used on a set affects the video image.

Introduction

An interior decorator working in a house selects the furniture, wall treatments, curtains and drapes, accent accessories, and many other design and visual elements to make the rooms appealing and to meet the needs of the homeowner. In the television industry, all of the design and visual elements chosen for a set are considered *set dressing*. The *set dresser*, also called *set decorator*, is responsible for selecting the furniture, wall and window coverings, accent accessories, and all the other design elements that complete a program's set. In making these decisions, the set dresser must consider the contrast ratio (see Chapter 14, *Image Display*) of the items chosen, as well as accurately create the director's vision of the set. This chapter discusses the various design and visual elements of set design, as well as related techniques and professional tips.

Creating the Set Design

set dressing: All the visual and design elements on a set, such as rugs, lamps, wall coverings, curtains, and room accent accessories.

set dresser: The person responsible for selecting the furniture, wall and window coverings, accent accessories, and all the other design elements that complete a program's set. Also called a *set decorator.*

set design: A scale drawing of the set, as viewed from above, that illustrates the location of furniture, walls, doors, and windows.

The *set design* is a sketch of the set, as viewed from above, drawn to scale, **Figure 18-1**. The set designer lays out the location of walls, doors, and windows on the set. Then, the set dresser adds the location of furniture and larger decorator items. The director uses the design when rehearsing the program with the actors and talent while the set is under construction. The set design notes the location of major pieces on the set, but does not necessarily indicate the placement of accent and decorative items.

PRODUCTION NOTE

To help the performers get accustomed to the amount of space available once the set is completed, the set design is used to mark the floor of the rehearsal space. Masking tape is commonly used to indicate where the walls and doors will be located on the finished set. Masking tape is inexpensive and readily available, but it leaves a sticky residue on the floor if left in place for very long. Set marking tape, or spiking tape, is also used to mark the floor of the rehearsal space and is available from any theatrical supply company. Set marking is different from masking tape in that it is brightly colored and leaves no sticky residue on the floor.

Furniture

When selecting furniture for a set, consider that the talent should be able to get into and out of furniture gracefully. The furniture needs to be solid and firm and the seat cannot be lower than the talent's knees, **Figure 18-2**. If the chair is too low, the talent must either bounce out of the seat or roll out on one side or the other. The problem is that the center of gravity is not correct and does not allow a fluid movement into and out of the chair.

Figure 18-1. A set design resembles the floor plan of a model home.

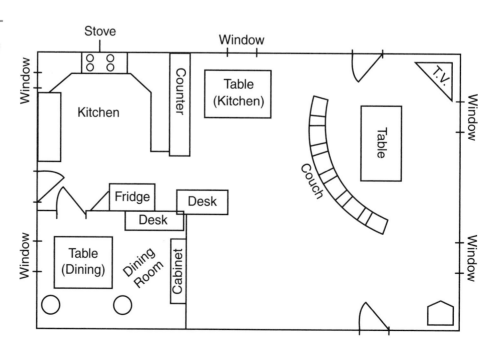

Figure 18-2. A chair that places the talent's midsection lower than their knees looks very unattractive and poses a problem in gracefully rising from the chair.

Talent is forced to slouch when sitting naturally in a low chair. Slouching is not only unattractive, but the talent's diaphragm is compressed. This makes it very difficult, if not impossible, to clearly project his or her voice. Also, the talent's eyes are positioned lower than the camera lens when seated in a low chair. This forces a high angle camera shot, looking down on the talent.

Never use chairs that swivel or rock on a set. The talent has a tendency to swivel back and forth or to rock. Talent moving in this manner creates a shot that is unpleasant to the audience. Chairs that swivel or rock also tend to squeak, which is a distracting, unintentional sound in a scene.

Be aware of any shiny surfaces, such as chrome or brass fittings, on furniture and other items on the set. Shiny surfaces may cause unattractive or distracting light hits, **Figure 18-3**. These reflections are very distracting to the audience and interrupt the viewer's attention from the message of the program.

Certain products can be used to keep reflections off shiny surfaces and can be removed with a damp cloth after the shoot.
- Lightly apply crème makeup on the surface to dull the shine.
- Spray the area with hair spray.
- Apply dulling spray (available at camera shops).

Figure 18-3. The shiny chrome arms on this chair will produce distracting light hits. Adding dulling spray, or another treatment, reduces or even eliminates light hits. Additionally, talent is likely to swivel back and forth while seated in a chair of this type. The swiveling motion is distracting to viewers.

VISUALIZE THIS

While light hits are usually considered to be negative program attributes, they can also be used artistically within a program. For example, hundreds of light hits are produced if you shoot the surface of a lake under bright sunlight. These light hits create the appearance of a sparkling, clear, inviting body of water. When multiple light hits are a desired effect, a star filter may be attached to the front of a camera lens to cause each light hit to become a star, rather than simply a bright white spot. A star filter was used for the image in **Figure 18-3**. Star filters are available at photo supply stores in many variations and sizes. Be sure to bring your camera to the store to verify the correct filter size for your camera.

Assistant Activity

You can create your own star filter using silver wire-mesh window screening from any hardware store. The screening must be made of shiny metal, not plastic, with a fine mesh.

1. Measure the diameter of the end of your camera lens.
2. Measure and mark a circle that is slightly larger than the camera lens diameter on a small piece of wire-mesh window screening.
3. Cut out the circle.
4. Carefully attach small, thin strips of duct tape in a few places on the mesh, as close to the edge of the circle as possible.
5. Attach the screen to the end of your camera lens by pressing the strips of duct tape onto the *outside* of the lens shroud.

WARNING: Do not allow any of the tape to touch the glass lens. Do not allow the mesh to touch the glass lens—it will likely scratch the lens.

Aim the camera at an object with many light hits, like a body of water. Experiment by rotating the mesh slightly to see what effect rotation has.

Placement

In most homes, it is very common to see the furniture that people sit on (sofas, loveseats, and chairs) placed against the walls of a room. Placing this type of furniture in the middle of a room looks odd, unless the room is rather large. Think of the sets in various situation comedy shows that portray the living room in a home. On these sets, furniture may be placed against a wall, but none of the characters sit on those pieces of furniture. The furniture used by talent is placed in the middle of the room. This arrangement of furniture appears so natural on the television screen that it probably never stood out to you while watching.

With furniture arranged against the wall of a set, it is not possible to backlight the talent seated on the furniture. The purpose of backlighting is to separate the talent from the background. Without the appropriate backlight, the talent, as well as the entire image, appears flat and unrealistic. Additionally, lighting the talent so close to a wall causes a shadow of the person to be cast on the wall. Multiple background lights are typically used, which creates several shadows. It is not likely that you see multiple shadows on a wall behind someone in your home. If these shadows are present on a set that the audience sees, the illusion of reality of the living room will be broken. A standard rule in set design is to place the furniture used by the talent at least six feet away from any wall of the set. Furniture not used by the talent is considered set dressing and may be placed wherever the designer likes.

Props

Props are any of the items handled by the performers during a production, other than furniture. Just as there are exceptions to spelling rules in English class, television production principles are equally loaded with exceptions. A simple piece of furniture may become a prop if it is used in a way, the audience assumes, that it was not manufactured to be used. Examples of this may be a couch that is single-handedly hoisted into the air by a character with super strength, a bed that collapses when the talent gets into it, or a bookcase whose shelves give way with the weight of a single book.

props: Any item handled by the performers during a production, other than furniture.

VISUALIZE THIS

To help clarify how furniture may be used as a prop, visualize the following scene. A desk in an office setting is piled with towers of papers and files. Sitting at the desk, barely visible behind the piles, is a man who looks overwhelmed. His boss approaches carrying a single sheet of paper, perhaps an interoffice memo, and places this single sheet of paper on one of the towers already on the desk. Without warning, the desk collapses from the additional weight of the single page.

Office desks are not constructed to collapse. This office desk was constructed of a particular material or assembled in such a way that it collapsed on cue. This action makes the desk a prop in the scene, not a piece of furniture.

When selecting or creating props, it is not always necessary to attend to every last detail. The television camera is more forgiving to smaller sized props. For example, the phasers used in the Star Trek television series were pieces of wood glued together and painted a dark gray with a few pieces of colored plastic attached. Before spending a great deal of time and money on props, consider the cardinal rule of television: "It does not have to be, it must only appear to be." However, high-definition digital cameras create images of such high quality and detail that much more attention must be paid when building sets, props, and scenery. With high-definition cameras, the original Star Trek phasers would look like what they actually were—wood with colored plastic pieces glued on.

Assistant Activity

On a classroom set, there is a coffee mug, a grade book, various papers, pens, a paper clip dispenser, and a stapler on the teacher's desk. Behind the desk is a chair. A large map and various pieces of student art hang on the wall. The performer (teacher) walks in, sits down on the chair, picks up a pen, and begins grading a paper. He records the grade in the grade book and takes a sip of coffee. In this scenario, which items are props? Which items are furniture? Which items are set dressing?

Flats, Curtains, and Backdrops

Scenery is whatever stops the distant view of the camera. In a studio setting, this includes flats, curtains, and backdrops. If a set is not supposed to reproduce a real-life environment, such as someone's living room, flats can be placed at odd angles, with gaps between them, or be painted in unusual colors and textures. The effect can be attractive and eye catching without upstaging the talent or subject matter of the program. Set designers must always consider contrast ratio and the limitations they place on other items in the picture when choosing a background color.

A background may be loose curtains, having the attractive folds found in living room curtains, or be an indistinct, solid color. A *cyclorama*, or *cyc* (pronounced "sike," rhymes with "hike"), is a solid-color background without any visual texture, **Figure 18-4**. A cyc may be created by stretching background curtains tight to cover the walls and curves of the studio, forming a solid background color. A cyc can also be constructed of a rigid material. The corners of a rigid cyc, as well as where the background meets the floor, are curved to create a completely indistinct background. A cyc is usually used for limbo shooting and chromakey shooting.

A cyc differs from a backdrop because a cyc is usually just one color and has no definition. On the other hand, a backdrop may have scenery painted on it. For a studio production set in London, as an example, someone may be contracted to paint a skyline of London on a backdrop that hangs behind the set. That way, if a shot ever moves off the set, the audience sees London in the distance. In modern studios, a backdrop is a rare thing indeed. Digital technology allows computer-generated backgrounds to be inserted into a picture that previously had no background at all.

cyclorama (cyc): An indistinct, solid-color background that is typically used for limbo shooting and chromakey shooting.

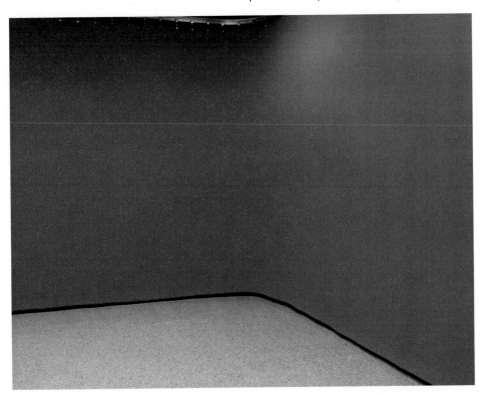

A

B

Figure 18-4. A cyc creates an indistinct background. A—A cyc can be a curtain pulled tight around a set. B—A rigid cyc is constructed of modular panels. *(Photo courtesy of Pro Cyc. Inc./ www.procyc.com)*

When computer-generated backgrounds are used, the only backdrop on the set is a chromakey wall behind the talent. (Chromakey is discussed in Chapter 23, *Electronic Special Effects*.)

Visual Design Considerations

While choosing items for the set, the set dresser must be conscious of other factors that affect the visual appeal and realism of the set. Both the placement of items on the set and the patterns on set items have a great impact on the video image.

The 3-D Effect

A television screen is a two-dimensional piece of glass. The creative use of light and shadow (previously discussed in Chapter 15, *Lighting*) creates a third dimension. Another technique in producing the illusion of three-dimensionality is to place items in layers on the set, **Figure 18-5**. Items on the set should be placed in the front area of the set (closest to the cameras), in the middle of the set, and at the back of the set (furthest away from the cameras). The sets of most modern sitcoms use this layout by placing some

Figure 18-5. Placing items in layers on a set helps create a three-dimensional feel to the picture.

item of furniture right in front of the camera. This may be a table, a chair with its back to the camera, a TV set, or any other item found in a home. With the talent placed in the middle of the set, objects layered in front of and behind the talent add great depth to the picture.

Patterns

Patterns are an issue in the areas of upholstery, wallpaper, curtains, and costumes. Television systems that use interlace technology create a picture from colored dots arranged in rows on the screen (discussed further in Chapter 25, *Getting Technical*). The rows, or lines, flicker on and off in such rapid succession that the human eye cannot detect the flicker. The odd lines light, then the even lines light, then the odd, then the even, and so on. As a result, any horizontal line or high-contrast patterns on the set appear, to the viewer's eye, to be jumping up and down. This effect may even cause a rainbow of colors to appear in the patterned area called *moiré* (pronounced "more-ray"), **Figure 18-6**. This is distracting to the viewers and can be avoided by carefully selecting the materials and patterns used on the set.

- Avoid bold horizontal and vertical lines.
- Avoid tightly woven, complex patterns of high-contrast lines, such as herringbone patterns.
- Avoid elaborately or thinly striped neckties and scarves.

Striking the Set

When a set is no longer needed, the crew must *strike* the set. To strike a set means to dismantle or tear it down. Large set pieces, like flats, doors, windows, and stairs, are usually salvaged for use on future sets. The number and types of pieces saved depends on the amount of storage space available.

moiré: An effect caused by shooting certain patterns usually frabrics, in which the television system reproduces the pattern with a rainbow of colors or moving lines displayed in the patterned area.

strike: To dismantle or tear down a set that is no longer needed.

Figure 18-6. The rainbow effect caused by high-contrast patterns in a picture is called moiré. Notice the distortions on the shoulders and around the collar of this shirt.

Figure 18-7. Various school news sets.

Fayetteville High School (Fayetteville, AR)

K-AHS/Austin High School (Austin, TX)

Keystone High School (LaGrange, OH)

Titan TV/Centennial High School (Frisco, TX)

EVTV/Eastview High School (Apple Valley, MN)

Anderson Districts I & II/Career and Technology Center (Williamston, SC)

Figure 18-7. *Continued.*

Lake Charles-Boston Academy of Learning
(Lake Charles, LA)

Oakleaf Junior High (Orange Park, FL)

Kaleidoscope TV News/King Career Center
(Anchorage, AK)

OC-TV/Ocean City High School (Ocean City, NJ)

ONW...Now/Olathe Northwest High School
(Olathe, KS)

VHSTV/Voorhees High School (Glen Gardner, NJ)

Wrapping Up

Most of the principles of set construction in theater apply to television production. The important exceptions to the theatrical principles are color and contrast ratio. These two concepts weigh heavily in most every aspect of television production. Unlike an interior decorator working in someone's home, a set dresser must be aware of how the completed set will appear to the camera and on the final video image. When designing and constructing sets, props, and set dressing, remember: "It doesn't have to be; it only has to appear to be." Review the gallery of news sets in **Figure 18-7** for the various set design elements discussed in this chapter.

Review Questions

Please answer the following questions on a separate sheet of paper. Do not write in this book.

1. What are some factors to consider when selecting furniture for a production set?
2. What is a *prop*?
3. How is a cyc different from a backdrop?
4. How does the set dresser contribute to creating the illusion of three-dimensionality on the television screen?
5. Compare the responsibilities of an interior designer to the responsibilities of a set dresser.
6. How does the crew strike a set?

Activities

1. Visit a local fabric store. Choose some fabrics that would cause moiré on a television image. Bring a few samples to class for discussion.
2. Create a basic set design for an afternoon talk show. The designs should include furniture, camera placement, decorative items, and a faux window.

STEM and Academic Activities

Science

1. How does your diaphragm assist in projecting your voice when speaking? How does slouching hinder your diaphragm's function when speaking?

Technology

2. Choose a movie or television program that largely used computer-generated backgrounds. Research the digital technology used to create the backgrounds and find "behind the scenes" information to discover how different the images recorded by the cameras are to what you saw on the screen.

3. Make a sketch of the set in your school's production studio. Include the location of walls, doors, windows, furniture, and other large items on the set. What improvements can be made to the current set design? How can those improvements be implemented?

4. Next time you watch your favorite entertainment program, pay attention to the placement of furniture on the set. How does the arrangement of furniture create dimension in the scene? Have you ever noticed the placement of furniture items on this program before now?

www.setdecorators.org The Set Decorators Society of America provides fellowship and networking opportunities for set decorators, crew members, and vendors in the industry. The organization also offers Associate and Student Memberships, internships, and other educational opportunities for students and apprentices hoping to enter the field.

Chapter 19

Production Staging and Interacting with Talent

Objectives

After completing this chapter, you will be able to:

- Identify the foreground, middle ground, and background on a set.
- Recall the function and importance of the vector line in camera staging.
- Explain the difference between a jump cut and an error in continuity.
- Illustrate the staging for both two-person and three-person studio interviews.
- Explain the difference between a dramatic aside and ad-libbing.
- Identify considerations and methods that production staff members should use when working with non-professional talent.

Introduction

This chapter discusses the placement of furniture, props, and talent in front of the camera. The arrangement of items in a shot is called *staging*. In theater production, staging refers to the movement instructions given to performers by the director. In television production, staging also applies to the placement and movement of cameras. This chapter presents guidelines and methods of effective staging for television production.

Professional Terms

ad-lib	hand shots
background	jump cut
camera line	middle ground
cheating out	staging
cross-camera shooting	swish pan
cutaway	teleprompter
dramatic aside	vector line
error in continuity	whip pan
foreground	

staging: The arrangement of items, such as furniture, props, and talent, in a shot.

foreground: The area between the talent and the camera.

middle ground: The area in which the action of a program typically takes place and where the most important items in a picture are usually positioned.

background: The material or object(s) that are placed behind the talent in a shot.

Areas on a Set

The television screen is a relatively flat piece of glass. All images displayed on the screen are two-dimensional. An important goal in production is to attempt to create the illusion of three dimensions in order to increase the realism of television images. Purposeful use of the areas on a set—foreground, middle ground, and background—is an effective method in creating three-dimensionality on the television screen.

The *foreground* is the area between the talent and the camera. Placing items in the foreground of a shot is a simple and effective way to create three-dimensionality, **Figure 19-1**. Novice camera operators often make the mistake of ignoring the foreground area of a set when framing a shot. While leaving the foreground of a set empty creates additional space for camera movement, it does not help in creating the illusion of a three-dimensional image on the flat television screen. This is an important area for staging to create a realistic image for the viewer.

The *middle ground* is the area in which the talent performs and the action of a program usually takes place. This is the area where important items in a picture are commonly positioned.

The *background* of a picture is the material or object(s) behind the talent in a shot. The distance between the talent and the background, if properly lit, greatly contributes to creating three-dimensionality.

Talk the Talk

While the terms *background* and *scenery* may seem very similar, they are two different elements on a production set. For example, a glass window on the back wall of a set is background. The painting or photograph depicting the outdoors placed behind the window is scenery because it stops the distant view of the camera.

Figure 19-1. The center of focus in this shot is the person at the desk. Notice that the person seated in the foreground helps create the illusion of three-dimensionality.

For a set depicting the interior of a living room, for example, a back wall and partial left and right walls would be constructed. A couch facing the cameras may be placed in the middle ground. Creative lighting in conjunction with the middle ground and background set creates an acceptable illusion of depth in the picture. However, placing a coffee table or chair in the foreground creates even more depth in the picture. Effective use of the foreground area helps heighten the impact of a dramatic program. Even most news programs have an anchor desk that separates the talent from the cameras. The desk is placed in the foreground, which enhances the three-dimensional illusion of the program.

Assistant Activity

Watch four sitcoms with scenes that take place in the home of one of the characters. List the foreground, middle ground, and background items that are common in the shots. Be prepared to discuss your findings in class.

Camera Staging

The placement and movement (such as dolly, truck, and arc) of cameras, particularly when using multiple cameras, is planned during the pre-production process of marking the script. Planning makes the most efficient use of both the talent and staff's time and the production budget.

If using multiple cameras for a production, each camera is assigned a number. The industry convention is to number the cameras from the camera operator's point of view (facing the set). Therefore, the camera on the operator's far left is camera 1 and the remaining cameras are numbered sequentially moving from left to right—camera 1 is on the left, camera 2 is in the middle, and camera 3 is on the right. The images from the cameras are displayed on monitors in the control room, which are also arranged from left to right. The monitor corresponding to the video from camera 1 is on the left in the control room and the monitor for camera 3 is on the right.

Stage directions, however, are given from the performer's point of view (facing the cameras), **Figure 19-2**. Imagine a weekly interview program that focuses on local musicians called "Musician's Corner." The studio segment of the program is a brief, informal interview with a different musician every week, one-on-one with the host of the program. The host of the current segment of the program, Michael, is speaking with a saxophonist named Katharine. Michael is seated on Katharine's right. In this example, Michael is stage right of Katharine. From the camera and audience's point of view, he is on the left of Katharine. When referring to the position of people or items in front of the camera, the direction is communicated as stage direction. When referring to the items and people behind the cameras, the directions are given from the perspective of someone looking at the set.

Figure 19-2. Proper stage direction terms.

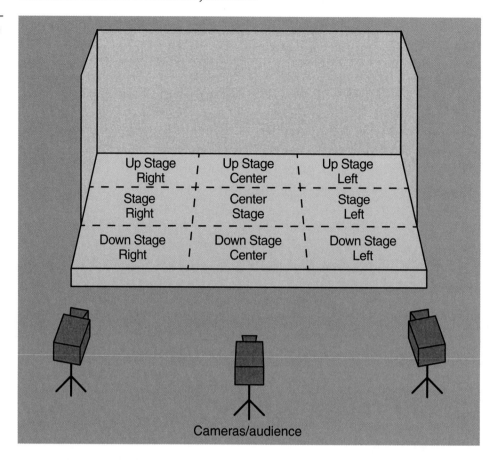

Cameras/audience

Talk the Talk

When directing talent, always place the word "stage" in front of the movement direction. For example, "Bill, please move two steps stage right." When referring to movement of items and people behind the cameras, directions are given using normal, everyday language. For example, "Camera 2, pan left to follow Bill as he moves."

cross-camera shooting: A two-camera shooting technique in which the camera on the left shoots the person on the right of the set and the camera on the right shoots the person on the left of the set.

cheating out: Positioning the talent's body to slightly face the camera to give the audience a better view.

To shoot this interview, it may seem most logical to have camera 1 shoot Michael and camera 3 shoot Katharine. Camera 2 provides a two shot of them both, **Figure 19-3**. This, however, is not the most appropriate camera placement. Because the interview is a conversation and the participants face each other, camera 1 and camera 3 would capture profile shots of Katharine and Michael. Profile shots create a very flat and confrontational feel in a program. Cross-camera shooting is the solution in this situation.

Cross-camera shooting is a technique where the camera on the left shoots the person on the right in the shot, and the camera on the right shoots the person on the left in the shot. See **Figure 19-4**. In the "Musician's Corner" interview example, camera 1 shoots Katharine and camera 3 shoots Michael. Camera 2 remains available and positioned for a two shot. In a two-person interview, the talent should angle their bodies slightly to place their upstage shoulders close together and their downstage shoulders are farther apart. Positioning the talent's body to slightly face the camera to give the audience a better view is called *cheating out*.

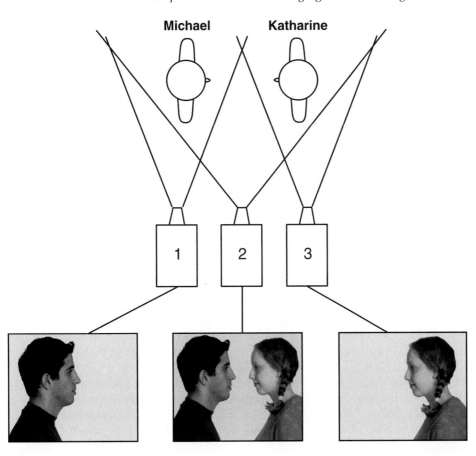

Figure 19-3. If a camera operator shoots the talent positioned directly in front of the camera, with the talent positioned face-to-face, the result is a flat, unattractive profile shot.

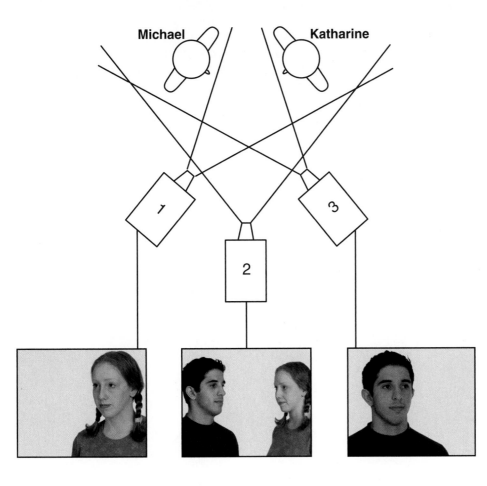

Figure 19-4. Changing the subject of cameras 1 and 3 to the talent on the opposite side of the set creates more realistic shots with dimension.

Vector Line

vector line: An imaginary line, parallel to the camera, that bisects a set into a foreground and a background. Also called a *camera line*.

The *vector line* is an imaginary, horizontal line that bisects a set into the foreground and background. In some facilities, the term *camera line* is used synonymously with vector line. The vector line extends all the way across the set from left to right. In the "Musicians Corner" interview example, imagine a line drawn horizontally across the set, through the noses of both Michael and Katharine. The line drawn is the vector line of the set. All cameras must remain on the same side of the vector line while shooting. If the program cuts to the image of a camera shooting on the opposite side of the vector line, all items in the picture are reversed, **Figure 19-5**. This is a grave production error called a jump cut.

In the "Musicians Corner" interview, Michael and Katharine are facing each other. Imagine that cameras 1 and 2 have two shots with Michael on the left of the screen and Katharine on the right. If camera 3 crosses the vector line, Katharine will be on the left of the screen. This creates a terrible jump cut because Michael and Katharine appear to "jump" to opposite sides of the screen when the director cuts from camera 2 to camera 3.

VISUALIZE THIS

While preparing to shoot a basketball game, you place a camera at the top of the stands on both sides of the court to make sure that you don't miss any of the action. Camera 1 is at center court on the home team's side and camera 2 is at center court on the opponent's side. During the game, a player makes a very long shot. The ball goes high in the air, traveling from screen left to screen right. Camera 1 has the shot. At the peak of the midair arc, you cut to the shot from camera 2. Because camera 2 is on the other side of the court—opposite side of the vector line for camera 1—the ball now appears to be traveling from screen right to screen left. On the viewer's television screen, the ball appears to have magically reversed direction in the middle of the shot. This image is a major error on the part of the director and could have been avoided.

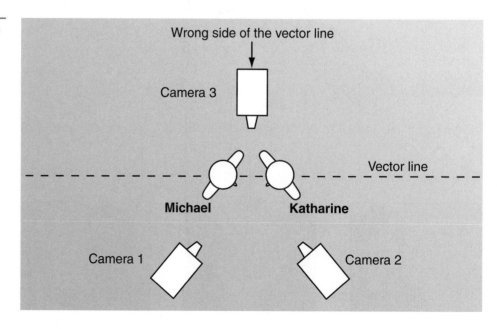

Figure 19-5. In the image captured by cameras 1 and 2, Michael is on the left side of the screen and Katharine is on the right. Camera 3, however, depicts Katharine on the left and Michael on the right. Because of this difference, cutting from camera 1 to 2 to 3 will create jump cuts.

A camera may cross the line *while it is hot* and, essentially, take the audience with it as it crosses the line. If the audience crosses the vector line with the camera, the camera reorients the vector line as it moves. Once it stops moving, the other cameras must be repositioned to be on the correct side of the new vector line. Because the placement of cameras on a set cannot be identical, each camera has its own vector line. The vector lines of all the cameras must complement each other so that none of the cameras are on the wrong side of any other camera's vector line.

Creative use of the vector line may result in savings in the production budget. Suppose two scenes need to be shot of a train moving across the prairie. In one scene, the train is on its way to Arizona from Oklahoma; moving from screen right to screen left. In the other scene, the train is returning to Oklahoma from Arizona; moving from screen left to screen right. Instead of setting up and shooting the scene twice, simply position a camera on either side of the railroad track. Make sure the two cameras do not shoot each other by hiding them or staggering them along the track. The train runs once and two scenes are accomplished with one take.

Cutaways and B-Roll

A *cutaway*, also called B-roll, is a shot that is not a key element in the action. It is usually a close-up of different items found on the set, such as some object or a person in the background. A cutaway may also be a shot that provides a visual of what is being discussed in the audio track. When a cutaway is used, the audience should not feel that the shot is jarring or out of place. Also, a cutaway shot should not include an integral moment or action in the scene. If the cutaway shot is not included in a scene, the audience should not feel as though some part of the scene is missing.

cutaway: A shot that is not a key element in the action. It is commonly used to bridge what would otherwise be a jump cut.

Talk the Talk

The term "cutaway" is more prevalent in television production environments, while "B-roll" is common in broadcast journalism environments.

B-roll footage is generally shot after the primary footage is shot, but may be shot prior to shooting the primary footage. The timing may vary from production to production, but it is important to remember that B-roll *must* be shot! To reshoot B-roll after leaving a location or striking a set requires that all the equipment be obtained again, all the necessary personnel be rehired, and the entire crew must return to the location or have the set rebuilt. All of this is a great inconvenience and expense for the production.

While shooting the primary footage, listen carefully to what the talent says. If any objects are specifically mentioned in the dialogue, make an effort to get cutaway footage of those specific objects after the primary video is shot.

Hand shots are a type of B-roll shot that features a close-up of hands, such as interlaced fingers, an index finger pointing, or counting "1, 2, 3, 4" by tapping the finger(s) of one hand into the palm of the other hand. Hand shots are related to nod shots (discussed in Chapter 8, *Scriptwriting*). Nod shots and hand shots are critically important when editing a panel

hand shots: A type of B-roll shot that features a close-up of someone's hands.

discussion program. Because camera operators never know which panelist will speak next, they often miss shooting the beginning of someone's comments. The camera operator rapidly pans the camera to find the person speaking and get the shot of that person. The rapid camera pan is called a *whip pan*, or *swish pan*. If the camera is hot, this shot is jarring and dizzying for viewers. In the editing room, B-roll footage, such as a nod shot or hand shot, can be inserted over the whip pan to eliminate the jarring shot.

One of the most common uses of cutaways is to bridge what would otherwise be a jump cut. For example, cutting a shot of a teacher writing on the board to a shot of the teacher standing over a student offering assistance creates the illusion that the teacher "jumped" from the blackboard to the student. Placing a cutaway of a student looking forward, then writing studiously between the two shots avoids a jump cut in the shot sequence.

Jump Cuts and Errors in Continuity

A *jump cut* is an error in the sequence of shots that occurs when cutting between similar sized camera shots of the same object. A jump cut occurs when two camera shots that contain the same object or character are cut together. Because the two shots are recorded from different angles or positions, the object seems to "jump" to a different location on the screen. This error is found far too often in television programs because it is a very easy mistake to make.

To illustrate a jump cut, imagine a three-shot has been set up in the studio, **Figure 19-6**. The staging has Chris, Renna, and Dave standing in a rough triangle facing each other. Chris is on screen left, Renna is in the center, and Dave is on screen right. Renna is the moderator interviewing both Chris and Dave.

- Camera 1 has a shot of Chris and Renna. In this shot, Chris is on screen left and Renna is on screen right.
- Camera 2 has a shot of all three people.
- Camera 3 has a shot of Renna and Dave. In this shot, Renna is on screen left and Dave is on screen right.

If the director cuts from camera 1 to camera 3, Renna appears to "jump" across the screen from the right to the left. This movement is very jarring to the audience.

An *error in continuity* occurs during the editing process and creates a finished product that contains physically impossible actions or items, **Figure 19-7**. A jump cut is sometimes, incorrectly, called an error in continuity; the terms are not interchangeable. Some examples of errors in continuity include a hat that disappears off someone's head from one shot to the next, wounds that look severe in one shot are almost healed the next shot, or a glass that is one-quarter full of soda in one camera angle is three-quarters full in another camera angle.

whip pan: An extremely fast camera pan. Also called a *swish pan*.

jump cut: A sequence of shots that constitutes an error in editing. This error can occur during production when cutting between similar sized camera shots of the same object or during post-production when shots are edited together. The result is an on-screen object or character that appears to jump from one position to another.

error in continuity: An error that occurs during editing where a sequence of shots in the finished product contains physically impossible actions or items.

ASSISTANT ACTIVITY

Errors in continuity constitute many of the mistakes found in films. Make a list of some errors in continuity you have noticed in films and television programs. Visit www.moviemistakes.com and investigate mistakes in some of your favorite movies. Be prepared to discuss your findings in class.

Figure 19-6. In a jump cut, a performer appears to jump across the screen when the image displayed cuts from one camera to another.

   The audience rarely sees the plates of food in a shot with performers dining. The camera shoots the talent eating, but the plate is not usually in the picture. Most often, the audience is shown before and after shots of the plate. If, for example, peas were on a plate, it would be very difficult to place each pea in the same position on the plate for every take. If the peas are not placed exactly as they were in the previous take, it would be comical to see them appear in another location on the plate.

Figure 19-7. The change in this woman's earrings from one shot to the next is an example of an error in continuity. A—In the first over-the-shoulder shot, the woman is wearing small, star-shaped earrings. B—In the next shot (a reverse angle of the same shot), the earrings are long and dangling.

In a scene that takes place in a restaurant, keeping track of the amount of liquid in each glass from take to take would be a very tedious job. For example, a woman in the scene is drinking soda from a glass that is one-quarter full. It would be very noticeable if the camera cut to a different angle of the woman drinking and the glass was three-quarters full of soda. This is an error in continuity that can easily occur when dealing with multiple takes of a scene.

On large shoots, a person in charge of continuity constantly snaps photographs of all the on-screen elements in a shot. The next time the scene is shot, everything is returned to its exact position.

Errors in continuity can occur with dialogue, as well. This happens when the dialogue in multiple takes of a scene does not match exactly. Picture a scene in which a man gets out of the car in his driveway and shouts a greeting to his next-door neighbor, who is watering the front yard lawn. He shuts the car door and walks over to talk to his neighbor. In the first take of this scene, the man opens the car door, stands up, and calls out to his neighbor. He then shuts the car door and walks over to his neighbor's front yard. In the second take of the scene, the man gets out of the car, shuts the door, and *then* calls out to his neighbor as he walks toward the neighbor's yard. When the editor tries to cut these two takes together, either the car door will be shut twice or the greeting will be called out twice. The performers in a scene must perform the dialogue and action exactly the same way in every take.

Both jump cuts and errors in continuity are not noticeable until the editing phase of post-production. At that point, it is often too late to reshoot a scene correctly. These editing issues can be avoided by carefully planning the production staging and camerawork before production and proper execution during production.

VISUALIZE THIS

Picture a scene of a man and a woman on a date. The woman is wearing a fancy scarf that is daintily arranged on her shoulders and clipped-on with an elaborate pin. The scene is shot near the end of the shooting day. The director sends everyone home and reviews the recording. He determines that he needs at least one other take of the same scene to provide greater variety of shot choices later in editing. Therefore, the second take of the scene occurs the next day. What are the chances that the scarf can be arranged exactly as it was 24 hours ago with the pin placed in the same position as the previous day? What if the woman is wearing different sunglasses? Imagine the audience's confusion if the woman gets into a car wearing black-rimmed sunglasses with dark lenses and the scene cuts to a shot of her turning the key in the ignition wearing white-rimmed sunglasses with mirrored lenses!

Production Equipment in the Shot

In dramatic programming, it is not acceptable to see production equipment in a shot. The program's director tries to simulate real life, and production equipment does not surround most of us in everyday life. Watchful viewers may, from time to time, see a microphone on a boom accidentally dipping into the picture from the top of the screen or catch the shadow of a boom on the background flats. Microphones and cameras should never be seen in a completed dramatic program.

There is, however, an exception. If the dramatic program is set in an environment that naturally includes production equipment, seeing these items in a shot is acceptable. A television drama about a television station would surely have equipment present on the set. The equipment seen on screen is, most likely, props or set dressing, not functioning equipment. If the audience sees production gear in non-dramatic programming, such as news programs, talk shows, and game shows, it is perfectly acceptable. Viewers are accustomed to seeing cameras, mics, and lights on these types of programs.

Talent Placement

All cultures have acceptable "bubbles" of personal space. In certain Far Eastern and some European countries, for example, having virtually no personal space is the cultural norm. Western cultures presume a larger "bubble" of personal space. We are very comfortable in our bubble of personal space and become uncomfortable when others enter this space uninvited. This space is smaller, or almost nonexistent, with close friendships and romances. Mere acquaintances, however, are not usually welcome within our personal space.

When performing on television, people must be positioned much closer to each other than is considered normal in Western culture. If performers were spaced apart as they are in real life, the distance would appear much greater on the television screen. Additionally, the empty space between the performers would be more prominent than the performers themselves. Therefore, all performers on television must adjust their personal space and understand they will be positioned very close to others, **Figure 19-8**. On any dramatic television program, notice just how close the performers actually are to each other. Because it appears so natural on the television screen, most viewers are unaware of it. The lack of personal space is a bit of a shock to an actor performing in front of television cameras for the first time because it does not seem natural to be so close to another person while speaking.

The ultimate goal is a good television picture. Everyone on the production team must contribute to this effort and realize that things that look or feel a bit unusual in real life, often make a good picture on camera.

Figure 19-8. Personal space must be reduced in order to obtain acceptable television staging. A—Actors are positioned to create the appearance of realistic distance for conversation. B—In reality, the actors are much closer than is commonly considered comfortable in Western culture.

PRODUCTION NOTE

The director must resist the urge to take performers by the hand and place them in the appropriate locations on the set. It is considered amateur behavior for a director to touch the talent in this manner. Because the director is often out of the talent's direct line of sight, using hand gestures is not effective. The director must maneuver the talent into position using only verbal direction.

VISUALIZE THIS

When positioning seated talent to capture a tight two shot, using only verbal commands can become challenging. The talent's knees seem to be in the way no matter how they move, **Figure A**. The solution is to remove personal space and have the talent tuck their inside feet under the chair behind their outside feet, **Figure B**. This allows their hips to move much closer together. The talent will look ridiculous in reality, but the audience will not know how contorted their bodies are in a medium close-up. The top half of **Figure B** is a common shot on television. When properly positioned, place a table in front of the talent with the corner of the table between them. This places the two participants on different, but adjoining, sides of the table, **Figure C**. When performers lean in and place their arms naturally on the table, they reduce the space between them even more. The overall appearance is normal. However, their feet are still contorted. The audience will not see what you do not show them on camera.

Figure A

Figure B

Figure C

Interviews

We usually think of an interview as a conversation or a question-and-answer session between two people. However, an interview can be performed with only the interviewee featured on-camera. The interview questions must be written in a way that prompts the interviewee to talk naturally about a topic. Additionally, the interviewee is asked to restate the question in the answer, which eliminates the need to see or hear the interviewer at all. This kind of interview requires a considerable amount of B-roll footage because the interviewee's responses will undoubtedly be edited and trimmed to keep the audio track interesting for viewers. Every edit must be "cushioned" with a B-roll shot to prevent jump cuts.

Staging for a Two-Person Interview

When positioning talent for a two-person interview, they should not be placed facing each other. The front of each chair should be slightly angled toward each other, **Figure 19-9**. If the angled lines of the chairs were extended, they would form an inverted "V." This staging opens the area downstage of the talent, which makes the audience feel like they are part of the conversation, **Figure 19-10**. If three people were standing and having a conversation in real life, the participants would be naturally positioned in some form of a triangle. This arrangement makes each person in the conversation feel involved.

Figure 19-9. The angled placement of on-screen talent creates an inverted "V."

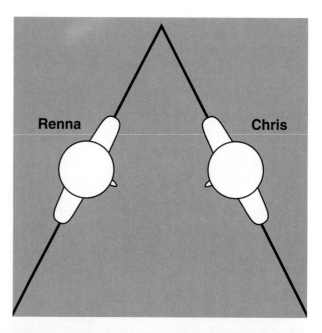

Figure 19-10. The inverted "V" staging leaves room for the audience to feel like they are a part of the conversation.

Think about almost every instance you have seen on television where people are seated around the dinner table at someone's home. To include the viewer as a participant at the table, all the actors are positioned around one half of the dinner table, **Figure 19-11**. This seating arrangement is unnatural in real life. However, leaving half of the table empty creates the illusion that the audience is seated at the open side of the table, downstage, and can see everyone at the table.

Staging for a Three-Person Interview

If a third person named Agnes were to join the interview with Renna and Chris, their chairs are shifted to place Chris at the apex of the triangle with Renna and Agnes on either side, **Figure 19-12**. The bottom of the

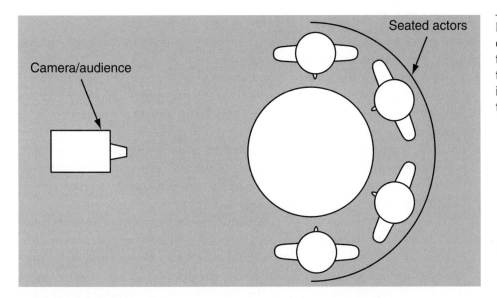

Figure 19-11. The downstage end of the table is left empty so that the camera (audience) is included as a participant at the table.

Figure 19-12. When three people are positioned in the inverted "V" arrangement, the middle person is usually at the apex of the triangle.

triangle remains open for the audience. Camera 2 has a one-shot of Chris and a three-shot of all the participants. Cameras 1 and 3 can alternate between one-shots of Renna or Agnes and two shots that include Chris.

Using the three-person interview scenario that includes Agnes, Renna, and Chris, cameras 1 and 3 can each capture a two shot. Cutting between the two shots, however, may create jump cuts because each of the two shots contains one common person—Chris. Cutting between the two shots results in Chris jumping from one side of the screen to the other. The solution is to assign either camera 1 or camera 2 a two shot that includes Chris, but not both camera 1 and camera 2. When marking the script in pre-production, camera shots must be thoroughly thought out to avoid jump cuts in the studio or in the editing room.

Dramatic Programming

In most dramatic programming, the talent cannot look directly at the camera. The exceptions to this are:

- If the camera is used as a subjective camera. Because the camera is shooting from the viewpoint of one of the program's characters, it is natural for other cast members to look directly at the character/camera.

- In the case of a dramatic aside. A *dramatic aside* occurs when a performer steps out of character, faces the audience, and directly addresses the audience. Dramatic asides are not regularly used in dramatic television programming.

Performers *ad-lib* when they deliver lines or perform actions that are not in the script or have not been rehearsed. Ad-libbing on a dramatic program can be a disaster. This may accidentally happen during a stage performance without great detriment to the production. A stage actor must eventually return to the script, so the other actors can proceed with their lines. On television, however, the script involves more than just the actor's dialogue. The technical director follows the script exactly and uses certain words as cues to cut to shots from different cameras. If a particular cue word or line is not delivered, the TD does not cut to the scripted shot, **Figure 19-13**. Ad-libbing is highly frowned upon.

dramatic aside: A performance technique; when a performer steps out of character and directly addresses the audience.

ad-lib: A performance technique; when talent speaks lines or performs actions that are not in the script or have not been rehearsed.

PRODUCTION NOTE

In television, the director *has* to know what is going to happen *before* it happens so a camera is positioned to shoot when it *does* happen. Likewise, the camera operators are not able to follow the shot sheets if a performer deviates from the script. Once a script has been finalized during the camera rehearsal or dry run, it is in the best interest of the entire production that *everyone* follow the script during the shoot.

Non-Dramatic Programming

In non-dramatic programming—such as game shows, news programs, documentaries, sports programs, and talk shows—talent may look at the

Figure 19-13. The technical director and camera operators are caught unprepared when actors suddenly start speaking lines not in the script. The crew cannot follow action that has not been planned and rehearsed.

camera at any time because addressing the audience is the very nature of the program. For example, a news anchor tells the audience about current events and, therefore, looks directly at the lens of the camera to speak to the audience. If the talent looks away from the camera in non-dramatic programs, the audience wonders what is happening and becomes distracted and frustrated.

VISUALIZE THIS

Imagine watching a network news broadcast and the anchor turns his attention to something on the ceiling of the studio. The news continues, but the camera does not tilt up to show you what he is looking at. Most likely, you will stop listening to the news being reported and wonder what the anchor is looking at.

In non-dramatic programming, the talent should directly address the audience/television camera. There are two exceptions to this rule:
- Talent may look down at their notes while addressing the audience.
- An anchor may look to the side at a co-anchor, but only if the co-anchor is included in a two shot that immediately follows.

Unlike the talent in dramatic programs, the on-screen participants of non-dramatic programs, such as news anchors and talk shows hosts, commonly read their lines from a teleprompter, **Figure 19-14**. A *teleprompter* is a computer screen positioned below the camera lens. A one-way, or half-silvered, mirror is placed in front of the camera lens at a 45° angle, pointed at the computer screen. The text on the computer screen is reflected in the mirror and read aloud by the talent, which allows the talent to look directly at the lens of the camera. The camera shoots right through the one-way mirror at the end of the lens, without seeing the dialogue displayed. If the camera zooms in too closely on the talent, the audience can see the talent's

teleprompter: A computer screen positioned in front of the camera lens that displays dialogue text in large type. This allows the talent to look directly at the lens of the camera and read the text.

Figure 19-14. A teleprompter allows the talent to look directly into the lens of the camera, at the viewing audience, while reading the program's script.

Figure 19-14. A teleprompter allows the talent to look directly into the lens of the camera, at the viewing audience, while reading the program's script.

eyes moving from left to right as they read the teleprompter. Because of this, the tightest shot captured of anchors on news programming is usually between a mid-shot and a medium close-up.

It is quite common to see a small pile of papers on the desk in front of news anchors. Those papers may actually be props that provide something for the talent to grasp so their hands do not move around. Many people use their hands when they speak, but this is very distracting to a television audience. The audience commonly assumes that the papers are a script, so it does not appear to be out of place.

Ad-libbing is more common in non-dramatic programming. In some formats, like talk shows, very few things are actually scripted. A comment from an audience member or show guest may provoke an unplanned course of discussion. This does not cause a great disruption, as in dramatic programming, because the talent is usually stationary. Additionally, an experienced TD can easily follow the conversation.

Staff and Talent Interaction

The interaction of production personnel and talent affects the success of the production process and is an important topic when learning television production. Every member of the production staff has the opportunity to interact with the program's talent at some point in the production process. Just as in other workplaces, professional behavior in a studio can be friendly, even jovial at times. When guests are present in the studio, however, more serious and professional behavior is necessary.

Managing Guest Talent

When guest talent enters a television studio, they are naturally uncomfortable. The environment is strange to them and they know very little about the activities of others around them. They see many strangers bustling around in semidarkness and probably have some anxiety about being placed under bright lights in front of untold numbers of people. A nervous guest will not look good on camera and does not contribute to a successful production. To help guest talent relax:

- Prepare guest talent for the experience before they arrive at the studio. Explain what they should expect when they arrive, provide suggestions for clothing and makeup selection, and offer some information on what is expected from them during the production process. Refer to the Talent Information Sheet in **Appendix B**.
- Designate a staff member to greet the guest upon arrival and be their friendly guide during the production process.
- The greeter should introduce the guest to the director, offer some refreshments, and keep the guest talking.
- A tour of the facility, including an explanation of the various activities, helps ease the guest's anxiety. For example, allowing guests to observe the editing process may help distract them from their nervousness.
- The greeter may take the time to introduce the guest to some of the crew members. Everyone on the production team is responsible for making the talent comfortable to produce a good program.
- The greeter should try to answer all of the guest's questions.
- Guest talent should not be placed under the studio lights until the program is ready to begin. Under the bright lights, they are not able to see anything in the studio. They can only hear the surrounding activity and will wonder what is happening. Keep the talent informed to ease their anxiety.
- Explain to the guest that no one will see mistakes; editors take out the errors. If mistakes occur, you can fix them.
- Schedule extra time for rehearsal so that guests do not feel rushed or overwhelmed—plan three times more rehearsal time than typical rehearsals.

Working with Non-Professional Talent

    Production personnel may often work with non-professional talent
and must be prepared to react and compensate for non-professional talent's actions. Camera operators must be especially alert when working
with non-professional talent. Non-professional talent may move unpredict-
ably, such as a sudden move left or right, unlike the scripted and rehearsed
movements of professional talent. If the camera operator is not alert or
the shot is too tight, the talent may completely leave the frame before the
camera operator has time to react, **Figure 19-15**. To compensate for this, a
camera operator should never have a shot tighter than a medium close-up
of non-professional talent standing in a shot.

    Professional talent always provides a cue to the camera operators
when they are about to move. When standing, professional talent shifts
their weight to one foot and turns their body in the direction they are about

Figure 19-15. Camera operators must always be alert to sudden movements made by non-professional talent while on screen. A—As long as the talent follows the script, the camera operator can maintain a good shot. B—If the talent makes an unplanned movement, like standing up quickly, the resulting image is worthless to the production.

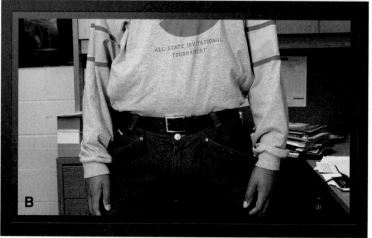

to move before actually moving. This gives the camera operator time to adjust and follow them. From a seated position, a professional performer leans forward, perhaps placing their hands on the desk to push up, and smoothly rises from the chair. Experienced talent will not abruptly spring from the chair, like a jack-in-the-box.

Headphone Etiquette

In the studio, the floor manager and camera operators wear headphones to communicate with the director and other control room personnel. During production, no one wearing headphones should laugh at any time. In a studio setting where the talent is not able to see the staff, sudden laughter from an unknown source is very unsettling to an already nervous guest. Their first thought is usually, "They are laughing at me." Even though this is probably not the case, the guest is already extremely self-conscious in an unfamiliar environment.

The volume of the headphones is also a consideration. Some operators have the volume of their headphones set so loud that others can actually hear what the director is saying. This sound is not usually picked up by the mics, but talent can hear it. The barely audible sound of someone speaking is very distracting to talent, especially to non-professional talent. Ensure that headphone volume controls are set appropriately for the studio environment.

Wrapping Up

A majority of the previous chapters focused on explaining the equipment used and the responsibilities of personnel on a production. This chapter addresses what to do with the people and objects in front of the camera. The placement of visual elements in the picture is called staging. Correctly staging a set adds to the visual appeal and realism of a program. Successfully managing guest talent and keeping them relaxed during production also improves the visual appeal of a program. When talent is nervous or anxious they may fidget, sweat, or shake while on the set. This is not the best portrayal of the talent on the television screen. Even if all other production guidelines are followed, neglecting staging techniques and guidelines will result in a program that resembles a common home video.

Review Questions

Please answer the following questions on a separate sheet of paper. Do not write in this book.

1. Define *foreground*, *middle ground*, and *background*.

2. In a dramatic program, production equipment should not be seen in a shot. What is the exception to this?

3. What is cross-camera shooting? What problem does it solve?

4. What problem is created when a camera shot crosses the vector line? What is the exception to this rule?

5. What is a cutaway? How is it most commonly used?

6. What is the difference between a jump cut and an error in continuity?

7. How does the concept of personal space change when performing on television?

8. Which production staff members are affected when a television actor ad-libs?

9. List six things that studio or production personnel can do to help guest talent relax before shooting begins.

10. What must camera operators be prepared to handle when working with non-professional talent?

Activities

1. Create an illustration that depicts the best placement of talent and equipment on a set for an interview program with an interviewer and two guests.

2. Research some of the differences between stage acting and television acting. Summarize some of the differences and be prepared to discuss them in class. Some topics may involve gestures, projection, movement, and memorization of lines.

3. While watching any half-hour sitcom, pay particular attention to jump cuts or errors in continuity. Note each instance that you find and be prepared to discuss them in class.

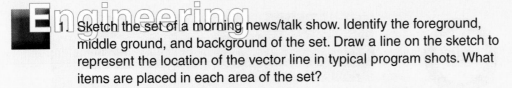

STEM and Academic Activities

Engineering

1. Sketch the set of a morning news/talk show. Identify the foreground, middle ground, and background of the set. Draw a line on the sketch to represent the location of the vector line in typical program shots. What items are placed in each area of the set?

Language Arts

2. Directors must maneuver talent into position using *only* verbal direction. Try positioning two friends or family members into the proper staging for a two-person interview using only your words—no hand gestures and no physical contact.

Social Science

3. Review a television programming schedule and make a note of the programs in which talent needs to ad-lib. Explain why the talent on these programs ad-libs.

www.sbe.org The Society of Broadcast Engineers provides a forum for the exchange of ideas and the sharing of information to help professionals keep pace with the rapid changes in the broadcast industry.

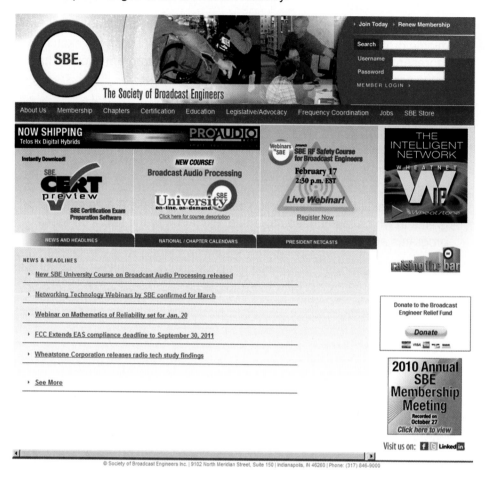

Chapter 20

Directing

Objectives

After completing this chapter, you will be able to:

- Recall the types of script breakdowns and identify the information included in each.
- Summarize the director's responsibilities in each phase of production.
- Explain the importance of marking the script when shooting on location.
- Identify some qualities common to effective directors.

Professional Terms

audition
camera rehearsal
cast breakdown by scene
countdown
cut
dry run
equipment breakdown
location breakdown
prop list

prop plot
scene breakdown by cast
script breakdown
shooting for the edit
shot log
slate
take
take log

Introduction

This chapter focuses on the director's activities and responsibilities during program production. Some of the topics presented apply to both studio and remote shoots, while others apply only to one type of shoot. Directing is often perceived as the most exciting, high-profile job in the television industry. It is also one of the most difficult jobs because the director is involved in each phase of production. Before any shooting can take place, a tremendous amount of pre-production work and planning must be completed. During production shooting, the director must coordinate the activities of the crew and talent, determine when sufficient takes have been recorded, and keep an organized record of the scenes and transitions. Even in the editing room, during post-production, the director works to ensure that the best possible program is produced.

In large-scale productions, many of the director's functions are actually assigned to several assistants working for the director. In small-scale and academic production environments, however, the director may be responsible for all the program production tasks presented in this chapter.

The Director's Role in Pre-Production

During the pre-production phase, the director's responsibilities include script breakdowns, marking the script, auditions, and pre-production meetings with the staff and crew. Organization is the absolute key to directing. It is not possible to be a successful director with poor organizational skills. A director must be willing and able to make things happen, either by doing things independently or by delegating tasks to competent coworkers. Program directors must understand that the responsibilities of an entire production cannot be accomplished by one person—taking on too much results in mediocrity. Television broadcasting is a team activity.

Script Breakdowns

Before script breakdowns can begin, several tasks must be completed:
- The program proposal must be approved by the executive producer.
- The outline must be approved.
- The script must be written.
- Locations must be scouted.

script breakdown: The process of analyzing a program's script from many different perspectives.

A *script breakdown* is the process of analyzing the script from many different perspectives. This process results in a production that is well organized and efficient. Once each breakdown is completed, the director can confidently answer production questions and develop a realistic production schedule. Professionals develop many different types of breakdowns as part of their pre-production procedure.

Prop List

prop list: A list of each prop needed for a production.

A *prop list* is developed by reading through the script and making a list of each prop that is referenced. The director should visualize how the shots will be staged and note any additional props necessary. The completed prop list becomes a shopping list for all the production's props, **Figure 20-1**.

Prop Plot

prop plot: A list of all the props used in a program sorted by scene.

As soon as the prop list is complete, the *prop plot* should be developed. The prop plot is a more involved list of all the props, sorted by each scene of the program, **Figure 20-2**. If a prop is needed for multiple scenes, it is listed separately for each applicable scene. The prop plot is longer than the prop list because items are listed more than once. The prop plot is used to access the props once they are obtained.

VISUALIZE THIS

To emphasize the importance of a prop plot, consider that all the props have been acquired for a large production and packed into two tractor-trailer trucks. The day for shooting scene 38 arrives. Props needed for the scene include general office supplies and a treasure map. The shoot location is four miles from the studio. It is not practical to drive two tractor-trailer trucks to the location, in addition to the necessary equipment, crew, talent, etc. By reviewing the prop plot, only the props needed for scene 38 can be gathered and transported.

Prop List

Pens
Paper
Coffee Mug
Date Book
Contract
Letter Opener
Desk Phone
Cell Phone
Place Setting (2)
Salt and Pepper Shakers
Napkins
Silverware Settings (2)
Small Vase with Flowers
Sugar and Cream Dispenser
Water Skis
Tow Rope
Life Jackets (4)
Cooler

Figure 20-1. A prop list is essentially a shopping list of props for the production.

Prop Plot

Scene 4

Pens
Paper
Coffee Mug
Date Book
Contract
Letter Opener
Desk Phone
Cell Phone

Scene 5

Contract
Cell Phone
Place Setting (2)
Salt and Pepper Shakers
Napkins
Silverware Settings (2)
Small Vase with Flowers
Sugar and Cream Dispenser

Scene 6

Cell Phone
Water Skis
Tow Rope
Life Jackets (4)
Cooler

Figure 20-2. A prop plot sorts the prop list by scene number.

Location Breakdown

location breakdown:
A list of each location
included in a program
with the corresponding
scene numbers that
take place at that
location.

A *location breakdown* is a list of each location included in the program. Listed next to each location are the scene numbers that take place at that location, **Figure 20-3**. Organizing the information in this way further assists the director and crew in scheduling resources and general time management.

Cast Breakdown by Scene

**cast breakdown by
scene:** A list of the
program's cast members
with the corresponding
scene numbers in which
they appear.

The *cast breakdown by scene* is very similar to a location breakdown, **Figure 20-4**. A cast breakdown by scene lists cast members in the left column on a page, with the scenes in which they appear in the right column of the page. The cast receives a copy of the cast breakdown by scene to use in conjunction with the production calendar. The production calendar indicates the scene numbers that will be shot each day. The cast members reference the cast breakdown by scene so they know which days they need to be on the set.

Scene Breakdown by Cast

**scene breakdown
by cast:** A list of each
scene number in a
program with all the
cast members needed
for each scene.

The *scene breakdown by cast* is a list of each scene number in the left-hand column of a page with all the cast members needed for that scene in the right-hand column, **Figure 20-5**. The production staff—primarily the

Figure 20-3. All of the scenes to be shot at each location are listed on a location breakdown.

Location Breakdown

Park:

Scenes 3, 9, 17, 30, 49

Apartment:

Scenes 5, 14, 28, 35, 36, 37

Office:

Scenes 1, 4, 10, 13, 20

Figure 20-4. A cast breakdown by scene lists each cast member with the scenes in which they appear.

Cast Breakdown by Scene

John: 2, 5, 6, 7, 12, 14

Mary: 2, 4, 5, 6, 9, 10

Eric: 1, 3, 15

Alex: 1, 3, 15

Susan: 4, 8, 10, 11

Extras: 13, 15

Scene Breakdown by Cast

1	–	Eric, Alex
2	–	John, Mary
3	–	Eric, Alex
4	–	Mary, Susan
5	–	John, Mary
6	–	John, Mary
7	–	John
8	–	Susan
9	–	Mary
10	–	Mary, Susan
11	–	Susan
12	–	John
13	–	Extras
14	–	John
15	–	Eric, Alex, Extras

Figure 20-5. A scene breakdown by cast is a listing of each scene number with all of the cast members appearing in that scene.

director, assistant director, makeup artist, and costumer—use the scene breakdown by cast to ensure all of the necessary cast is present when shooting each scene. The production assistant typically uses the scene breakdown by cast to contact performers to remind them when and where to be for the next day's shoot.

Equipment Breakdown

An *equipment breakdown* lists each scene with all the equipment needed to shoot that scene, **Figure 20-6**. The equipment breakdown benefits the production in several ways:

- With the equipment organized in this manner, it is very unlikely that something will be forgotten when shooting outside of the studio. Gathering equipment without a checklist increases the chances of crucial items being left behind.
- An equipment breakdown checklist alleviates the confusion that occurs when more than one person packs the gear. Each person may assume that the other packed a certain item or two. While setting up equipment on location, 50 miles from the studio, is an unfortunate time to discover that both people thought the other had packed an essential piece of equipment, such as the camera. No matter how many people help pack the gear, use one checklist and mark off each item as it is packed.
- An equipment breakdown ensures that unnecessary equipment is not transported to a location, and that excess crew members are not scheduled for a shoot.

equipment breakdown: A list of each scene in a program with all the equipment needed to shoot each scene.

Figure 20-6. Listed in this breakdown is all the equipment needed for a two-camera, film style shoot.

Equipment Breakdown for Scene 12:

- ☑ 2 Camcorders
- ☑ 2 Tripods
- ☐ 2 Power supplies (for cameras)
- ☐ 8 Mic cables
- ☐ 2 Lapel mics
- ☐ 2 Boundary mics
- ☐ 4-input Mic mixer
- ☑ 4 Light kits
- ☑ 4 Extension power cables
- ☑ 2 Multiple outlet strips
- ☑ 2 Monitors
- ☑ 2 Male BNC to male BNC cables
- ☐ 2 Double male RCA to male RCA cables
- ☑ Videotape
- ☑ Duct tape
- ☑ Tool kit
- ☑ Lens cleaning paper
- ☑ 4 Camera batteries
- ☑ 2 Battery chargers

PRODUCTION NOTE

Students frequently ask if all these breakdowns are really necessary.

- Preparing all the breakdowns forces you to look at the script analytically. You become extremely familiar with the production, which enhances your leadership abilities.
- When directing your first production, the value of these breakdowns will become crystal clear. Without them, you will be directing a disorganized mess and you will risk losing the confidence of your crew. The breakdowns are part of the meticulous organization required by the director of any production.
- In the professional world, time is money. Regularly using breakdowns saves time and, therefore, money.

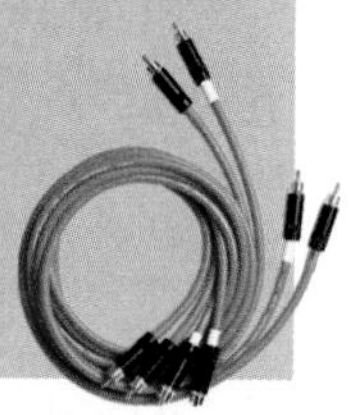

Marking the Script

If a production is planned using film-style shooting (see Chapter 17, *Remote Shooting*), the director must carefully mark the script with scene and shot numbers to help later in the editing room. For example, if scene 3 of a program is set for shooting, each planned shot of the scene needs to be labeled 3a, 3b, 3c, 3d, and so on. These labels are written into the shooting script prior to shooting the scene. As each shot is recorded, it is checked off the list. This ensures that a shot is not accidentally omitted, whether shooting in the studio or on location. Returning to the set days or weeks later to record a shot that was accidentally omitted may be unreasonably expensive.

Camera Shots

The director, instead of the scriptwriter, often enters all the camera shots on the shooting script. The script is clearly marked with abbreviations understood by all the crew members. A detailed and complete camera script aids in a smooth production shoot. (See script samples in Chapter 8, *Scriptwriting*.)

Shots should vary considerably, using both horizontal and vertical angles and a great variety of shot sizes. Novice directors often place cameras at uninteresting angles, such as directly in front of talent at eye level. Using a wide variety of angles and sizes keeps the audience interested, even if the program's subject matter is relatively unexciting. Cutaways and reaction shots should be noted on the camera script, in addition to primary action shots.

Shooting for the Edit

A director has very clearly defined duties in planning the production of a program. All the time and work spent planning each shot allows the program to be edited efficiently and effectively. Planning is a critical element in creating a professional video program.

PRODUCTION NOTE

Inexperienced directors often cut corners in the planning process. As a result, the majority of editing in their productions involves removing bad things that happened while shooting, such as the bad takes, bad shots, bad performers' mistakes, bad errors in continuity, etc. In this case, the director must settle for a sequence of the best of many bad takes—whatever is left over becomes the program.

Experienced directors plan things obsessively with several backup plans. As a result, there are many, many choices of good shots to use while editing the program. There may actually be enough footage to put together a sequence several different ways. With a variety of good shots, a director has the freedom to truly create a program.

The director must plan exactly how the scene being shot will transition from the scene that immediately precedes and follows it. This process is called *shooting for the edit*. Before shooting scene 9, for example, the director needs to know how scene 8 ends (fade, cut, dissolve, or other effect), as well as the direction of the action at the end of scene 8. This information dictates the beginning of the shoot for scene 9. The two scenes must be edited together later without jarring the audience. The director must also be concerned with how scene 9 leads into scene 10. Planning these transitions would be easily accomplished if scenes were shot in the order they appear in a finished program. This, however, is not the reality of television production shooting.

shooting for the edit: The process in which a director plans exactly how each scene of a program will transition from the scenes that immediately precede it and to the scenes that follow it. Production shooting then follows the director's plan.

It is very likely that scenes will not be shot in the order the audience sees in the final program. Therefore, before the first day of shooting, the director needs to mark the entire script noting how each scene transitions out of the previous scene and into the next scene. The screen direction of the action and camerawork should also be noted at the beginning and end of each scene. With this information marked for every scene, the director can confidently shoot the scenes in any order and know that, if the script notations are followed, the scene will be smoothly edited into the program.

PRODUCTION NOTE

The director may try making a change in a scene during production, as long as the written script is followed. The director has thoroughly thought through the "as written" version of the script and knows if the change will work in the scene. Shoot the scene as written in the script, then shoot the scene with the director's change. If the changed scene does not work out, the original version of the scene has already been recorded.

Another important aspect of shooting for the edit is to consider the four basic rules of action on the screen. Breaking any of these rules is guaranteed to jar the audience, which is *usually* a negative program attribute, depending on the nature of the program. The following rules apply to the processes in the editing room:

- A stationary camera shot should not immediately precede a moving camera shot.
- A moving camera shot should not immediately precede a stationary camera shot.
- An edit can be made from a stationary camera shot to another stationary camera shot.
- An edit can be made from a moving camera shot to another moving camera shot.

Talk the Talk

The phrase "moving camera shot" refers to both a shot where there is action on the screen and a shot where the camera itself is actually moving. The phrase "stationary camera shot" refers to both a shot where there is no action on the screen and a shot where the camera itself is not moving.

Set Design

After both the director and producer approve the set design, it should be studied in detail, **Figure 20-7**. The director visualizes where the cameras can be placed and where the action will take place. Camera placement needs to be determined before set construction begins. Many sets are built with removable walls to allow room for specific camera positioning. If alterations to the set are necessary, it is easier and more efficient to make changes to the plan than to make changes after a set has been built.

Because individual designers for the production may not be in communication with each other, the director is the final authority on the colors used on the set. The director must consider the colors with respect to makeup, costumes, set pieces, furniture, and lighting. Each of these elements factor into the contrast ratio and contribute to the visual quality of a production.

Auditions

Whether the program is a drama, comedy, game show, documentary, or a news program, the director needs to decide who to put in front of the camera. This process is called the *audition*. During an audition, the director watches and listens to a performer and decides if the performer is capable of portraying the role needed in the program. The director must try to be extremely objective when auditioning talent.

For the initial audition, directors often listen to the quality of the talent's voice without visually observing the talent. The talent may be asked to read an excerpt from the script, a sample news story, or a prepared monologue. During a visual audition, the director may test the talent's ability to follow stage directions, in addition to another reading from the script (**Figure 20-8**). Unlike performance theater, the director and talent should not be in the same room for an audition. The director should watch the aspiring cast member on a monitor in the control room—much the same way that television viewers will see the performer. The performer's "look" on screen is extremely important to the director.

audition: The process in which a director makes casting decisions for a program by watching and listening to prospective performers.

PRODUCTION NOTE

One of the biggest mistakes made in the audition process is casting a good friend. A friend is less likely to follow direction because of the familiar relationship outside of the studio. This may lead to arguments, which could result in a poor finished program or the loss of a friendship.

Directors sometimes forget that they are "stuck" with their casting decisions. Once shooting begins, a cast member can be fired, if necessary. However, depending on the program type, each scene that the cast member was in may need to be reshot with a replacement performer.

When casting decisions are made, the producer must obtain a talent release for each person appearing on camera, unless they are included in a news program, are in the background of a shot, or in a crowd. (For details on talent releases, see Chapter 12, *Legalities: Releases, Copyright, and Forums*.) Always obtain a talent release *before* recording video. If a talent release is not signed, the footage cannot be used.

Pre-Production Meeting

A pre-production meeting includes every member of the staff. At this meeting, the expectations for each member of the staff are discussed and everyone collaborates to develop a production schedule/calendar. Because so many different personal and professional schedules are combined in the production process, a commitment to the production calendar is taken very seriously. Neglecting a scheduled commitment is irresponsible—the negative impact on the entire production team will leave a lasting impression.

The equipment needed for shoots must be scheduled. The director must make arrangements for the necessary equipment to be available and transported to the location, if necessary, at the scheduled time. This requires that the equipment be reserved well ahead of time. Most studios allow equipment to be reserved up to two weeks before the shoot date.

Depending on the type of program, the director may hold rehearsals with the cast before production begins. As the cast becomes proficient with the script, the crew begins attending rehearsals. The cast and crew need to know what is going to happen in a scene before it happens to ensure a camera is positioned to record the shot. This requires tremendous coordination between the director, talent, technical director, and camera operators.

The Director's Role in Production

Before a studio shoot begins, the camera script is reviewed and final camera directions are noted in the left column of the script. Transitional devices for the beginning of scenes and the end of scenes are indicated, as well as a variety of shots from many angles.

The *dry run*, or *camera rehearsal*, includes only the director, talent, technical director, audio engineer, and camera operators. Costumes and makeup are not worn and nothing is recorded. To save money and keep the heat down, the studio lighting instruments are not turned on. The talent goes through the scenes and the camera operators are given directions. All shots are rehearsed, along with the audio cues, while the technical director practices the camera switching, **Figure 20-9**. After the director is satisfied with everyone's performance, he calls for the actual shoot to begin. While the performers get into costume and makeup, the crew readies the set, lights, and other equipment.

During the shoot, it is important that the director always use correct terminology. The correct use of terms fosters the crew's confidence in the director. Using correct terminology is also the most efficient method of communicating. Many directors memorize a start-up sequence of commands to get the program started smoothly, **Figure 20-10**. Aspiring directors should memorize a standard start-up dialogue. On the day of a shoot, novice directors often get nervous or stressed and may forget a crucial command, such as "begin recording." Committing a start-up sequence to memory decreases the likelihood of this happening.

dry run: A program rehearsal session that includes the talent, technical director, audio engineer, camera operators, and director. Also called a *camera rehearsal*.

Figure 20-9. A dry run, or camera rehearsal, is necessary to coordinate the technical production staff with the movements of the performers.

Figure 20-10. An example of a director's start-up sequence of commands.

Start-up Sequence

DIRECTOR: Studio ready?
FLOOR MANAGER: Ready.
DIRECTOR: Standby.
FLOOR MANAGER: Standby.
DIRECTOR: Audio ready?
AUDIO ENGINEER: Ready.
DIRECTOR: Video ready?
VIDEO OPERATOR: Ready.
DIRECTOR: CG ready?
CG OPERATOR: Ready.
DIRECTOR: Standby control room.
DIRECTOR: Begin recording.
VIDEO OPERATOR: (begins recording) Recording.
DIRECTOR: Countdown.
FLOOR MANAGER: 10, 9, 8, 7, 6, (5–0 are counted off with his fingers. His hand is in front of the talent with the first line).
(When the FM gets to 2, the Director speaks again).
DIRECTOR: Fade up on (camera) 1.
DIRECTOR: Audio up.
FLOOR MANAGER: (gesture for zero).
DIRECTOR: Bring in title.
DIRECTOR: Lose title.

Multiple Takes

take: Each recording of an individual scene. The take number increases by *1* each time an individual scene is shot.

cut: A command given by the director that indicates all production activty, talent performances and crew activities, should stop immediately.

Each time a scene is shot, it is called a *take*. Multiple takes of scenes may be planned to capture different angles. Multiple takes may also be necessary due to a mistake made by the talent or crew, which causes the director to yell "*Cut!*" The floor manager relays the director's commands to the studio and the performances and crew activity stop immediately.Each scene should be reshot until at least three "good" takes are recorded. The camera can then be moved to capture a couple more good takes from a different angle. All the extra footage *will* be needed in the editing process. It is better to have the choice between several good takes than to come up short in the editing room. Never move on to another scene until takes of the current scene are acceptable. When shooting a scene is complete, review the shots and takes of the scene. This additional time is well justified when compared to the alternative: gathering all the crew, talent, sets, props, and equipment for a re-shoot weeks later.

slate: A board or page that is held in front of the camera to note the scene number, the take number, and several other pieces of information about the scene being shot.

The *slate* is a board or page that is held in front of the camera to note the scene number, take number, and several other pieces of information about the scene being shot. Recording the slate provides a visual shot of the title of the scene at the beginning of the recorded footage. In post-production, the editor uses the shot of the slate to quickly locate the correct take of a scene to use in the final program. The camera should shoot the slate for at least 10 seconds, but not more than 15 seconds, before every scene. The slate is part of the head recorded for the scene. (Heads and tails are discussed in Chapter 5, *Videotape, Video Media, and Video Recorders.*) A slate can be an elaborate clapboard or a simple piece of paper, **Figure 20-11**. When multiple takes of a single scene are shot, the slate becomes particularly important. For example, the sixth take of scene 7 is slated as "Scene 7, Take 6."

After the slate is recorded, the countdown begins. *Countdown* refers to the way the floor manager initiates action on the set and cues the performers. The director gives the floor manager the "Countdown" command over the headsets. The floor manager calls, "Quiet on the set" and all sound in the studio ceases. The floor manager stands in the line of sight of the performer who has the first words or actions of the scene, and begins to countdown the seconds from "10" in a clear, loud voice. When the countdown reaches "5," the floor manager stops speaking aloud and holds up his hand with five fingers spread apart. The remaining seconds are silently counted down using the floor manager's fingers. When the countdown reaches zero, the floor manager quickly throws his arm forward and points to the first performer, who immediately begins the performance. The first five numbers of a countdown are audible so everyone in the studio can hear the rhythm of the countdown and keep pace during the last five seconds in silence. Silence during the countdown allows the audio of a scene to begin cleanly from silence.

A *take log*, or *shot log*, is a written list of each scene and take number shot and recorded on a particular tape or other recording medium, **Figure 20-12**. The director circles the acceptable takes of each scene on the log. Later in the production process, the director can scan directly to the slate shot at the beginning of the good take instead of reviewing each preceding

countdown: The procedure used by a production's floor manager to initiate action on the set and cue the performers by counting down from "10."

take log: A written list of each scene and take number that have been shot and recorded on a particular tape, disc, or other recording medium. Also called a *shot log.*

Figure 20-11. The slate can be a simple notation indicating scene and take numbers. A—Scene information can be written with a dry-erase marker on a clapboard. B—Scene and take numbers noted on a piece of paper also makes a simple and effective slate.

A   B

Figure 20-12. The director develops a take log to decrease the amount of time spent in the editing room.

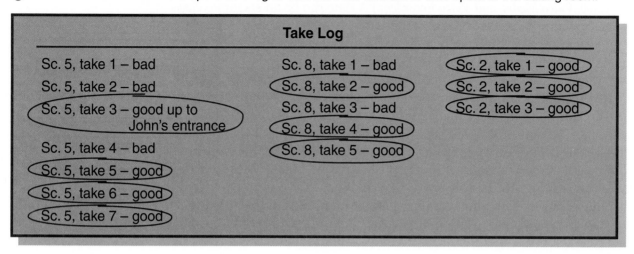

bad take. By reviewing only the good footage, the director can save hours of editing time.

PRODUCTION NOTE

Always clearly label all tapes and other video media with identifying information, such as the program title, director, and scene numbers. Imagine a full tape of footage is recorded after an entire day of shooting. The gear and equipment are shut down and stored until the next day. When shooting resumes, the tape with the first half of the program's footage is used to record the next day's shoot. Because the tape was not labeled, the operator thought it was blank and a full tape of footage was recorded over.

The Director's Role in Post-Production

After the shooting is complete, post-production begins. The director's role in post-production varies from program to program and from one production environment to another. Processes during post-production include editing, adding background music, scene transitions, sound effects, some special effects, and titles.

Talk the Talk

Post-production is most commonly referred to as "post."

Keep in mind that very few aspects of a program can be satisfactorily fixed in post. Accepting a substandard shot while shooting a scene is a terrible mistake because fixing substandard shots in post is often nearly impossible. While digital processes offer more tools that can be used in the effort to fix a shot, the task may become too time consuming and, therefore, prohibitively expensive for the production. The best practice is to plan and shoot the scenes correctly during production.

Being an Effective Director

Good directors commonly possess certain characteristics that contribute to quality and successful productions.

- A good director is not the dictator of the production.
- A good director fosters a team mentality among the cast and crew, so that everyone works toward a common goal.
- A good director does not wait to make decisions—a decision is made and the director confidently moves on.
- A good director takes initiative to do whatever is necessary to successfully complete the program.
- A good director knows the capabilities of the equipment and strives to make the most of available resources, instead of complaining about what is lacking or not available.
- A good director maintains an even temper in front of the cast or crew. Showing anger or bursts of emotion causes others to lose faith in the director's abilities.
- A good director is a "people person" and understands the benefits of saying "please" and "thank you" when communicating with the cast and crew.
- A good director gives only constructive criticism when instructing the cast and crew.
- A director is part artist and part technician. The best directors work their way up through the production team and know the job responsibilities of each production crew member. This knowledge and experience allows the director to develop effective relationships with the production staff.
- A good director delegates tasks, rather than trying to attend to everything personally.
- A good director is highly organized and efficient, almost to a fault.
- A good director takes responsibility for making final decisions.

Starting Something You Can Actually Finish

When given the opportunity, children will often put more food on their plate than they can actually eat; we say "their eyes are bigger than their stomach." The same theory applies to inexperienced directors. Novice directors may have excellent ideas for complex and involved programs they want to produce as their first show. However, the best way to achieve success is to work within the boundaries of a production. Do not attempt a program that requires more funding, equipment, people, time, and experience than is available. The following are some tips to help you start a production you can actually finish.

- Keep the program short; dynamic 5–7 minutes vs. boring 30 minutes.
- Keep it simple. The more complex a program is, the greater the chance for mechanical failure or human error.

- A small crew reduces complications. When more people are involved, the variables for errors increase, which equals a higher probability of failure.
- Have a realistic vision of the program that is proportional to the budget. Scale the production for success, not disappointment.
- Be a professional. Treat people with respect, provide plenty of schedule reminders, and have maps and phone numbers available for everyone.
- Be organized and do not waste anyone's time. Have lists of everything and breakdowns of everything—including props, locations, camera shots, equipment, eating locations, and restrooms.
- Keep a record of contact information for clients and each member of the cast and crew to facilitate quick and efficient communication. This information should include address, home phone, business phone, cell phone, and e-mail address.
- Keep an eye on the big picture. Do not spend an excessive amount of time getting one small scene perfect, while sacrificing time needed to complete the entire program.

Wrapping Up

The best directors have come up through the ranks and have held almost every production staff position on the way. All of that experience is called upon throughout the production process. Professional directors work in the business for years before directing their first program.

Student directors attend courses on the psychology of presentation, which address methods in making the audience "feel" things and the responsibilities related to using those methods. A director can influence attitudes, emotions, and actions of the audience by using aspects and principles of visual media. Classes in the ethics of visual media instruct student directors on how to be unbiased in the presentation of information, as well as how to be biased and manipulative. Unbiased presentation of information is the very cornerstone of professional broadcast journalism. Having a well-rounded education in geography, history, and political science is also beneficial for a director.

Throughout your education, watch television programs and film productions to analyze what professionals do and how they do it. Examine how a director makes a particular scene exciting. Turning down the audio will help you stay attentive to the production aspects of the program, without becoming involved in the plot of the program.

As a broadcasting student, you will begin to develop your own style as you direct more programs. Meanwhile, work as often as possible in all of the technical positions. Experience in each position is beneficial for success with future projects and responsibilities.

Review Questions

Please answer the following questions on a separate sheet of paper. Do not write in this book.

1. List some of the director's responsibilities for each phase of production: pre-production, production, and post-production.
2. What steps must be completed before script breakdowns can be prepared?
3. List each type of script breakdown and note the information contained in each.
4. What is the director's role in the audition process?
5. Which members of a production are involved in the camera rehearsal?
6. What is "shooting for the edit?"
7. Identify five characteristics of effective directors.

Activities

1. Record an episode of your favorite television show. Create a prop plot for one scene in the program. Include a description of the events and activities in the scene.
2. A slate can be as elaborate or as simple as the available resources allow. Create a reusable slate, using only items found in your home.

3. Research the Directors Guild of America. Summarize important facts and initiatives of the organization. Be prepared to share your findings in class.

STEM and Academic Activities

1. Research the methods and tools used for set design. What resources and tools used by set designers today were *not* available 25 years ago?

2. During a pre-production meeting, the expectations and responsibilities for each member of the staff are discussed and a plan is established. A detailed planning phase for any project improves organization and efficiency. Identify the "pre-production" planning you performed for another project or task. Explain the responsibilities of each participant and detail the plan of action that was created.

3. A shot of the slate is recorded for 10 seconds at the beginning of every take of every scene. In a production with 25 scenes and 3 takes of each scene, how much footage (in terms of minutes) is dedicated to recording the slate?

4. Research your favorite television or movie director. Summarize the director's experience and methods. What do crew members and talent say about the director's abilities? What qualities of this director stand out?

Chapter 21

Makeup Application and Costume Considerations

Professional Terms

base
blending
character makeup
crème makeup
foundation
highlight

pancake makeup
prosthetics
shadow
spirit gum
straight makeup

Objectives

After completing this chapter, you will be able to:

- Explain why the use of makeup is necessary on television.
- Recognize the differences between character makeup and straight makeup.
- Identify the materials and products used for each layer of makeup application.
- Recall common considerations when selecting the costumes for a production.

Introduction

Makeup is the collection of various cosmetics and materials that are applied to the skin. Makeup is necessary for all studio productions on both male and female talent—wearing makeup is the norm for television. Even news anchors, though not considered actors, wear makeup when on camera. This chapter introduces different kinds of makeup products, the uses of each, and some application tips.

Why Is Makeup Necessary?

Students in television broadcasting classes often have some exposure to stage makeup in drama classes or theater production. Stage makeup is worn for three reasons:

- To make the actor look attractive under very bright stage lights.
- To help the actor portray a character by creating a "look" that is more appropriate for that character.
- To add three-dimensionality to the actor's face by replacing the natural shadows removed by the bright lights.

Stage makeup is usually applied rather heavily so the audience can see the actors' exaggerated facial expressions from anywhere in the theater. If the audience were to get a close-up view of the actor, the makeup would appear garish and overdone, **Figure 21-1**.

Television makeup is used for the same three reasons as theatrical stage makeup. The application for television, however, is much more subtle. All of the aspects and techniques of stage makeup are used for television, but the makeup is not applied as heavily. The talent on television should, most often, not appear to be wearing makeup at all. The goal is to create a natural appearance from a distance of 8 to 12 feet, which is the average distance between a television set and viewers at home.

Television performers placed under bright 32K white lights look gray, pasty, flat, and unattractive without makeup. Digital video technology creates images so sharp that even the slightest skin imperfection is greatly magnified on television. Features that may be insignificant in real life—such as blemishes, dark circles under the eyes, acne, rashes, bruises, five o'clock shadows, and wrinkles—appear magnified on television.

While many performers may resist wearing makeup, they need to understand its necessity. A capable lighting designer lights performers

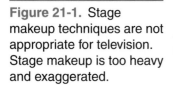

Figure 21-1. Stage makeup techniques are not appropriate for television. Stage makeup is too heavy and exaggerated.

evenly and brightly. Being evenly lit means that practically all shadows are removed from performers' appearance because light comes from all directions, **Figure 21-2**. A face without shadows does not have any depth on screen; it looks flat. When the image is displayed on a television screen, the performers appear blemished and flat and seem to be a part of a horror film, rather than a professional television production. Most people performing on television, whether portraying a character or being themselves, want to be as attractive as possible. Sometimes, the simplest way to convince a performer to wear makeup is to record a bit of footage and let them see how they look without it. Any resistance usually melts after seeing themselves on camera without makeup.

Figure 21-2. The even, bright white light in a television studio causes a face without makeup to lose all depth and dimension.

Makeup Styles

Fundamentally, there are two styles of makeup: painted and natural. Both styles are perfectly acceptable, depending on the type of program and the director's goals for the program.

When makeup is applied in a painted style, the audience can clearly see that the performer is wearing makeup, **Figure 21-3**. The painted style of makeup application is not meant to look natural. For example, an actress playing a woman in the 1960s would be made-up with dark eyeliner around the eyes, long dark eyelashes, and layers of blue eye shadow. This was the trend in fashion makeup during that era and is appropriate for a realistic portrayal of the character. But, eyes do not naturally have dark lines around them and the color blue is not a pigment found in healthy human skin. However, some shades of blue may appear in damaged or bruised skin.

Makeup applied in a natural style simply enhances a person's facial features, but does not draw attention to the makeup applied. The only colors

Figure 21-3. The painted look for television is an accepted style of makeup design. A character wearing the painted look obviously appears to be wearing makeup.

used are those found naturally in human skin, such as various shades of cream, rose, tan, and brown. Female performers often do their makeup for television in a completely natural style and add cosmetics based on their character's wardrobe, personality, or situation. It may seem odd to spend time applying natural makeup if the purpose is to appear as though no makeup was used at all. However, if a natural style of makeup is not applied, the television image presents an unhealthy, unattractive, and unnatural picture.

Applications

Character makeup is used to make a performer look like someone or something other than the performer's own persona, **Figure 21-4**. For example, the performer can be made to appear older or younger, as a different ethnicity, or as an alien from another world. A myriad of special effects makeup and injuries are categorized with character makeup, including cuts, bruises, scars, and warts.

Special-effect makeup for television creates or exaggerates physical features on performers, such as noses, wounds, swelling, and warts. These items are prosthetic devices that are added to a performer's appearance based on the character portrayed or the action in a scene. A *prosthetic* is an appliance, usually made of foam, latex, or putty, that is glued to the skin with special adhesives. The adhesive most commonly used to apply prosthetic items is *spirit gum*. Spirit gum is as thin as water and is applied with a brush. To attach a prosthetic using spirit gum:

1. Brush the spirit gum onto the skin in the area where the prosthetic is to be attached. Let the adhesive set for a few moments.

2. When the spirit gum is no longer shiny, gently tap it with your fingertip. If it is tacky and strings of adhesive attach to your finger, the gum is ready for bonding.

character makeup: Makeup application technique used to make a performer look like someone or something other than the performer's own persona.

prosthetic: A cosmetic appliance, usually made of foam, latex, or putty, that can be glued to the skin with special adhesives.

spirit gum: A type of adhesive commonly used to apply prosthetic items.

Figure 21-4. Character makeup is designed to make a person look like someone or something other than the performer's own persona.

3. Clean any adhesive off your fingers with spirit gum remover.
4. Attach the prosthetic with gentle, even pressure.

Safety Note

Do not use spirit gum around the eyes. Irritation, rash, or inflammation can occur when the chemicals in the adhesive come into contact with the sensitive skin around the eyes.

Use spirit gum remover to remove prosthetics from the skin. The remover chemically dissolves spirit gum on contact. To apply spirit gum remover, dip a cotton ball or makeup brush into the remover and gently work it into the edges of the prosthetic. This dissolves the gum and allows the prosthetic to be removed from the skin with ease.

Straight makeup is applied to make people look like themselves under the bright television lights, **Figure 21-5**. This application technique corrects or hides blemishes, makes the complexion more even, and improves general attractiveness on television.

straight makeup:
Makeup application technique used to correct or hide blemishes, make the complexion more even, and generally help people look attractive and like themselves under bright television lights.

Makeup Products

In general, there are two types of makeup to choose from: theatrical and over-the-counter. If economy is a concern, theatrical makeup is less expensive than over-the-counter, consumer cosmetics. Over-the-counter makeup is packaged in smaller quantities than theatrical makeup, so it will need to be purchased more frequently. Additionally, the amount of powder and fragrances used in over-the-counter cosmetics may irritate and dry the

Figure 21-5. The natural look enhances facial features under the bright lights without drawing attention to the fact that the performer is wearing makeup.

skin. For the sake of convenience, however, the corner store is more accessible for some people than a professional makeup supply center.

Makeup is available in two forms: crème makeup and pancake makeup, **Figure 21-6**.

Crème Makeup

crème makeup: An oil-based makeup product that easily blends with other colors.

Crème makeup is an oil-based product that easily blends with other colors. As additional layers of makeup are applied, colors can be mixed together to create a natural progression from one shade to another. Today's oil-based makeup is far removed from the greasy products of yesteryear. There is a stigma that oil-based makeup compounds problems with skin that is already oily. On the contrary, the natural oil in skin mixes with the oil in crème makeup and does not clog the pores if removed properly.

Figure 21-6. The makeup applied to on-screen talent is available in two forms: pancake and crème. Pancake makeup is a powder that must be applied with a moistened sponge. Crème makeup can be applied lightly with a sponge.

Pancake makeup   **Crème makeup**

Pancake Makeup

Pancake makeup is a water-soluble, pressed powder makeup foundation that is pressed into a compact container. Pancake makeup once was the most common type of makeup used by most television and film performers, but that is no longer the case. Pancake makeup can clog the pores of the skin and cause breakouts. Additionally, pancake makeup does not blend well with other colors. Once the color is applied, the makeup remains in the same place and at the same intensity until it is removed. Therefore, a very light touch is required to apply pancake makeup. Any additional layers of color must also be powder-based. When pancake and crème products are mixed, a sticky goop results.

pancake makeup: A powder makeup foundation that is water-soluble.

Makeup Application

Several layers of makeup are required to create a realistic appearance on camera. This is true for both character and straight makeup applications. Makeup should be applied under lighting conditions that closely match the lighting on the studio set or shoot location.

First Makeup Layer

The first layer of makeup applied usually covers the entire face, neck, ears, back of the hands, and bald spot (if applicable). This layer is called *base*, or *foundation*, **Figure 21-7**. Base is best applied with a slightly moistened cosmetic sponge. These dense foam sponges are in the shape of a wedge and available at most retail outlets in the cosmetics section. Cosmetic sponges are very effective for applying base on everyone except men who are old enough to shave. The beard stubble on a man's face shreds the sponges and leaves bits of foam all over the face. Use a different kind of sponge on adult men, such as a natural sea sponge or one made of polyester. If base is applied properly, performers should not feel the makeup on their face at all. If performers can feel this layer of makeup, it was applied too heavily.

base: The first layer of makeup applied—usually covers the entire face, neck, ears, back of the hands, and bald spot (if applicable). Also called *foundation*.

PRODUCTION NOTE

To simplify the process, use the same kind of sponge to apply makeup on everyone. This way, there is no need to keep a supply of both cosmetic and sea sponges or polyester sponges on hand for each production. However, each performer should have a dedicated set of sponges. Polyester and sea sponges should be cleaned after each use with soap and hot water.

Second Makeup Layer

Highlight and shadow are applied after base makeup. Bright studio lights remove all of the shadows on the face, and makeup is necessary to replace these shadows.

Shadow makeup is three or four shades darker than the surrounding area. Anywhere shadow is placed makes that area appear to sink into the

shadow: Makeup that is three or four shades darker than the area to which it is applied.

Figure 21-7. These photos illustrate the difference between a bare face and a performer wearing a layer of base. Notice that the entire face has a completely even color when the base layer is applied.

No makeup

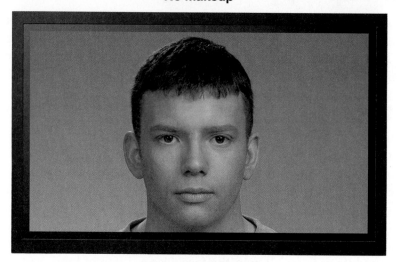

Base layer applied

plane of the face. Shadow is most commonly applied to the following areas of a performer's face:

- Above the eyes, but below the brow.
- Below or to one side of the nose.
- Below the chin.
- In the temple area for an aged appearance.
- On the cheek, below the cheekbone.

Cheek shadow, sometimes called rouge, should not be placed directly on the cheekbones. This makes the cheekbones appear crushed or damaged. To determine the proper placement of cheek shadow, follow these steps:

1. While looking in a mirror, use a light-colored eyebrow pencil to place a tiny dot halfway between the outside corner of the eye and the opening of the ear.

2. Look straight ahead directly into the mirror and place your fingers in the center of the cheek.

3. Gently press your fingers against the cheek and slowly walk your fingers up the cheek until you feel a bone. This is the bottom edge of the cheekbone.

4. Continue walking your fingers up the face to the top of the cheek-bone, which is located just below the depression of the eye socket.

5. Look straight ahead and place a dot one-quarter of the way up from the bottom of the cheekbone and position it directly in line with the pupil of the eye.

6. Imagine a line between the pupil and the second dot on the cheek-bone. Continue that line down until it is even with the outside corner of the mouth. Place the third dot in this location.

8. Connect the three dots to create a rough triangle shape with a slight curve on the top-side, **Figure 21-8**. This is the area, and roughly the shape, to apply the cheek shadow.

Highlight is makeup that is three or four shades lighter than the surrounding area, **Figure 21-9**. Highlight is usually applied to the following areas of a performer's face:

- The bridge of the nose.
- The bone just above the eye, below the eyebrow.
- Above the eyebrow, on the bony ridge across the forehead.
- The chin and the jaw line, which gives the appearance of strength (usually used for men).
- The cheekbones.

Applying highlight to the bone just above the eye, below the eyebrow, is necessary on both men and women. When highlight is applied to this area on men, it should be subtle and particularly well blended. If applied too heavily or improperly blended, it is apparent that the talent is wearing eye makeup.

Highlight should be applied to the bony ridge across the eyebrows on men who do not have a strong brow line. Accentuating this area with highlight gives the appearance of a stronger, more masculine individual. If a man already has a prominent brow ridge, accentuating this area makes him appear brutish and savage. Therefore, pay particular attention when

highlight: Makeup that is three or four shades lighter than the area to which it is applied.

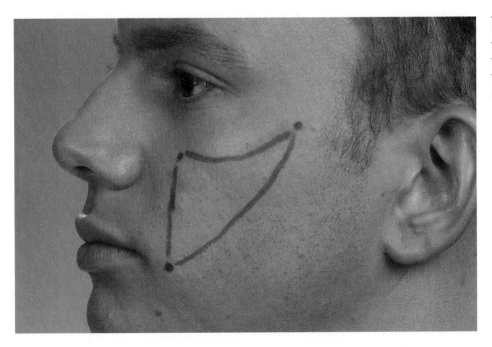

Figure 21-8. Connecting the dots creates a triangular placement template for cheek shadow.

Figure 21-9. Highlight is placed on areas that should stand out on the face.

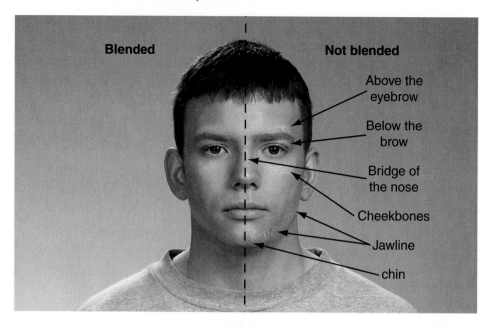

Blended   Not blended

Above the eyebrow

Below the brow

Bridge of the nose

Cheekbones

Jawline

chin

blending: Incorporating a layer of makeup into the areas surrounding it by brushing the makeup with the fingers or a brush.

deciding to highlight this area. Adding highlight to the bony ridge across the eyebrows on women creates a rather unattractive appearance. Always keep in mind that makeup should help a performer portray exactly the type of character necessary for the program.

When applied, highlight and shadow should be blended. *Blending* involves brushing the makeup with the fingers or a brush until the makeup applied seems to merge into the surrounding areas, **Figure 21-10**. There should be no definitive line or separation between areas with only base makeup and areas with additional colors or layers applied.

Third Makeup Layer

After the crèmes of various shades have been applied, the entire face must be powdered. The object of powdering crème makeup is to dull its

Figure 21-10. Cheek shadow is blended into the surrounding areas of base makeup. There are no visible lines of separation between the base and shadow.

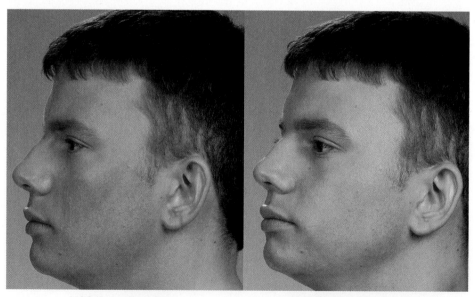

Initial application   **Blended into the area**

shine and to set the makeup. Remember that crème makeup is oil based, so it naturally has a slight shine. Setting makeup with translucent powder keeps it from easily smearing. The powder used must be translucent makeup powder. Never use baby or talcum powder—these powders are white and will change all the shadows and highlights strategically placed on the face.

To apply the translucent makeup powder, perform the following:

1. Place a powder puff on the powder and work some powder into the puff with your fingers.

2. Gently pat the powder on the face, **Figure 21-11**. Do not rub the powder onto the face.

3. Use a makeup powder brush to gently brush off any excess.

PRODUCTION NOTE

A bald spot or a completely bald head usually needs to be powdered to avoid reflecting large light hits.

Fourth Makeup Layer

Lipstick should be applied after powdering. All men and women need lipstick. Without it, a person's lips do not seem to exist at all when under bright lights. For a character who should not appear to be wearing lipstick, choose a ruddy-brownish color. If lightly applied, the lips appear quite natural. Glossy lipstick should not be used because it causes light hits that are very distracting. Talent wearing glossy lipstick look like they have sequins glued to their lips.

Eyeliner is rarely used on men or children. Women should wear eyeliner only if it should be obvious that they are wearing makeup. Television amplifies the intensity of eyeliner, which does not look natural.

Mascara thickens and lengthens eyelashes. Men should never use mascara unless they are extremely blonde with eyelashes that are not visible without mascara.

Figure 21-11. The powder is applied with a powder puff and the excess is removed with a powder brush.

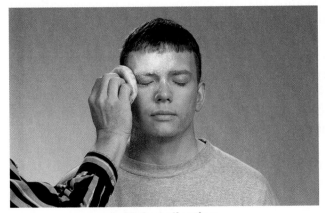

Initial application

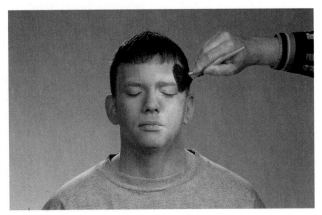

Remove excess

Safety Note

A single tube of mascara should never be used on multiple people. Mascara should never be shared with another person. A tube of mascara is a fertile breeding ground for bacteria. Eye diseases such as conjunctivitis, or "pinkeye," are highly contagious and could affect the entire cast if mascara is shared. Also, once mascara is opened, it can go bad after a period of time making it unhealthy to use. It is recommended that mascara be discarded one month after the tube is opened. Even if only one person has used the mascara, the product should be discarded and replaced once a month.

A finished makeup job gives the face depth and a three-dimensional appearance when viewed on a flat television screen, **Figure 21-12**. The talent's complexion is completely even, without varied tones on the face, and the shadows and highlights applied add dimension.

Figure 21-12. With all the layers of makeup applied, the talent's face has dimension and even tone.

No makeup applied

Four layers of makeup applied

Makeup Removal

At one time, cold cream was the most commonly used product to remove makeup. Cold cream is still used, but much less commonly. The disadvantage of using cold cream is that it is very messy—it must be slathered on the face and wiped off with many tissues. The face must then be washed with hot water and soap. Neglecting to wash with hot water and soap practically guarantees a skin breakout by the next day. Makeup remover is an alternative that accomplishes the same task, but often with the same degree of messiness. Makeup remover is an oil-based product that is rather expensive at cosmetic stores. Presently, most people in the performing arts remove makeup with baby wipes, which are readily available, easy to use, and inexpensive (**Figure 21-13**). Additionally, some brands offer fragrance-free wipes and many contain moisturizers, like aloe. Using two or three wipes removes all traces of makeup without the mess of other products. Regardless of the removal product used, talent should always wash their face with soap and warm water after removing the layers of makeup.

Makeup Application Considerations

Each production brings a unique set of challenges for a makeup artist. The action in the production, the characters involved, and the characteristics of individual performers are all considerations for a makeup artist.

Eyes

Before applying makeup on another person, always ask if they wear contact lenses and if the lenses are "hard" or "soft." The response determines which application techniques to use and the ease with which makeup can be applied around their eyes.

If the person wears hard lenses, it is recommended that a novice makeup artist not apply their makeup. Extra care needs to be taken to prevent hurting the person's eyes. Hard contact lenses are just that—hard. If a novice makeup artist presses too hard when applying makeup to the eyelids, the edge of the contact lens can scratch the cornea of the eye underneath the lid. In this situation, there are two alternatives:

- Coach the person on applying their own makeup, **Figure 21-14**.
- Obtain assistance from another makeup artist who has experience with applying makeup on talent wearing hard lenses.

Soft contact lenses, on the other hand, are soft and do not have hard edges. Performers who wear soft lenses pose no additional challenge in makeup application.

When applying makeup, you will work near the talent's eyes. Those who wear contacts are more comfortable when others work around their eyes, because they deal with their contacts daily. These people are well aware that working around the eyes is not a source of pain or discomfort.

People who do not wear contacts are more likely to be squeamish when others work near their eyes. Female performers may be more comfortable, as they are more likely to apply their own eye makeup on a regular basis. Men and children are not used to anything being near their eyes and may react with unintentional resistance. They may blink, tear, or move away when you try to apply makeup near their eyes. The best solution is to be

Figure 21-14. In some cases, it may be easier to coach someone in applying their own makeup, rather than trying to apply their makeup yourself.

patient and rely on your personality. An effective technique is to distract them by talking constantly with them. Because it is more difficult to trust a stranger, another option is to coach talent on applying their own makeup.

Talent who wear glasses may present unique difficulties. The lenses, and sometimes the frames, can produce unattractive light hits. The simplest solution is to have the talent remove their glasses. If this is not possible due to poor vision or a specific trait of a character, the glasses can be tilted slightly downward. This adjustment directs the light reflection to the floor. Another option is to move or re-aim the lighting instruments in the studio, which is a more time consuming task.

Skin Sensitivity or Allergies

Because the Food and Drug Administration has not set standards or definitions for products labeled as "hypoallergenic," it is not accurate to rely on the "hypoallergenic" label on some products. Commonly, products labeled "hypoallergenic" contain all the same ingredients as those without the label. The hypoallergenic version of consumer products typically omit the fragrance—a common allergen. Theatrical makeup does not contain fragrance. Performers who are allergic to lanolin or wool should not wear crème makeup—they must use pancake makeup. However, people without lanolin or wool allergies can typically wear theatrical crème makeup without problems.

PRODUCTION NOTE

Consider that every actor and actress wears theatrical makeup for every performance and public appearance. Actors and actresses make their living with their face and would not risk damaging the skin on their face. Understand that theatrical makeup is safe for the skin and necessary for an attractive appearance on camera.

Men and Makeup

Men frequently resist the very idea of wearing makeup, as it is not generally viewed as a masculine attribute in our society. This is ironic because most men the public views to be the most masculine wear makeup. Actors, politicians, and sports figures all wear makeup on television, when in a studio environment. Recall the appearance of any professional athlete in a locker room or on the sidelines speaking with a newscaster after coming off of the playing field—the athlete's face is typically wet with perspiration and their complexion is flushed, blotchy, and uneven. In this shot, the athlete is not wearing makeup—this is their natural face under the bright television lights. The same athlete looks drastically different when seen in a studio interview or promotional product spots. This is because studio makeup has been applied to ensure the best quality camera image of the athlete.

Costume Selection

Selecting the costumes for a program depends on many existing factors, such as plot, setting, set dressing, program format, lighting arrangement. The costume design for a production cannot be determined without accounting for the contributing factors. In dramatic programs, for example, the plot and setting dictate the costuming for the program. The type and style of costumes used in a Western set in the late 1800s would be very different from the costuming used in a science fiction program set in the year 3005.

The program set and set dressing must also be considered when selecting the costumes. The patterns and colors used on a set and in the set dressing coordinate with the talent's costuming. Both striking clashes and perfectly matching colors and patterns should be avoided. Clashing colors or patterns can be so distracting to the audience that the message of the program is completely lost. When matching colors or patterns are used on both the set and the costumes, the talent and background may blend together causing the talent to disappear from the shot.

Lighting is another consideration in costume selection. If the lighting designer uses colored gels in the lighting instruments, for example, the colored light will change the appearance of the costume. The lighting designer's plans should be included in the costume selection process to avoid problems and ensure the costuming meets all the other needs of the program.

The following items should be considered when selecting costumes for a production:

- Avoid extreme contrasts between individual items of clothing, between clothing and skin tone, and between clothing and the background. Always remember the limitations of contrast ratio.

- Avoid the color white unless the set and other costumes are light in color.
- Avoid the color black unless the set and other costumes are dark in color.
- Costume colors should not match the background color. When these colors match, the visual depth created between the performer and the background is lost.
- Avoid the color red, if possible. Red is the most difficult color to accurately process for both the television camera and the home viewer's television set.
- Be aware of the chromakey color (Chapter 23, *Electronic Special Effects*), if used in a production. The talent should not wear a color that matches or is similar to the chromakey color.
- Avoid flashy jewelry because the pieces of jewelry produce distracting light hits.
- Avoid vertical or horizontal thin stripes and small, busy patterns, such as herringbone. These patterns create a distracting moiré pattern on screen.

Appropriate Clothing

Selecting appropriate costumes for a program also includes evaluating the decency of clothing items. Clothing with any obscene graphics or phrases are not acceptable costuming. Television broadcast stations, cable television, and satellite television are all regulated by the Federal Communications Commission (FCC). The FCC has established decency standards for television programming and enforces the standards through their licensing process and violation penalties. For example, several stations had to pay millions of dollars in fines following the airing of Janet Jackson's "wardrobe malfunction" during the 2004 Super Bowl half-time show. Severe fines and legal penalties are levied against obscene or indecent videos that are aired.

In an educational environment, schools have policies regulating the use of obscene, indecent, and vulgar language on school grounds. Student journalists are guilty of breaking school policy if they air a news story over the school's broadcast system that depicts talent using obscene, indecent, or vulgar language. The school's same decency policy is violated if the news story aired contains an article of clothing with text or graphics considered to be obscene or indecent. In both cases, either audio or video brought indecent content into the school environment. Even if the video was recorded <u>off</u> school grounds, in an environment where the content may not be considered obscene or vulgar, when the story airs <u>on</u> school grounds, the student producers are guilty of breaking school policy (*not* the talent who wore the indecent clothing or used vulgar language). In an educational environment, there is never a valid reason to air obscenities or indecent content over the school's television broadcast system. School policies cannot be broken in the process of doing a story.

Planning for Productions

If a scene is scheduled for shooting over a period of a few days, both the costuming and makeup on the talent must not change from one day of shooting to the next. All of the footage shot for a single scene is likely to be edited down to a scene that lasts mere minutes. To ensure consistency in costuming and makeup, take the following steps:

1. Take photographs of talent on the first day of shooting, after their makeup and costuming is complete.

2. For each performer, write the brands and colors of makeup used on a makeup chart, **Figure 21-15**. The makeup chart should also include placement and application techniques for the products used.

3. Create a chart that records the articles of clothing worn by each performer and the scenes in which the clothing is worn.

Even with proper planning, mistakes do occur. For instance, a large wound on an actor's face appears smaller and shaped differently from the first shot to second shot in the same scene, but returns to its original size and shape in the third shot. This error in continuity may occur because the first and third shots were taken on a different day than the second shot. The makeup artist did not accurately duplicate the wound on both days. A makeup chart helps prevent errors in makeup continuity. On each consecutive day of shooting, the makeup artist uses the photographs and the makeup chart to duplicate the makeup application. This also applies to costumes used in a production. If a scene began with a male actor's shirttail tucked into his pants, but the scene ends with the shirttail untucked, the shirttail must be retucked if the scene is shot a second time. Costuming records and photographs are reviewed before additional takes of the same scene are shot.

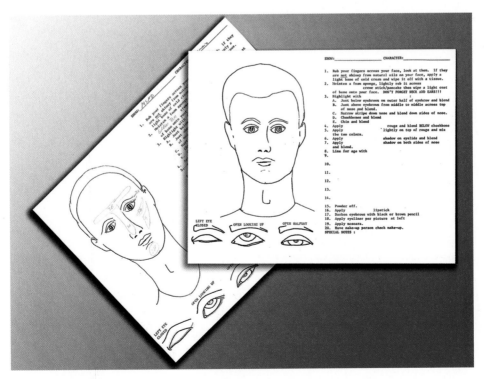

Figure 21-15. A makeup chart ensures that the talent's makeup is applied the same way for every day of shooting. This avoids makeup-related errors in continuity.

Makeup, Clothing, and the News

When it comes to news programs, the viewing public has certain expectations of what news people should look like. The audience expects news people to dress in professional, business-like attire. Men should wear a coat and tie and women should dress in business-appropriate, modest attire. Large or flashy jewelry is not appropriate on news programs. News people should not wear heavy "evening style" makeup.

PRODUCTION NOTE

Business-like attire is the norm in mainstream news programs. However, more casual attire is often seen on entertainment and pop-culture news programs.

Appropriate dress and grooming are important factors in fostering the audience's respect and trust in news people. Any aspect of a news person's general appearance that diverts attention from the words of a story is not acceptable. Many station managers require an "approval" clause in the contract with on-air news personnel. This clause gives management control over the wardrobe, makeup, and hair style (including color) of the reporters and anchors.

Wrapping Up

Much practice is necessary to become competent with makeup application and design techniques. Television makeup is almost identical to theatrical makeup, except in the amount and intensity of application—subtle application is the rule. To gain some practical experience, volunteer to help with makeup for the next theatrical production at your school. A variety of theatrical makeup books are also available through local libraries and bookstores.

Review Questions

Please answer the following questions on a separate sheet of paper. Do not write in this book.

1. Why is makeup necessary on television talent?
2. What is the one distinct difference between stage makeup and television makeup?
3. What are the advantages of crème makeup over pancake makeup?
4. List the cosmetic products used in each of the four layers of makeup application.
5. What products are available for makeup removal? Which product is recommended?
6. What are some of the considerations a makeup artist must weigh for each production?
7. List some of the cautions related to costume selection.
8. What regulates decency standards in an educational environment?
9. How can errors in continuity related to costuming and makeup application be avoided in a production?
10. How does the appearance of news people affect the perception of viewers?

Activities

1. Create a spreadsheet comparing the prices of comparable professional makeup and over-the-counter retail makeup products. Compare items such as base/foundation, cheek shadow, eye makeup, and brushed-on powders. Include the product type and brand in the spreadsheet data.
2. Make a pictorial display of images depicting both character and straight makeup applications using ads and layouts in current periodicals and local publications.

STEM and Academic Activities

1. What chemicals or other substances are commonly omitted from products labeled as "hypoallergenic"? Why do these chemicals or substances cause irritation?

2. How has high-definition television technology affected the use and application of makeup on talent? Have you noticed a difference in the appearance of makeup on newscasters or other on-screen talent?

3. Review the standards for decency on television established by the Federal Communications Commission. How do these standards affect student productions? How do these standards affect programs you watch regularly?

Chapter 22

Video Switchers and Special Effects Generators

Professional Terms

bank	fader lever
black video	glitch
bus	key
cut	lap dissolve
cut bar	mix bank
cut button	production switcher
delegation control	soft cut
dissolve	special effects generator (SEG)
effects bank	take
fade	technical director (TD)
fade in	up from black
fade out	video switcher
fade to black	wipe
fader bar	
fader handle	

Objectives

After completing this chapter, you will be able to:

- Explain the main function of a video switcher.
- Identify some of the effects possible when using a special effects generator (SEG).
- Understand the function of a bus and a bank in relation to a SEG.
- Recall the steps to use the cut bar on a SEG to cut between different camera shots.

Introduction

When entering the control room of a professional television studio, the most interesting and intimidating piece of equipment in the room is the video switcher. Large video switchers have row after row of lighted buttons, levers, and knobs that look terribly confusing. The video switcher is the "brains" of the operation. A special effects generator (SEG) is the next generation of a basic switcher, with basic switcher functions in addition to the capability of producing many video effects commonly seen in television programs.

This chapter discusses switchers, beginning with the simplest types. All modern switchers perform the same basic functions. Switchers that are capable of more complex tasks with many features and options are more expensive than simple switchers. The definition of a video switcher has evolved significantly from the early days of television to the modern digital age.

The Video Switcher

All of the video sources in the studio are connected to the *video switcher*, or *switcher* (**Figure 22-1**). This piece of equipment allows the operator to select one signal from various video sources connected to the switcher—such as camera 1, camera 2, camera 3, CG, or VCR 1—and output the selected signal to the video recorder, **Figure 22-2**. Simple video switchers can be inexpensive—less than $50 in many consumer electronics stores. A switcher operates much the same as many common household items:

- The remote control for a consumer television allows the operator to select from all the inputs going to the television set, such as the various channels of programming, VCR, DVD player, and video game console.
- Most cars have radios with buttons that are programmed with preset stations. When different buttons are selected, the radio tunes in the corresponding radio station. Pressing two buttons at the same time, however, does not allow the listener to hear two stations at the same time.
- The receiver component of most home sound systems has a knob or individual buttons labeled with various audio sources, such as AUX, TUNER, DVD, CD, and DOCK. Selection allows the signal from one component of the system to be heard through the speakers. The song playing on a radio station and a song on a CD cannot be heard simultaneously on an individual home sound system.

Video Switcher Controls

A video switcher performs one function at a time like the household items in the previous examples. All the video inputs are plugged into the switcher. At least one row of buttons on the switcher control panel is marked with numbers representing each of the inputs. A word or abbreviation may also be used to mark buttons on the control panel, **Figure 22-3**. Some switchers come with extra,

Figure 22-1. Initially, a large video switcher can be very intimidating. It works just like a small switcher, except it allows more video inputs.

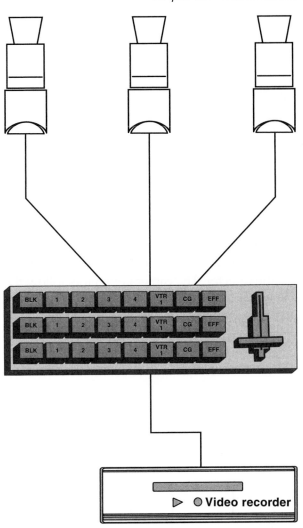

Figure 22-2. Any of the various inputs connected to a video switcher may be selected and sent to the video recorder.

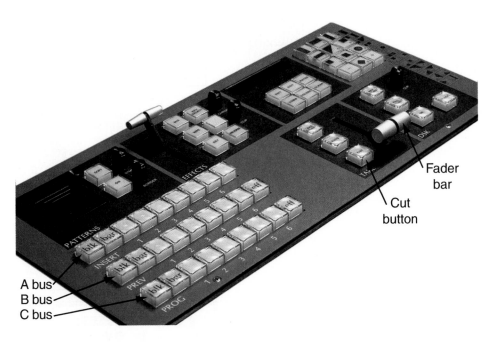

Figure 22-3. Each button on the control panel of a video switcher is marked to indicate the corresponding signal or function.

preprinted buttons that can be placed on the panel after removing a numbered button. For example, a small studio with 3 cameras and 1 CG may remove the camera "4" button and replace it with a "CG" button. When the CG button is selected, the video recorder receives the output of the CG. Component-specific labels simplify the selections for the operator.

One cable connected to the output of the switcher runs to the video input of the recorder. When the operator selects button "2" on the control panel, for example, the signal coming from camera 2 is sent to the recorder for recording. When the button for camera 3 is pressed, the image switches from camera 2 to the image from camera 3.

Video Switcher Operation

Operating the switcher is the primary job function of the *technical director*, or *TD*. The TD follows the camera script (prepared by the director) or verbal commands directly from the director. The camera operators follow the orders of the director and the TD selects the camera shots by skillfully operating the switcher. In smaller operations, the director may also function as the TD.

Connected to the switcher are a video monitor for each input signal and a monitor for the output signal, **Figure 22-4**. Each monitor is labeled with the signal it displays. The TD watches the monitors and constantly evaluates the

technical director (TD): A member of the production team whose primary job function is to follow the camera script or verbal commands from the director and operate the video switcher accordingly.

Figure 22-4. The control room system has a camera monitor for each camera signal, as well as a program monitor that displays the signal being recorded.

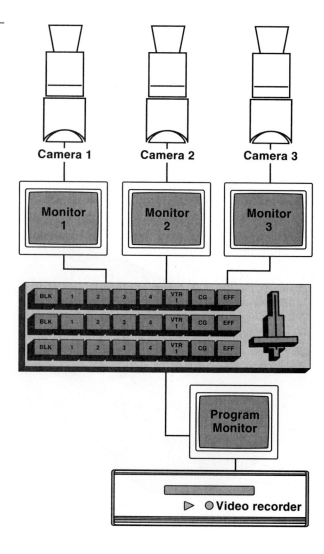

images from each camera in determining which signal to send to the recorder. The video switcher only switches from one input to another input. The result on the screen is an instantaneous picture change called a *cut*, or *take*. Sometimes, a cut is accompanied by a momentary "trashing" of the video signal (such as a roll, tear, or briefly appearing noise) called a *glitch*. Glitches are not acceptable in professionally produced programs and usually indicate that there is a technical problem in the digital video system. The problem could be in many different pieces of equipment or in the cables.

cut: An instantaneous picture change that occurs on screen. Used synonymously with *take* in this context.

glitch: A momentary "trashing" of the video signal (such as a roll, tear, or briefly appearing video noise).

Talk the Talk

In this context, "take" is synonymous with "cut." The director may say either, "Cut to Camera 2" or "Take 2."

PRODUCTION NOTE

Resolving a glitch is a task for a crew member with plenty of troubleshooting experience. It is easy to make the problem worse if you do not know exactly what you are doing. Contact a video engineer, technician, or repairman to address the problem. Remember as many details about the problem as possible because you will need to recreate the scenario exactly in order for the expert to offer a solution.

The output of the switcher is called the "program" and is viewed on the program monitor. The program signal goes to the video recorder for recording. The image on the program monitor is always identical to the image on one of the video input monitors. The video input monitors are generally placed in a horizontal row, reading from left to right. The program monitor is usually placed on top of the row of video input monitors. By properly operating the switcher, the TD creates a program with varying shots that keeps the audience's interest.

VISUALIZE THIS

To understand the importance of correctly using a switcher, consider that many people fall asleep when they are bored. When students become disinterested in a classroom activity or discussion, for example, they begin to feel drowsy and their eyelids get heavy. To help themselves stay alert, they look at different elements in the environment to provide stimulus to their brains. If history class gets boring during a lecture, you might cast your eyes around the room looking at your book, fellow students, your fingernails, a pen, doodles in your notes, or the teacher's hairstyle. Looking at anything different changes the picture in your mind, or switches to a different camera image. The audio in your mind remains the same, but you make the visual images more interesting. This process is the same as production switching. The moment the audience decides that a program does not provide sufficient visual stimulus, they turn away from the television screen and the director fails in his mission. The TD uses the switcher to vary the shots in a program and keep the audience's eyes from wandering away from the screen.

Production Switcher

A *production switcher* is a video switcher that is used in the control room to cut between live camera shots while a program is being recorded. While a video switcher can be used many places in a production facility— such as the editing room, control room, master control room, or on a duplication system—a production switcher will only be found in the control room. All production switchers are video switchers, but all video switchers are *not necessarily* production switchers. A production switcher is a video switcher used exclusively for switching the live recording of programs in the studio or on location. For example, at the Super Bowl, multiple cameras are used to shoot the game and related activity. A production switcher is used to cut between shots from the various cameras and broadcast the event live.

The Special Effects Generator (SEG)

A *special effects generator*, or *SEG*, is a piece of production equipment based on the principles of the video switcher, with the additional capability of producing various video effects. A SEG can either be a stand-alone piece of equipment or a computer program. As a stand-alone piece of equipment, a SEG has a control panel that sits flat on a countertop and is operated by the TD. As a computer program, the SEG interface displayed on the computer monitor looks like a stand-alone SEG. The TD may perform operations using a traditional computer mouse and keyboard. The SEG interface may incorporate a touch screen, which allows the TD to tap a finger on the interactive control panel to perform the SEG functions.

Talk the Talk

The term *SEG* is pronounced like "egg" with an *S* at the beginning. This piece of equipment is also commonly referred to by simply saying the three letters "S-E-G." Either way of referring to a SEG is perfectly acceptable.

Like a video switcher, a SEG allows the operator to select one signal from the various video inputs available. In addition, a SEG performs clean cuts, fades, dissolves, wipes, and various keys. To accomplish these effects, the SEG allows the TD to mix video signals from multiple sources. Compared to the simple cut created by a switcher, these effects are much more appealing to program viewers.
- *Fade.* The video image slowly appears from a solid-colored screen. Or, the video image slowly disintegrates from an image to a solid-colored screen. The color of a solid screen is usually black, but the director can choose any color the SEG is capable of creating. Some SEGs only offer black as a screen color for a fade.
- *Fade In.* The viewer sees a black screen and the image appears, **Figure 22-5**. This effect is usually used for the beginning of a program

Figure 22-5. A totally dark picture transitions into a fully visible picture when using a fade in. A fade out is the reverse.

or scene, and corresponds to the concept of opening the curtain on a stage performance. A fade in may also be used when a character wakes up, while shooting with the subjective camera technique. The appearance is almost as if the set was completely unlit, and the lighting instruments are gradually brought up. An *up from black* effect is the same as the fade in.

- *Fade Out.* As a scene in a program ends, the image gradually goes to a black screen. When used at the end of a scene, this effect corresponds to the concept of closing the curtain on a stage performance. Another use of fade out may be when a character goes to sleep, passes out, or dies while using subjective camera. A *fade to black* effect is the same as a fade out.

- *Dissolve.* One picture gradually disintegrates while another gradually appears, **Figure 22-6**. During a dissolve, the screen is never black (or any other solid color). A dissolve is a visual statement to the audience, saying "you are now seeing another place during the same time frame," or "meanwhile." It is also used to show a time passage in the program. For example, a cook puts a cake in the oven and sets the timer for 35 minutes. The audience does not want to stare at an oven timer for 35 minutes. Therefore, they see a dissolve to the same oven timer 35 minutes later. In this case, 35 "real time" minutes have passed in 3 seconds of "screen time." This effect is called a *lap dissolve* in the film industry because, originally, two pieces of film would have to overlap to produce a dissolve. If the fader level is moved very quickly when performing a dissolve, the result almost appears to be a cut, but is not as sharp as a cut. This fast dissolve is called a *soft cut*.

fade out: A video effect where the image slowly goes to a black screen as a scene of a program ends. Also called *fade to black*.

dissolve: A video effect where one picture slowly disintegrates while another image slowly appears. At no time is the screen solid black (or any other solid color). Also called a *lap dissolve*.

soft cut: A fast dissolve that appears to be a cut, but is not as sharp as a cut.

Figure 22-6. In a dissolve, one image slowly becomes another image. At no time is the screen black.

VISUALIZE THIS

Many students confuse the terms "fade" and "dissolve," using them interchangeably. To help differentiate the two terms, consider that the original color of cotton is white. White cotton is dyed blue before making blue jeans. As blue jeans are worn and washed, the blue color slowly washes out, or *fades*, revealing the original white color of the cotton. You would not say that an old pair of jeans had dissolved; you would say they had faded. When the applied color fades, the natural color begins to show through. The "natural" color of a television screen is black. Therefore, when colorful images on the television are removed, the screen returns to black. When a television image is removed from the screen, or is washed out, the screen fades to black.

wipe: A video effect where a line, or multiple lines, moves across the screen and replaces one picture with another.

key: A video effect where a portion of the picture is electronically removed and replaced with another image.

bus: A row of buttons on the control panel of a SEG that access different inputs and functions.

- *Wipe.* A line, or multiple lines, moves across the screen, replacing one picture with another. If the line stops at some point while moving across the screen, the screen is divided into two or more sections, also known as a split screen (discussed in Chapter 23, *Electronic Special Effects*).
- *Key.* A portion of the picture is electronically removed and replaced with another image (discussed in Chapter 23, *Electronic Special Effects*).

Components of a SEG

A *bus* is a row of buttons on the control panel of a SEG. Most SEGs have at least three buses, with a minimum of five buttons per bus. Some SEGs, however, may have more than 30 buttons on a bus. Each button accesses a different video input and is clearly labeled. If there are ten cameras, for example, ten of the buttons are labeled 1 through 10. Other buttons may be available for additional video playback units, CG, and BLACK. The more video inputs on a SEG, the more buttons are required to access each at any time. The buses are identical, with the possible exception of an additional button on the right end, **Figure 22-7.**

PRODUCTION NOTE

Do not let the great number of buttons on larger SEGs intimidate you. A SEG that has additional video inputs requires more buttons to access those inputs. The additional buttons are simply more of the same buttons and functions you will become familiar with on a smaller SEG.

Figure 22-7. Each bus is identical to the others, with BLK, 1, 2, 3, 4, VTR 1, CG, and EFF buttons.

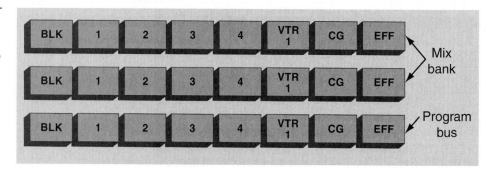

Most commonly, the lowest bus displayed on a computer interface or the bus closest to the TD on a stand-alone SEG control panel is labeled *Program Bus*. Any signal button pressed on the program bus goes directly to the output of the SEG and to the video recorder. The other buses are usually organized in pairs, which are electronically connected to each other. To create a dissolve, for example, two pictures must be on screen at the same time. Since multiple buttons cannot be pressed simultaneously on a single bus, two buses are connected to allow the display of two images. When two buses are electronically connected to each other, they are called a *bank*.

Next to each bank is a *fader lever*, also referred to as a *fader bar* or *fader handle*, **Figure 22-8**. A fader lever is usually a T-shaped handle that can be moved forward and backward. The further the bar is moved in one direction, the stronger the signal coming from the bus in that direction. Using the fader lever allows the signal from each bus in the bank to be manipulated. To begin a dissolve, for example, one signal can be at 90% strength with the other signal at 10%. To perform the dissolve, gradually move the fader lever (80/20%, 70/30%, 60/40%, 50/50%, and so on) until complete.

To differentiate between the buses, the industry convention is to assign a letter of the alphabet to each bus. The bus displayed above the others on a computer interface or the bus farthest away from the TD on a stand-alone SEG control panel is called A bus, **Figure 22-9**. Each bus that follows is sequentially assigned a letter. For example, the program bus on a SEG with seven buses would be G bus. Buses are referred to by the letter assigned.

bank: Two buses on the control panel of a SEG that are electronically connected to each other.

fader lever: A control on a SEG, usually a T-shaped handle, that controls the strength of that signal coming from each bus. Also called a *fader bar* or *fader handle*.

Assistant Activity

Review the explanation of how letters are assigned when labeling buses. Using **Figure 22-3** as a guide, sketch the control panel of a SEG and close the book when you are finished. Label each of the buses and identify the program bus and mix bank. Check your work using the previous sections in the text and the corresponding figures.

Figure 22-8. The fader lever can be moved forward and backward. As it moves, it diminishes one signal and increases another signal.

Figure 22-9. Each bus is assigned a letter, beginning with the row farthest from the TD.

To illustrate a dissolve:

1. Imagine that button 3 has been pressed on the program bus (C bus in this case) and the audience is watching the image coming from camera 3.

2. To set up a dissolve, the fader lever is pushed all the way forward, so the handle is pointing at A bus.

3. Press 1 on A bus and 2 on B bus. The SEG control panel looks like **Figure 22-9**.

4. Press the EFF (effects) button on the far right of C bus and the SEG output jumps up to the mix bank. The SEG control panel looks like **Figure 22-10**. When the EFF button on the program bus (C bus) is pressed, the output of the SEG changes from C bus to the effects bank (combination of A bus and B bus). Since the fader lever is pushed all the way forward to A bus, 100% of the output from the SEG is the image selected on A bus (camera 1, in this case).

5. The audience watching the image from camera 3 simply sees a cut to camera 1. The audience doesn't know they are now watching a signal coming from the mix bank.

Figure 22-10. If camera 1 is selected, or "punched up," on A bus and camera 2 is punched up on B bus, what happens when the fader lever is pulled down to the lower position?

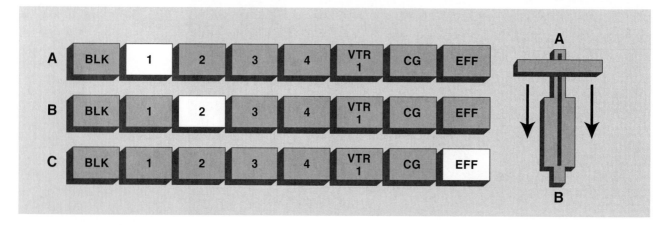

6. Dissolve from camera 1 to camera 2 by smoothly moving the fader lever down from the forward position. When the dissolve is complete, the handle of the fader lever is pointing at B bus.

If the fader lever is stopped halfway through a dissolve, the images from both camera 1 and camera 2 are displayed on the screen at 50% strength. As a result, both images are semitransparent and look "ghostly." This effect is called a superimposition. The image in the middle of the series in **Figure 22-6** is a superimposition. If the BLK (black) button were selected on B bus in step 3 (Figure 22-9), a dissolve to black would have been produced. From the audience's perspective, a dissolve to black appears the same as a fade to black.

VISUALIZE THIS

To relate the concept of a dissolve to an audio example, turn on your sound system. Move the balance control so that all the right channel sound comes out of the right speaker. Now, move the control to the left side so you only hear the left channel of audio. In adjusting the balance control from the right channel to the left channel, you have just "dissolved" the sound from the right channel to the left channel. If you stop the balance control halfway between the right channel and left channel, both channels are equally "superimposed."

Banks

Cut, fade, and dissolve are the simplest effects a SEG can produce. All three of these effects are contained in a single bank called the *mix bank*. Like the ingredients combined in a kitchen mixer, the inputs on the SEG cannot be separated and processed individually once they have been mixed. An *effects bank*, however, allows each signal to be processed individually. The TD can set up various keys and wipes on the effects bank (discussed in Chapter 23, *Electronic Special Effects*).

A SEG with both a mix bank and an effects bank has at least five buses; A and B may be the mix bank, with C and D as the effects bank, and E as the program bus, **Figure 23-11**. Large SEGs may have four mix banks and four effects banks, totaling 17 buses. Even these large SEGs are nothing more than multiples of the same functions found on a three-bus SEG.

The most economical SEGs have only three buses, but are still capable of both mixing and producing effects. A small SEG accomplishes this by allowing bus A and B to be *either* a mix bus, an effects bus, a key bank, or a DVE bank (discussed in Chapter 23, *Electronics Special Effects*). A switch or button, called a *delegation control*, simply changes the functionality of the selected bank, **Figure 22-12**. This delegation control is similar to the one on your home stereo system that allows you to switch from AM to FM or from CD to MP3 player.

mix bank: A bank on an SEG that contains the cut, fade, and dissolve effects.

effects bank: A bank on a SEG that allows each signal to be processed individually.

delegation control: A switch or button on the control panel of a SEG that changes a bank from one function to another. For example, the delegation control can switch the selected bank from operating as a mix bank to an effects bank.

Figure 22-11. A SEG with both an effects bank (A bus and B bus) a mix bank (C bus and D bus). Each bank has a dedicated fader bar. The program bus (E bus) has both MIX and EFF buttons, but only one of these buttons can be selected at a time.

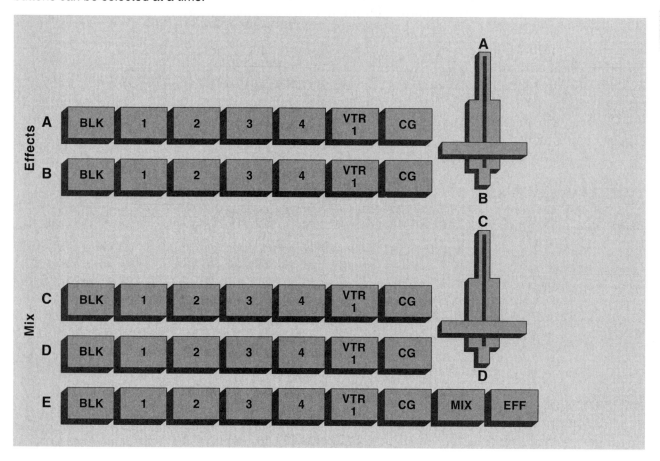

Figure 22-12. A delegation switch or button allows the A/B bank to perform either mix or effects. Once "effects" is selected as the function of A/B, some switchers present the option of indicating which kind of effect should be processed.

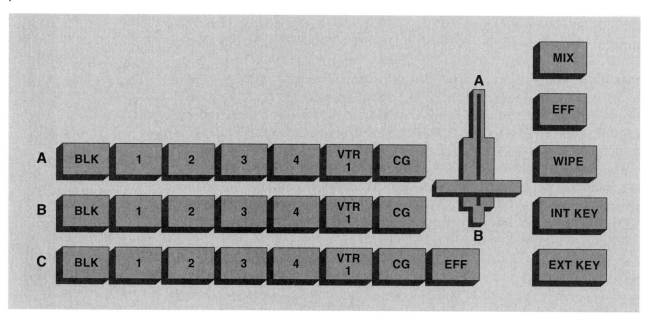

Black Video

Black video is a bona fide video signal that has no lightness. When the BLK button on a SEG is selected, the output is a purely black signal. The BLK button is not an input on the SEG. Black is generated by the SEG itself. The black output may be used at the beginning or end of a program to make either end neater than displaying video noise. Black may also be used within a program. If, for example, a character loses consciousness, selecting this button represents the world going "dark."

Only when light is applied to the screen is an image visible, because the natural color of the television screen is black. When the SEG outputs a black signal, it is an actual signal with no light. During a program, the director may want the screen to go black. To record a clean black screen, the black must be created on the SEG. Black video is very important to the successful operation of a television production facility. To record a black signal on a camcorder, merely place the lens cap on the camera or close the iris all the way and press record.

black video: A bona-fide video signal that has no lightness.

VISUALIZE THIS

A human's eyes are like a video camera and the signal going up the optic nerve to the brain is like the video signal going to a television screen. Imagine that you are exploring a deep cave that extends far into the earth. You turn off your flashlight and are suddenly in total darkness. Does this mean that your eyes have stopped functioning and that you are blind? Of course not. Your eyes are functioning perfectly, but there is no light for your eyes to detect an image to send to your brain. Likewise, black video is a perfectly good video signal that simply lacks any light.

Preview Monitor

The preview monitor is connected to the output of the mix bank, **Figure 22-13**. The preview monitor allows the TD to set up an effect on the mix bank (A bus and B bus). The TD can see the effect on the preview monitor before pressing the EFF button on C bus, which sends the effect out to the viewers. While the audience watches camera 3 on the C bus, for example, the TD sets up a split screen between cameras 1 and 2 and watches it on the preview monitor. When the TD is satisfied with the completed split screen effect, the EFF button on C bus is pressed and the audience instantly sees the split screen already created.

Cut Bar

A *cut bar*, or *cut button*, is a convenient accessory on many SEGs. A cut bar is a selection on a SEG that provides a quick way of cutting between the input selected on one bus and the input selected on another bus in the bank. For example, when shooting a 2-person interview with two cameras, each camera captures a different person while the program is recording. The entire interview involves cutting from one camera to the other as the two people speak to each other.

cut bar: A feature on a SEG that provides a quick way of cutting between the input selected on one bus and the input selected on another bus in the bank. Also called a *cut button*.

Figure 22-13. A preview monitor allows the director to see an effect before it is sent to the program output of the SEG.

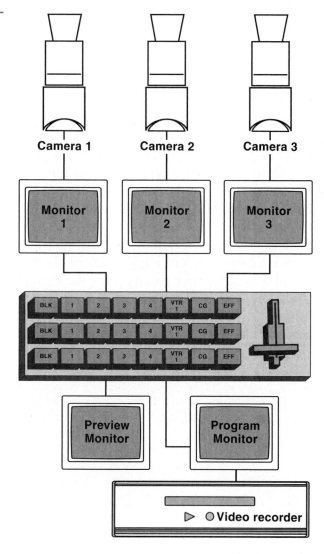

Using the SEG controls, **Figure 22-14**, the TD may simply alternate pressing button 1 and button 2 on C bus. This motion can become monotonous after a short time, which may lead to mistakes. Pressing the wrong button may send the audience the interviewer's face as the interviewee speaks, which is probably not the desired image. The cut bar eliminates this mistake. To use the cut bar in this scenario:

1. Select the EFF button on C bus.
2. Press camera 1 on A bus and camera 2 on B bus.
3. Begin the interview on B bus. The fader lever should be down, pointing at B bus.
4. To change cameras, simply press the cut bar or CUT button.
5. The first time it is pressed, the output of the SEG switches from B bus to A bus. This creates a cut from camera 2 to camera 1.
6. Press the cut bar again to cut the output from camera 1 back to camera 2.
7. Repeat steps 5 and 6 as many times as necessary during the program.

Figure 22-14. The CUT button on a SEG allows a simple switch from one camera to another and back again, by pressing just one button instead of two.

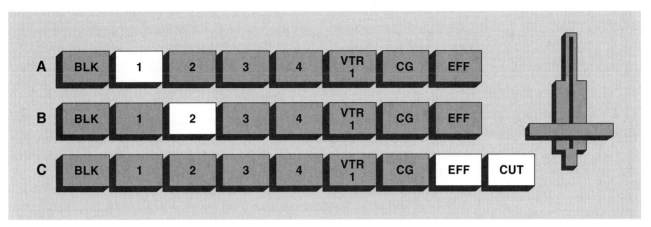

Wrapping Up

A switcher is the central component in a television studio environment to which all video sources in a studio are connected. Many terms are commonly used to refer to the switcher, including video switcher, switcher, production switcher, special effects generator, and SEG. All these terms are acceptable, but proper use depends on the capabilities of the equipment—all SEGs are switchers, but not all switchers are SEGs.

Whether capable of only basic functions or equipped with many advanced features, almost all switchers operate on the same basic principles. A switcher allows the operator to select one or more signals from various video inputs, layer/combine the signals, and output the resulting signal to the video recorder. Most TDs switch when instructed to do so by the director. Over time and with experience, many TDs develop a talent for the sequences and timing of images. They can often predict the director's instructions.

Review Questions

Please answer the following questions on a separate sheet of paper. Do not write in this book.

1. What is the difference between a video switcher and a SEG?
2. How do you differentiate between a fade and a dissolve?
3. What is a bus?
4. What is the function of a bank?
5. What is the difference between a mix bank and an effects bank?
6. How is black video used in a production?
7. How is the preview monitor used with a SEG?

Activities

1. Record about five minutes of television and watch the recording with the sound turned off. Notice the camera switches. Determine why the switch was made at each point and make note of the reason.
2. Research various brands and models of SEGs. Summarize the options available on five different models and compare the purchase price of each.

STEM and Academic Activities

Science

1. To create an image on a television screen, light must be applied. The same is true of human vision. Explain how the human eye processes light. How is eye function impaired in someone who has nyctalopia?

Technology

2. Describe the differences in functionality between a basic video switcher and a SEG. Identify the advances in technology that led to the increased functionality found in a SEG.

Mathematics

3. Research the prices and features of various SEGs. Create a chart comparing the prices and features of six different models. Explain any correlation you find in the price and the features of the models compared.

Social Science

4. A switcher allows the operator to select one signal from various sources connected to the switcher. Are there any pieces of equipment or other items you use in your daily life that operate like a switcher?

www.iatse-intl.org The International Alliance of Theatrical Stage Employees, Moving Picture Technicians, Artists and Allied Crafts supports people working in various branches of the entertainment industry, including motion picture and television production, product demonstration and industrial shows, conventions, facility maintenance, casinos, audio visual, and computer graphics.

Chapter 23

Electronic Special Effects

Objectives

After completing this chapter, you will be able to:

- Explain how effects are used as transitional devices.
- Differentiate between a superimposition and a key.
- Recall the importance of pixels to DVEs.

Introduction

Students learning about television broadcasting and production are often tempted to add "cool" effects that serve no real purpose to the overall goal of their program. Special effects in a program that do not serve a purpose and do not support the program become nothing more than gimmicks and distractions, and are a big mistake in a production. The audience stops paying attention and the message of the program is lost. Using special effects is very seductive, but they should only be used when a special effect supports the program and its message.

Professional Terms

chromakey	key level control
chrominance	lower third
circle wipe	lower third key
clip control	lower third super
corner insert	over-the-shoulder graphic
corner wipe	pixel
digital video effect (DVE)	soft wipe
fill camera	split screen
fill source	superimposition
horizontal wipe	super
key	transitional device
key camera	vertical wipe

SEG Effects

A SEG has many effects built into it. Some SEGs even allow users to program custom-designed effects. The most common electronic special effects include wipes, mixes, and keys.

Wipes

A wipe is an effect consisting of a line that moves across the screen, replacing one picture with another. The line that moves across the screen during a wipe reveals a picture that is "behind" the image being wiped off the screen. In a simple wipe, no part of either image is moved in any way; each is either removed or revealed by the line. Common types of wipes are *horizontal wipes*, a vertical line that moves across the screen horizontally, and *vertical wipes*, a horizontal line that moves up or down the screen. A vertical wipe looks similar to a window shade being raised or lowered. A wipe can take many different forms, depending on the wipe effects built into the SEG, **Figure 23-1**. The line may be sharp, may have a colored border, or may be slightly out of focus. Using a line that is slightly out of focus in a wipe effect is called a *soft wipe*.

To accomplish a wipe, one camera signal is selected on one bus and a different camera signal is selected on another bus. The wipe pattern is selected from the control panel of the SEG. Moving the fader handle controls the progress of the line across the screen.

horizontal wipe: A video effect where a vertical line moves across the screen horizontally, replacing one picture with another.

vertical wipe: A video effect where a horizontal line moves up or down the screen vertically, replacing one picture with another.

soft wipe: Any wipe effect in which the wipe line is out of focus.

PRODUCTION NOTE

It is important to note that a simple wipe involves a line moving across the screen that reveals an image "behind" the image being removed from the screen. Imagine a teacher erasing the writing on a blackboard. As the words are erased, they are not moved or pushed off the blackboard. The chalk writing is removed by the eraser and the blackboard surface is revealed. A simple wipe is much like this example because no part of either image is moved in any way. The images are either removed or revealed by the moving line.

transitional device: An effect that is used as a means of getting from one scene to another.

split screen: A wipe that is stopped part of the way through its move, dividing the screen into two or more parts.

circle wipe: A video effect where a circle grows or shrinks, replacing one picture with another.

Wipes are usually used as a method of getting from one scene to another, or as a *transitional device*. A wipe that is stopped part of the way through its move and divides the screen into two or more parts is called a *split screen*. This effect is most often used to portray a telephone conversation, **Figure 23-2**. Each person is shown on one-half of the screen as they speak to each other on the telephone. A split screen is also used for interviews on news programs when the interview participants are in different locations. Showing both interviewees on the screen allows the audience to see the conversational interaction during the interview.

A *circle wipe* uses a circle that grows or shrinks to replace one picture with another. A SEG capable of a circle wipe often has a joystick to control the position of the circle on the screen.

A *corner wipe* is seen on many local news broadcasts, where an image related to a story is displayed over the studio anchor's left or right shoulder. Achieving this effect requires two cameras or sources of video:

- Camera A shoots the story and positions the important part of the picture in the upper left or right (as appropriate) quarter of the viewfinder of the camera.
- Camera B shoots the anchor leaving an empty space above the anchor's shoulder where the image of camera A can be placed.

corner wipe: A video effect where a small image is positioned in any corner, usually the upper right or left corner, of the full screen image. Two edges of the smaller image touch the frame of the larger picture.

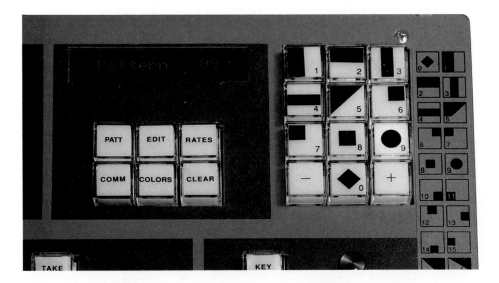

Figure 23-1. Some of the most common wipe patterns are indicated on the buttons and illustrations on this control panel.

Figure 23-2. A wipe is a line that moves across the screen, replacing one picture with another. If the wipe line stops at some point on the screen, a split screen is created.

When the wipe is performed, the image from camera A appears over the newscaster's shoulder. In a corner wipe, two sides of the smaller picture touch the actual frame of the larger picture, **Figure 23-3**. The corner wipe appears only in the upper corner of the full screen image shot by camera B. This is an important consideration when framing a picture for a corner wipe. The only portion of the picture shot by camera A that is seen by the audience is the part included in the wipe. The remainder of the picture from camera A may actually be poorly framed—the main concern is the portion of the image used for the wipe. To properly frame the shots, the camera operator must know which corner is going to be wiped before the image is shot.

A *corner insert* is a different kind of wipe that is similar to the corner wipe. All four sides of a corner insert image rest fully inside the frame of the larger picture, **Figure 23-4**. In news programs, a corner insert is often referred to as an *over-the-shoulder graphic*. This resembles the "picture-in-picture" feature on some television sets.

corner insert: A video effect where a small image is positioned in any corner, usually the upper right or left corner, of the full screen image. All four sides of the smaller image rest inside the frame of the larger picture. Also called an *over-the-shoulder graphic* in news programs.

Figure 23-3. In a corner wipe, two edges of the wiped image touch two edges of the frame of the background image.

Shadow and Edge

Edge and shadow are additional effects that can be added to wipes. The edge feature places a border around the image. The SEG operator can adjust the thickness and color of the edge. The shadow feature places a drop shadow around two sides of the image. Both edge and shadow provide a degree of separation between the wipe image and the background image, **Figure 23-5**.

Mixes

Mixes include fades, dissolves, and superimpositions. (Fades and dissolves were discussed in Chapter 22, *Video Switchers and Special Effects Generators*.) A *superimposition*, or *super*, places the output from two different

superimposition: A video effect where two images, or the output from two different sources, are placed on screen at the same time. Each image is less than 100% of its original intensity, but when combined, the result is 100% of the intensity possible on the television screen. Also called a *super*.

video sources on the full screen at the same time. Each image is less than 100% of its original intensity, but when combined, the result is 100% of the intensity possible on the television screen. For example, one image is the picture of an athlete and the other is the athlete's name created on a character generator (CG). The director wants the athlete's name to appear on the screen, so the audience can read his name as they hear what he has to say, **Figure 23-6**. When a name or other text is placed in the lower third of the screen using a superimposition, the effect is called a *lower third super*.

Suppose that the athlete's jersey is red and the CG letters are white. Using a superimposition, both images are less than 100% of their original

lower third super: A video effect created when an image or text is displayed in the lower third of the screen using a superimposition.

Figure 23-4. None of the four edges of the keyed image touch any of the edges of the frame of the background image in a corner insert.

Figure 23-5. Edge and shadow modes further separate a keyed image from the background image. This separation helps the viewer to more clearly understand the total image.

Figure 23-6. Both images mixed in a superimposition are somewhat transparent.

intensity and have a ghostly, see-through appearance. The image of the athlete is not quite as bright as it would be without the superimposition, because less of the image displayed. The white CG letters are less intense than pure white and allow some of the red color from the jersey to bleed through. The final picture on the screen is an athlete wearing a red jersey with pink letters that spell out his name. This is probably not the image the director intended. Because colors "bleed" into each other when using superimpositions, the lower third super has largely been replaced by the lower third key.

Keys

A *key* is an effect similar to a superimposition, except that both images are displayed at 100% of their original intensity. This solves the compromised color issues created when using a lower third super. A key visually cuts a hole in the original background picture and fills the hole with the desired text or image. When a key is positioned in the lower third of the screen, it is called a *lower third key* or *lower third*. A lower third key is the most widely used method of identifying individuals featured on-screen in programs, **Figure 23-7.**

key: A video effect where two images, or the output from two different sources, are displayed on screen at the same time. Each image is displayed with 100% of its original intensity.

lower third key: A video effect created when an image or text is displayed in the lower third of the screen using a key effect. Also called a *lower third.*

Figure 23-7. A key permits both the background image and the keyed image to be visible at 100% intensity.

VISUALIZE THIS

Using the athlete example, imagine cutting the letters spelling the athlete's name right out of a still photograph of the athlete—literally cutting holes in the shape of each letter right out of the photograph. If you place the photograph over a piece of white paper, the letters can clearly be seen. Each image, the player and the white letters, is 100% of the possible intensity. Visually, this is the result of a key.

Talk the Talk

The process of performing keys may also be called *compositing*.

At least two video sources must be in operation when a key is activated. The *fill source* is the video source that provides the background image. If the fill source is a live image coming from a camera, the camera shooting the fill source image is called the *fill camera*. The second video source is the *key camera*, which provides the shape of the hole cut into the background picture. The relative strength of the key camera's signal is adjusted with the *key level control*, sometimes called a *clip control*, located on the key module of the SEG (**Figure 23-8**). The signal from any camera selected as the key camera passes through the clip control. This control allows the operator to adjust the luminance, or brightness, of the key image.

fill source: The video source that provides the background image in a key effect; the image may be live, pre-recorded, or computer generated.

fill camera: A video camera shooting the live image used as the fill source in a key effect.

key camera: The video source (camera) that provides the shape to be cut into the background image in a key effect. The portions removed from the key camera image (usually areas of blue or green) will be replaced by the fill source image.

clip control: A knob on the control panel of a SEG that adjusts the amount of luminance in a video signal that is sent to the SEG. Also called *key level control*.

Figure 23-8. The clip control allows the operator to adjust the amount of luminance sent to the SEG.

Chromakeys

chromakey: A type of key effect where a specific color is blocked from the key camera's input.

chrominance: The color portion of the video signal, which includes the hue and color saturation.

A *chromakey* is a key effect that operates with specific colors, or *chrominance*, and can be set for any color. Chrominance is the color portion of the video signal, where luminance is the brightness quality of the signal. Chromakeys are frequently used on television, most often during the weather report on the local evening news. The audience sees the weathercaster standing in front of various maps, pointing to current and future weather activity. In the studio, however, the crew sees the weathercaster standing in front of a completely blue or green background—there is no map behind him.

Chromakey circuitry allows the SEG to tell itself not to "see" a particular color. In the case of the weathercaster, the blue or green of the background is filtered out of the key camera's image. Therefore, the SEG uses everything except items that are blue or green. The SEG sends an image, eliminating everything that is blue or green, to the effects circuitry. The second video source, either the fill source or fill camera, provides a new background to replace everything that is cut out of the key camera's image. In the weathercaster example, the fill image is a computer-generated weather map. The chromakey circuit places the fill image into the key source camera image wherever there is a hole in the key camera's image. So, the blue or green background behind the weathercaster is completely replaced by the weather map, **Figure 23-9.**

If a weathercaster has blue eyes or stripes in his tie that match the background, the viewers at home see right "through" his eyes or tie and see the map. This effect is eliminated with colored contact lenses and wardrobe personnel who are on their toes.

Blue is the most commonly used chromakey color because no race of humans on the planet has blue pigment in their skin. Therefore, people look natural on the mixed picture. Green is the second most common color for chromakey. However, people with olive complexions do not look very healthy if all the green is removed from their skin tone. This problem can be corrected with makeup.

The SEG can be tuned or adjusted to key any color desired by the production company. However, the most common colors used for keying are blue or green. More expensive chromakey modules can fine-tune the color

Figure 23-9. Using a keyed image allows a fill image to be inserted behind the talent. This creates a background without constructing an elaborate backdrop.

Key Camera Image   **Fill Image**   **Combined Image**

to very specific shades of blue or green. Less expensive chromakey modules may key only broad shades of blue or green that encompass many hues.

Lighting the background image used in chromakey is critical. The background image should be very evenly lit, so that no area is brighter or darker than other areas. If the chromakey color is tightly tuned to a specific shade of blue, for example, uneven lighting on the background image will create many different shades of blue. Without even lighting, the image will not key cleanly.

PRODUCTION NOTE

32K fluorescent instruments work much better than 32K incandescent instruments for lighting a chromakey background because the fluorescent instruments are soft lights that provide an even wash of light. Incandescent instruments tend to create "hot spots" of brighter light on areas closest to the instruments, which make it difficult to produce an even wash of light on the chromakey background.

Pixels and Digital Video

A picture printed in the newspaper is composed of little dots. A glossy magazine picture has many, many more dots that are much smaller and packed more closely together. A greater number of dots packed closely together results in a sharper picture. Therefore, a printed picture in a magazine is much sharper than a printed picture in a newspaper.

The picture on a high-definition television screen is also made up of millions of little dots called *pixels*, **Figure 23-10.** ("Pixel" is the shortened form of two words—"picture" and "elements.") Digital television systems use at least twice the number of pixels to create the television picture than analog television systems. This is why the picture on digital televisions is tremendously sharper than on analog televisions.

pixel: One of the millions of little dots that make up the picture on a television screen, in a photograph, and any other type of digital image display medium.

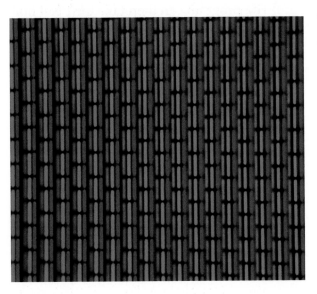

Figure 23-10. The television picture is made up of an incredible number of little dots, or pixels.

To understand how the digital television picture is made, recall exercises in math class on graphing points on the x-axis and y-axis. Each point graphed has coordinates that designate its location on the graph. Each pixel on the television screen also has coordinates (like an address) on an x- and y-axis. It is actually possible to pick a point on the television screen and carefully count columns and rows to determine the exact address of that point. With a method to address and control each pixel individually, the entire picture can be controlled.

VISUALIZE THIS

A mosaic is created of tiny pieces of material, such as colored tiles. An artist painstakingly places the small tiles close together and, eventually, a picture emerges from all the correctly positioned tiles. The smaller the tiles, the sharper the picture. A digital television picture is created in much the same way. The picture on a high-definition television screen is comprised of over 2 million pixels! Just the calculations necessary to create a new picture 30 times every second make digital televisions astounding.

Digital Video Effects

digital video effect (DVE): Video effects that are created using digital technology; based on the ability to alter an image by manipulating each individual pixel.

The special effects created by a SEG that uses digital technology are called *digital video effects*, or *DVE*. A DVE requires the ability to alter an image by manipulating each individual pixel. Each pixel of every image can be located and controlled, which allows us to control the image completely. DVE technology can produce nearly limitless spectacular effects. The simple example that follows illustrates the creation of a digital special effect.

Imagine a black capital letter *T*, almost as tall as the television screen, against a white background, **Figure 23-11**. (The numbers used in this example are not accurate. They are simple representative values to help illustrate the concept.) The address of the pixel on the very tip of the left arm of the *T* is (50, 500). This means that the pixel is in the 50th column of pixels from the left edge of the screen, and 500 rows up from the bottom of the screen. Knowing the exact address allows a computer to take that pixel and place it, for example, at (50, 25). The program can then take the pixel normally at (50, 25), and place it at (50, 500). This move essentially replaces a

Figure 23-11. Controlling each pixel of a picture allows individual pixels to be moved around, which can affect the entire picture.

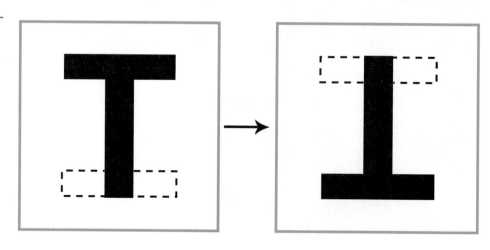

black dot with its exact white counterpart from the bottom half of the screen. Performing this move repeatedly with nearby pixels, would move the upper left arm of the *T* to the bottom of the television screen. One pixel at a time, the left and right arms of the *T* can be moved to the bottom of the screen and the white pixels from the bottom of the screen are moved up. Millions of mathematical calculations are performed in order to turn the *T* upside-down, but computers complete this operation in a fraction of a second. On screen, it appears as a smooth "morph" from a right-side-up *T* to an upside-down *T*. Morphing is one of the most well-known examples of DVE. Because morphing is such a commonly used DVE, an extra program feature that morphs images from one into the next either comes installed on many new computers or is available as a free download. When this technology was first used in music videos, the software cost hundreds of thousands of dollars.

Using DVE, an image on the screen can be stretched, twisted, scrunched, rolled up into a ball, and spun out of the frame of the picture, **Figure 23-12**. It can be torn, rotated, wiggled, changed into a runny liquid,

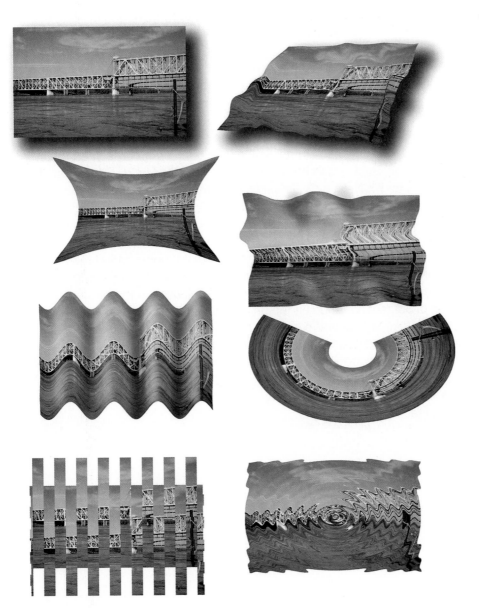

Figure 23-12. There are an infinite number of possibilities for the use of DVEs, depending on the program's budget.

made to slither across the screen like a snake, or any other motion that the budget and imagination allows. Standard SEG effects, such as wipes, can also be performed with DVE technology. A DVE wipe can make an image "move." For example, one picture can literally push another image off the screen or an image may appear to lift up and turn like a page being turned in a book.

DVEs are purchased in packages or bundles and installed on digital SEGs. The number of effects available on a digital SEG system depends entirely on the budget for purchasing the DVE equipment and software. If a desired effect does not already exist, someone can be paid to create it.

PRODUCTION NOTE

Digital video effects have had a profound effect on the film industry. Many special effects are far too expensive or impossible to create using the film medium. However, DVE technology provides complete control of all the pixels that make up a picture, making literally anything possible in digital video. Because of this, the film industry has begun utilizing the capabilities of high-definition digital video when creating visual effects.

The first major film to successfully use digital video effects to a notable extent was *Terminator 2: Judgment Day* in 1991. The realism created on the screen stunned many film and television industry professionals. In this film, viewers saw images that they knew could not be real, yet the images did not appear fake or unrealistic in any way. This movie marked the beginning of a trend in film making where anything the director envisions, no matter how fantastic or impossible, can become an on-screen reality.

George Lucas was one of the first high-profile film directors that chose to work exclusively in high-definition digital video. He waited to shoot *Star Wars–Episode I, The Phantom Menace* (1999) until the available technology could transform the images in his mind into realistic computer generated effects.

Video technology now exists to take film images, convert them to video, combine reality and fantasy seamlessly into a digital composite image, and then convert it back to film for projection in movie theaters. This, of course, depends on the production budget and the ability to hire digital effects artists and computer graphics specialists to carry out the director's vision. The future of digital video technology holds great possibilities. Viewers can be certain that the effects they see in films and television programs will continue to blur the line between reality and computer-generated fantasy as digital technology evolves.

Wrapping Up

Even some technical experts cannot easily determine if an image is real or has been generated by a computer. This is a factor in why "eyewitness" video is considered very suspect in most courtrooms. 3D televisions are already in the marketplace, but viewers must wear special glasses while watching the screen to see the 3D effect. The next step in this technology may be 3D televisions that do not require the use of 3D glasses. There is no way to predict what we will be watching on television at home in the future.

Special effects are amazing and great fun to experiment with. However, all video professionals must remember that, if not used to reinforce the message or plot of the program, special effects should not be used at all. If special effects are overused, they become mundane annoyances and the trademark of inept and amateur videographers.

Review Questions

Please answer the following questions on a separate sheet of paper. Do not write in this book.

1. What is a wipe? How is a wipe accomplished?
2. What is the difference between a corner wipe and a corner insert?
3. What are the challenges in using superimpositions?
4. How does a chromakey operate?
5. Explain how pixels relate to the picture on digital televisions.
6. What benefits does DVE technology offer when producing a program?

Activities

1. While watching television, make note of each time you notice the use of a key, the information or image displayed using the key, and the type of program that made use of keys. Be prepared to share your findings in class.
2. What effects can you apply to a still image using your computer at home? Bring some examples into class.

STEM and Academic Activities

1. Other than the weather forecast on a news program, where have you seen chromakey used? How was this effect used in the program?

2. Digital video effects have changed the way programs are made. What do you think the next advancement in digital video technology will be? How would this advancement affect the television production industry?

Social Science

3. Because computer generated video footage can be manipulated to look convincingly real, courtrooms now look at "eye witness" video with suspicion. Discuss other social impacts, positive or negative, that digital video effects have had.

Chapter 24

Video Editing

Objectives

After completing this chapter, you will be able to:

- Identify the difference between linear editing and non-linear editing processes.
- Summarize the creation and use of an edit decision list.
- Explain the considerations related to editing and action.
- Recall the application of edit transitions.
- Summarize the steps involved in non-linear editing.

Important Terms

audio delay edit
bin
capture
clip
cut rate
digitize
distribution amplifier (DA)
dub
dup
edit decision list (EDL)
edit point
edit through black
edit transition
editing
editor
export

kiss black
linear editing system
matched cut
matched dissolve
non-linear editing system
  (NLE)
pace
processing amplifier
  (proc amp)
split
time base corrector
  (TBC)
time coding
trimming
video delay edit

Introduction

Because television program scenes are not shot in order, they must be rearranged (or edited) into the correct order. Additionally, errors must be removed and any footage that is unwanted must be edited out. Editing is a very complex process with important ethical issues. In broadcast journalism, it is unethical to fundamentally change the intended message of a program or person. Advances in editing technology have made it relatively simple to manipulate the spoken word. A video editor has an awesome responsibility and power over the thoughts of the viewers. This chapter presents the basic concepts and skills necessary to begin using professional non-linear editing systems (NLE).

Editing Systems

editing: The process of selecting the best portions of raw video footage and combining them into a coherent, sequential, and complete program. Editing also includes post-production additions of music and sound effects, as well as effects used as scene transitions.

editor: A collective term that refers to the systems and equipment used to edit program footage.

linear editing system: Videotape-based editing equipment in which the best takes from the raw footage are copied in the order the audience will see them to a tape in the record VCR.

Editing is the process of selecting the best portions of raw video footage and combining them into a coherent, sequential, and complete program. Editing also includes the post-production addition of music and sound effects, as well as video effects used as scene transitions. All the editing processes are performed using pieces of equipment collectively called an *editor*.

Linear editing systems predate the non-linear editing systems of today. Non-linear editing systems are currently the primary tool for editing video.

Linear Editing Systems

Linear editing systems are based on videotape. Raw footage is placed in a playback VCR, or source VCR, with a blank videotape in the record VCR. Good takes of the program footage are copied to the record VCR in the order the audience will see them. Using this process, the program is assembled in a straight line, or "linear" fashion. Simple linear editing systems only perform cuts while editing the footage.

Linear editing systems have been in use for several decades. These systems require a minimum of five pieces of equipment: a source VCR, a record VCR, an edit controller to operate both VCRs, a monitor for the source VCR, and a monitor for the record VCR.

Videotape Generation Losses

A generation is each videotaped duplication of original camera footage. Generations are noted sequentially. Raw footage straight from the camera to tape is the first generation. If the footage is placed in an editor and edited, the edited version is second generation because it is a copy of the first generation. If the edited tape is placed into a duplication system to make 20 copies, each of the copies made is a third generation because they are all copies of a second generation tape. Using a duplication system to make many copies of a master tape is called dubbing, or duping. A copy of the master recording is a *dub*, or a *dup*.

dub: A copy of the master program recording. Also called *dup*.

Talk the Talk
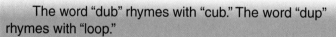
The word "dub" rhymes with "cub." The word "dup" rhymes with "loop."

The quality of the picture decreases with each generation of videotape. While most types of videotape experience generational losses, the losses are greatest with VHS. Currently, only digital tapes can be duplicated without significant deterioration in quality. The speed at which a tape is recorded is a factor in duplication. The slower the tape moves, the greater the loss in picture quality on the copy, **Figure 24-1**. Other factors that affect the quality of the copied tape include:

- The quality of the videotape.
- The quality of the VCRs.
- Use of a time base corrector, processing amplifier, distribution amplifier, or any combination of these.

Approximate Video Generation Signal Loss		
Copy From Speed	**Copy To Speed**	**Picture Quality Loss**
VHS SP (2 hr. speed)	VHS SP	10%
VHS SP	VHS LP (4 hr. speed)	25%
VHS SP	VHS EP/SLP (6 hr. speed)	50%

Figure 24-1. The approximate video generation loss expected when copying tapes at different speeds.

A *time base corrector (TBC)* is a machine that corrects mechanical errors, due to age or use, related to the operation of a VCR. A TBC strips any quality-related imperfections out of the signal and leaves only pure audio, video, and sync in the signal. This piece of equipment compensates for any deterioration in VCR functionality by giving the signal a virtual facelift!

A *processing amplifier (proc amp)* corrects some color and brightness problems in the video signal as it passes through. It is commonly used in videotape recording and duplication systems.

A *distribution amplifier (DA)* is used when a signal must be split and sent to multiple outputs. A DA amplifies the signal before it is split, so each output receives nearly 100% of the original signal. If the original signal was simply split and sent, each output would receive only a portion of the signal. For example, if a signal is split and sent to five outputs, each may receive only 20% of the original signal (100% ÷ 5 = 20%).

Non-Linear Editing Systems

Non-linear editing systems (NLE) are software applications that make use of digital technology and high-capacity computer hard drives to store and process video and audio, **Figure 24-2**. Raw footage is converted to a digital format and copied to the computer's hard drive. Scenes can then be properly arranged, with special effects and transitions added during the process. Once a program is complete, the video and audio can be recorded onto a blank videotape, DVD, or other media format.

Arranging scenes on an NLE is similar to the cut and paste or drag and drop functions of a computer word processing program. Instead of moving text within a document, scenes are arranged on a timeline. There are many different brands of non-linear editing systems on the market, each offering different features. The basic functionality of each brand is the same, but the processing options vary. Each NLE brand or model has a different appearance on the computer screen—the workspace is laid out differently and different brands call the same feature or operation by different names. All of the options available are the manufacturer's effort to provide for almost every conceivable editing situation. However, all of the options and features, in addition to the differences between brand interfaces, creates a learning curve for NLE users. A new NLE operator looking to perform a simple task, such as a dissolve without any other fancy effects, may be a bit overwhelmed. Fortunately, the basic features of an NLE are usually easy to find on the interface, which allows simple editing to begin fairly quickly and gives users time to become familiar with the interface.

time base corrector (TBC): A machine that corrects any quality-related imperfections in the video and audio signals caused by mechanical errors associated with the VCR's functionality.

processing amplifier (proc amp): A machine commonly used when recording or duplicating videotapes that corrects some color and brightness problems in the video signal passing through it.

distribution amplifier (DA): A machine used to amplify a signal before it is split and sent to multiple outputs.

non-linear editing system (NLE): Video editing equipment that is based on digital technology and uses high-capacity computer hard drives to store and process video and audio. The raw footage is converted to a digital format, copied to a computer's hard drive, and may then be arranged and otherwise manipulated.

Figure 24-2. The capability of a non-linear editing system depends on the hard drive capacity of the computer used. The camera footage is uploaded onto the hard drive, all the program components are manipulated, and the completed program is output to the desired format.

PRODUCTION NOTE

Because each NLE system has its own operational commands and procedures, review the manuals and training materials provided for the system by the manufacturer. Most NLE manufacturers also offer resources for private or group training, phone support, and many helpful forums online. Be sure to thoroughly explore the NLE manufacturer's website for usage tips.

Program Editing Basics

Some editing processes and concepts are the same on every brand of NLE system. Understanding and effectively utilizing these concepts, such as screen time vs. real-time and editing action sequences, adds to the production values of a program, which strengthens the delivery of the program's message.

Previewing the Raw Footage

The editing process begins with previewing all the footage shot. While previewing the raw footage, note the specific takes of each scene that should be used in the finished product and where those takes are located on the recording media. This list of scene and take numbers becomes an *edit decision list (EDL)*. To indicate the location of "good" takes, make note of the time code at the beginning and end of each good take while reviewing the footage, **Figure 24-3**. Creating an accurate EDL saves hours of expensive time in an editing suite because the good takes of each scene can be quickly located.

The take log (discussed in Chapter 20, *Directing*) that was created while shooting notes the number of takes recorded per scene and can be used as the foundation of an EDL. An accurate and detailed take log saves time when creating the EDL. Remember that the head recorded before each take displays the slate and countdown. Even during high-speed scanning, the slate is noticeable on the monitor and assists in locating the good takes of each scene recorded.

edit decision list (EDL): A list noting which take of each scene should be used in the final program and the location of each take on the raw footage tape.

Edit Decision List (EDL)		
In Point	**Out Point**	**Scene Number or Description**
00:05:10:21	00:07:22:10	Scene 7, Take 3
00:12:13:24	00:14:03:12	Scene 3, Take 2
00:21:17:02	00:25:09:14	Scene 6, Take 4

Figure 24-3. An EDL notes the location and a brief description of the good takes on the raw footage.

Time Code

Time coding is a system of assigning each frame of a video a specific number, like an address, **Figure 24-4**. To use time code editing, the camera must have a circuit that records time code while shooting. The time code is recorded on an area of the footage that does not affect the picture. The editing system must also have a circuit that reads time code. Time code editing is very precise. If all the necessary equipment is enabled with time code capability, the editor can make an edit within 1/30th of a second of where the edit is actually desired.

time coding: A system of assigning each frame of a video a specific number, like an address.

PRODUCTION NOTE

Many professional production companies, as well as schools, require that an EDL be prepared before using an available video editor. This ensures the efficient use of costly editing equipment for editing purposes only—not for reviewing raw footage. To save a great deal of time spent on the editors, view the raw footage on a regular digital playback unit with time code reading circuitry. The digital playback unit should display the time code to be noted in your EDL.

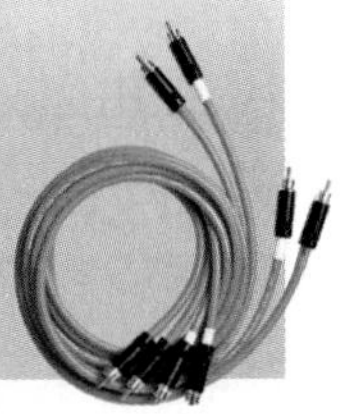

Screen Time

An important goal of successful editing is to create the illusion that the audience has not missed anything in the sequence of action. In editing a program, the "real time" of the footage shot must be edited to reflect the available "screen time" in order to keep the audience's attention. Consider the host of a cooking program demonstrating how to make chocolate chip cookies. The cookies go into the oven and the camera shoots them baking through the window of the oven for the entire 10 minutes of cooking time. This "real

Time Code

01:17:32:12

A shot marked with this time code is located exactly one hour, seventeen minutes, thirty-two seconds, and twelve frames from the beginning of all recordings on the media.

Figure 24-4. The time code indicates the exact location of a shot on the recording media.

time" footage does not contain action that keeps an audience's interest. Too often, directors feel that every frame of video shot is "too good to lose" and are unwilling to part with anything. The result can be 20 minutes of torturous programming that would have been quite interesting if it had been cut to five minutes. In the cooking program example, 10 minutes of real time can be cut to 10 seconds of screen time by editing the shots as follows:

- Shot of cookies in the oven.
- Cut to a shot of a timer set to 10 minutes.
- Dissolve to the timer's alarm sounding 10 minutes later.
- Cut to a shot of the baked cookies being removed from the oven.

Editing and Action

An edit must occur between two shots within a scene and between two scenes in a program in a way that connects them together. Carefully consider where and when an edit should be performed. All shots have a 10 second head recorded, which includes the countdown and slate. At what point during the head should the edit be made? All shots also have a 10 second tail. At what point during the tail should the edit be made? Some guidelines for video cuts and editing include (**Figure 24-5**):

- If shot A includes a moving object/action or the shot is taken by a camera that is moving or zooming, shot B should also include an action shot or camera movement for continuity.
- An edit should not take place between a still (non-action) shot and an action (moving) shot.
- Edits can occur between two still (non-action) shots.

Figure 24-5. Guidelines for video cuts and editing between action and still shots.

Editing Shots		
	Yes	**No**
Action to Action	✔	
Action to Still		✔
Still to Still	✔	
Still to Action		✔

The director of a program may, however, actually want the audience to be jarred, startled, or surprised by the edit. This reaction may help communicate the program's message or the feeling the director is trying to convey. To create a jarring edit, perform a cut or edit in a manner that contradicts the guidelines provided. There should always be a purposeful reason that supports the program and its message when conventional rule is broken.

Matched Cuts and Matched Dissolves

matched cut: A type of edit in which a similar action, concept, item, or a combination of these is placed on either side of a cut.

A *matched cut* is a creative type of edit that places a similar action, concept, item, or a combination of these on either side of a cut. The following are examples of match cuts:

- A historical drama portrays a prisoner about to have his head removed on a guillotine. The camera watches the blade fall in a close-up and the program cuts to a butcher swinging a meat cleaver onto a piece of beef. The same kind of action occurs on both sides of the edit.
- A man leaves his home, gets in a car, and drives off screen right. The program cuts to the car driving into the frame from screen left, as the man arrives at his workplace 15 miles from home. The same item (man and car) and same action (driving) occur on either side of the edit.

A *matched dissolve* is another type of creative edit that uses a similar action, concept, item, or combination of these to transition from one scene to the next. Instead of a cut between the scenes, a dissolve ties the scenes together. Cooking shows regularly use matched dissolves:

- A cake is placed in an oven.
- The oven closes and the camera zooms to a clock that reads "1:15 p.m."
- The shot dissolves to same clock that now reads "1:35 p.m." (20 minutes later).

The same concept (time) and same item (a clock) occur on either side of the dissolve.

matched dissolve: A type of edit in which a similar action, concept, item, or a combination of these is placed on either side of a dissolve.

Editing and Audio

In addition to video, there is also audio on both sides of an edit. The audio must also be considered when determining the best timing for an edit.

- If the audio track is the primary focus, pauses between the talent's lines must be a natural length of time.
- When the response to an interview question runs long and is more than can be used, the response is edited. This edit may create a jump cut. To correct the jump cut, keep the audio flowing and insert a nod shot or cutaway of the interviewer or a shot of relevant B-roll footage.
- All of the background sounds in each take of a scene must match to ensure continuity in the edits. Just as there must be video continuity within a scene, there must also be audio continuity within a scene. This is where taking extra time on location to record a few minutes of room tone and background sound can save a program in the editing room. The room tone and background sound can fill audio gaps so there are no periods of "dead" air in the program.
- The level of the accompanying music and sound effects tracks must not compete with the primary audio of the scene. The volume level of the background sound or music should be no higher than one-quarter the level of the primary sound (usually dialogue).

Breaking these rules for audio and video is acceptable only if the producer or director intends to jar the audience. Any time a conventional rule is broken, there should be a purposeful reason for doing so that supports the program and its message.

Editing Transitions

In English class, a "poor transition" means the ending of one paragraph does not flow well into the beginning of the next paragraph. Video editing is like writing a paper—each scene must transition into the next scene. An *edit transition* refers to the way one scene ends and the next scene begins, such as fading scene 6 out while fading scene 7 in. An edit transition can be a

edit transition: The way in which one scene is edited to end and the next scene is edited to begin.

fade, dissolve, wipe, special effect, or a digital video effect. Regardless of the method used for an edit transition, the action, plot, and theme must all transition smoothly from one scene to the next. If these elements do not transition smoothly, the audience is left confused. In the television industry, a confused audience will likely reach for the remote and change the channel.

Even on the simplest editor, called a "cuts-only" editor, all editing transitions do not necessarily look like cuts. It is possible, for example, to edit a fade out followed by a fade in if the script is marked appropriately before shooting. If the director indicates on the script that scene 6 should fade out and scene 7 fades in, the scene can be shot with the applicable fade out or fade in. Some camcorders have an automatic fade feature. If this feature is not available, shoot a fade out by smoothly closing the iris of the camera. Since the end of scene 6 fades to black and the beginning of scene 7 fades in, the edit between the scenes can be made while the screen is black. A cut from black to black is not noticeable to the audience. This kind of edit is called an *edit through black*, or *kiss black*.

To create the illusion of moving in and out of a flashback, a shot can be brought into and out of focus. While the picture is out of focus, the cut to another out of focus shot is nearly invisible to the audience. For this type of transition, the scenes must be shot using rack focus on the camera.

The frequency of cuts or edits during a program is the *pace*, or *cut rate*, and is usually expressed as a "per minute" rate. The pace of most prime time television programs averages one cut every 7 seconds. Always strive to keep a program moving to retain the audience's interest.

edit through black: An edit in which a cut is made during the period of black on screen between a fade out and a fade in. Also called *kiss black*.

pace: The frequency of cuts or edits per minute during a program. Also called *cut rate*.

PRODUCTION NOTE

While most NLEs offer hundreds of editing transitions, it is not appropriate to use the transitions just because they are available. Programs with gimmicky, flashy transitions are often assumed to be the work of amateur producers. The most common edit used, by far, is a simple cut. Dissolves and fades trail closely behind simple cuts.

ASSISTANT ACTIVITY

Watch one hour of prime time network television, not including the spots that appear during the hour of programming. One hour of prime time programming is only 42–44 minutes of program time when the spots are removed. With an average of 7 seconds between each edit, there are approximately 370 edits during an hour of programming. Watch the programming carefully and note the number of edits that are NOT a cut, dissolve, or fade. You'll notice a very small number of other edits used. Professionals do not use many edit transitions other than dissolves, cuts, and fades.

Cutaways and B-Roll

The importance of recording cutaway and B-roll footage has been discussed in previous chapters (see Chapter 8, *Scriptwriting* and Chapter 19, *Production Staging and Interacting with Talent*) and was presented as a way to bridge jump cuts. In the process of editing a program, cutaways and B-roll have several other uses.

Talk the Talk

In this context, the terms *cutaway* and *B-roll* mean the same thing. The only difference is the industry that most commonly uses each term—"cutaway" is primarily used in television production and "B-roll" is primarily used in broadcast journalism.

Cutaways can be used to pick up the pace of the program and create more visual interest. In a slow moving program with little action and few cuts between cameras, for example, cutaways can be inserted throughout the program to break up shots with little or no movement.

Cutaways can be added to a program to provide reaction shots. For example, the audience sees a drill sergeant yelling at a recruit. While the drill sergeant is yelling, the program cuts to the recruit's face so the audience can see the recruit's reaction to the yelling.

Cutaways may be used to support the message of the program. During a lecture or speech program, for example, the editor may insert cutaways of charts, graphs, or presentation slides with information related to the lecture or speech topic.

Cutaways may also be used to cover an audio edit of a long-winded speaker.

Video and Audio Delay Edits

While editing a program, it is possible to separate the video and the audio signals and edit each at different times. Consider the following scene:

Maria and Janet are sitting in Janet's office. Maria asks, "What time is your husband flying in?" Janet responds, "He'll be landing at five o'clock this afternoon." Cut to a shot of the airplane landing at the airport.

A *video delay edit* cuts to the audio portion of the next scene before the corresponding video is seen by the audience, **Figure 24-6**. Using a video delay edit, the sound of the airplane landing is audible when Janet says the word "landing" and video of the airplane is cut in after the word "afternoon." An *audio delay edit* cuts to the video portion of the next scene before the corresponding audio is heard by the audience. Using the example above, video of an airplane is cut in on the word "landing" as we hear Janet complete her line, "…at five o'clock this afternoon." The audio of the plane landing then cuts into the scene.

video delay edit: An edit that cuts to the audio portion of the next scene before the corresponding video of the new scene is seen by the audience.

audio delay edit: An edit that cuts to the video portion of the next scene before the corresponding audio of the new scene is heard by the audience.

Non-Linear Editing

One of the greatest advantages of using a non-linear editing (NLE) system is the efficiency that digital technology offers. Once the EDL is finalized and all the necessary raw footage is accessible, editing with a non-linear editing system can begin. The following are steps in the NLE editing process:

1. Capture, or digitize, the recorded footage.
2. Split the footage into clips.
3. Separate audio from video, as needed, to be edited individually.
4. Color correction.
5. Audio correction.
6. Create a timeline of scenes.
7. Trim the clips.
8. Apply audio and video effects and transitions, including music.
9. Insert titles.
10. Output the completed program.

Figure 24-6. A video delay edit or an audio delay edit can make an otherwise plain cut appear more interesting.

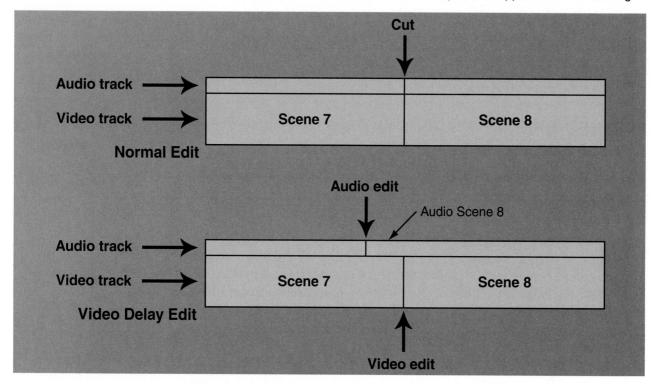

PRODUCTION NOTE

The duration of the editing process is dependent on the editor's knowledge of the NLE processes, the rendering capabilities of the system, and, of course the number and complexity of the needed edits. The best way to become proficient in operating an NLE is to practice on the system as often as possible.

capture: Non-linear editing process of copying all the good program footage to a computer hard drive or to a server. Also called *digitize.*

Capturing Recorded Footage

The first step in NLE is to copy all the good footage onto a computer with a high-capacity hard drive or, if available, a video server. This copying process is called *capturing*, or *digitizing*, the footage. For footage recorded onto digital videotape, capturing the video and audio is a real-time copying process. Footage recorded directly to a portable hard drive or other tapeless recording format, however, can be captured in seconds by connecting the portable hard drive or other tapeless recording format directly to the NLE computer. The footage files can be copied to the NLE computer or to a video server.

Talk the Talk

In a professional production environment, the terms *capturing* and *digitizing* are used interchangeably. Neither of these terms is used more commonly than the other.

PRODUCTION NOTE

If the footage is copied to a server, all the NLE editors can access the footage from any workstation. An editor can work on one editing station today and on a different editing station tomorrow, and still have access to the centrally-located footage. With footage stored on a server, it is even possible for several editors to work on different parts of the same program simultaneously.

Depending on the size and type of production, there may be a large amount of program footage to capture. While servers usually have sufficient space to store the footage, the computers used with NLE systems must have high-capacity hard drives with a great amount of available memory to store all the program footage. Using an EDL reduces the amount of hard drive space required to store the program footage and reduces the capture time, as well. The editor can upload and capture only the "good" takes with the corresponding heads and tails by referring to the EDL, instead of uploading all the program footage.

By converting the audio and video footage to a digital format, individual images, scenes, and audio tracks can be manipulated and edited using the features and technology available on an NLE system. Images can be viewed as they are transferred to the hard drive, if the capturing process is a real-time transfer. As the beginning of each new scene is displayed, the NLE operator presses a button to *split* the footage into individual clips. A *clip* is a captured scene or piece of video that can be used when compiling the completed program. If the footage is captured from portable hard drive or other tapeless recording format, splitting the footage is a separate operation that must be performed after the footage is placed on the NLE. Many newer NLE systems can detect when the incoming video footage contains a break (a record stop/start) and will automatically begin a new clip without any NLE operator action necessary.

As footage is loaded onto the NLE, a frame of each video clip—usually the first frame containing the slate—is displayed on the computer screen as a thumbnail icon. This is why it is so important that the slate be recorded during the head of every take while shooting. All of the thumbnail icons for the footage are contained in a folder called the *bin*, which may be viewed in a separate window on the computer screen (**Figure 24-7**). If there are 45

split: Non-linear editing operation in which program footage is separated into individual clips.

clip: A captured scene or piece of video that can be used when compiling the completed program.

bin: A folder on a non-linear editing computer that contains all of the captured footage for a program. A thumbnail icon of the first frame of each video clip contained in the bin may be viewed in a window on the computer monitor.

Figure 24-7. The first frame of each scene is displayed to help identify the footage contained in the bin.

scenes or video clips of a program loaded onto the hard drive of an NLE computer, the bin will contain 45 thumbnail icons. On some NLEs, both bins and clips can be given a short name that describes the contents. Bins may be named for the corresponding program, and clips may be named to reflect the scene footage, such as "Scene 4, Take 3."

At this point, the editor does not spend time on perfectly editing the clips. The clips are rough cuts and may still contain complete heads and tails. Precise editing occurs later in the process during trimming.

Timeline Creation

The captured rough scenes are arranged along a timeline in the order they will appear in the finished program, **Figure 24-8**. NLE systems offer the ease of click-and-drag scene organization—click on the footage file or clip for a scene, drag it to the desired place on the timeline, and drop the scene into the sequence. To place a scene on the timeline with some NLE systems, the NLE operator simply double clicks the footage icon. If another scene or effect is inserted at a previous point on the timeline, the existing material shifts forward along the timeline to make room; previous work is not overwritten. This process is very similar to inserting a sentence into a paragraph using a word processing program—all the exiting text after the inserted sentence moves down on the page to accommodate the new sentence.

Trimming

Trimming is the process of determining the exact place an edit should occur and cutting the clip to remove unnecessary footage. The exact place a scene begins or ends is called an *edit point*. The edit point at the end of a scene is called an edit "out point" and the edit point at the beginning of

trimming: Non-linear editing process of determining the exact place an edit should occur and cutting the clip to remove unnecessary footage.

edit point: The exact location in the footage where an edit should occur. The edit "out point" is the edit point at the end of a scene. The edit "in point" is the edit point at the beginning of a scene.

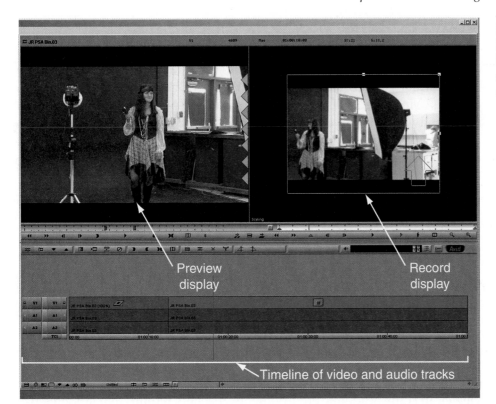

Preview display

Record display

Timeline of video and audio tracks

Figure 24-8. Using a non-linear editing system, the timeline can be rearranged as easily as clicking and dragging the scene icon.

a scene is called an edit "in point." The head and tail recorded for each scene are removed by the trimming process. However, the recorded heads and tails provide the editor with more flexibility in deciding when a scene should start. During the 15-second head, the slate is recorded for the first 10 seconds. This leaves 5 seconds before the first "important" action or dialogue begins in the scene. The editor may choose to allow an emphatic pause before the main action of the scene begins, or the editor may jump directly into the scene. Editing is all about choices, and the recorded heads and tails give the editor more choices.

When the scenes are placed in order, they are trimmed to flow naturally. Depending on the NLE system, precision trimming can occur between fields – 1/60th of a second. (The term *field* is discussed in Chapter 25, *Getting Technical*.) The order of the scenes, effects, or audio can be adjusted, previewed, and moved again until correctly placed.

Correction

Minor discrepancies sometimes occur between two scenes. For example, the color of an item is not identical in both scenes. Most NLE systems have color correction features to address this type of discrepancy. These features can help match scenes and correct the image to some degree. Color correction features on an NLE are considered quite advanced and may not satisfactorily correct the images. Shoot the scene right the first time and do not regard "fixing it in post" as a viable option.

Audio correction features are also available on most NLEs and are a bit more user-friendly than color correction features. However, audio correction features are not designed for a novice editor either. Again, shoot the scene right the first time.

Audio Editing

Once the clips of a program are trimmed and the program flows smoothly, the audio mixing, music, nat sound, and sound effects can be applied. Many NLEs offer an abundance of tracks of audio to use while editing. Using these tracks, each type of audio is placed on a different track to be easily mixed with all the other tracks. Although many tracks of audio are available to an editor, the primary audio (usually dialogue) should never be difficult for the audience to hear. Typically, other audio tracks should be no higher than 1/4th the level of the primary audio.

Effects and Transitions

NLEs are equipped with built-in special effects, DVEs, and editing transitions (**Figure 24-9**). A transition may be selected from a menu and set to last for a specific amount of time. The default edit type on an NLE is a simple cut. To perform any other transition type—from a dissolve to spinning an image then shrinking it, or curling the edges into a circle, inflating it into a 3D ball, and making it fly off the screen into infinity—requires additional menu and button selections. Before deciding to use a transition other than a cut, decide if the transition contributes to the overall message/purpose of the program or if it just looks cool. What looks "cool" to an editor may be "cheesy" to the viewer. In some academic environments, teachers require that students use only cuts in the first few projects they undertake.

Figure 24-9. Selecting the transition from one shot to another is as simple as clicking an option from an on-screen menu.

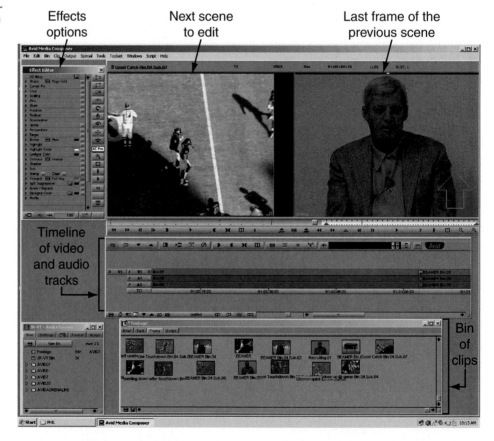

Effects options   Next scene to edit   Last frame of the previous scene

Timeline of video and audio tracks

Bin of clips

Titles

Many NLEs offer built-in titling devices, **Figure 24-10**. The opening and closing titles may be created and added to the finished program or keyed over existing video, if desired.

Exporting

Once a program is complete, it can be *exported* for duplication and distribution in the necessary format, such as videotape or DVD, or the program may be aired directly from the hard drive without any loss in quality. Exporting an edited program to tape is usually a real-time process. Exporting to DVD is a much faster process, but the speed of the export depends on the capabilities of the equipment being used.

export: The process of copying the completed program for duplication and distribution.

Preview display   Record display

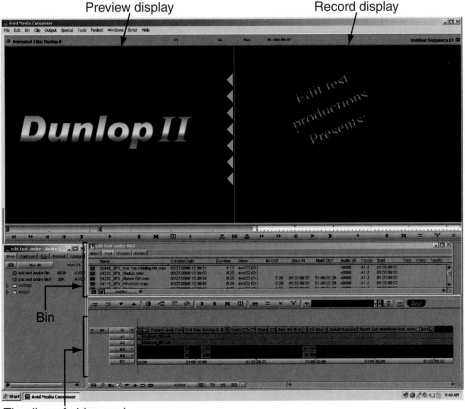

Bin

Timeline of video and
audio tracks

Figure 24-10. The character generator module built into an NLE has a display screen with various options available, such as bringing up blank pages, colorizing backgrounds, creating letters, applying edges, and creating motion. Text is entered once all the settings have been selected.

Producing Quality Programs

The goal of every production operation should be to produce the best program possible—from pre-production all the way through the process to duplication and distribution. Unfortunately, there are so many things involved in creating a really good program that sometimes the simple steps and details get overlooked. Failing to remember the simple things can often ruin an otherwise great program. **Appendix A** contains a comprehensive review of production values and technical considerations for maximum quality programs.

Wrapping Up

Editing is one of the most exciting positions on the production team. Novice editors should practice with cuts only, regardless of the type of editing system available. Learn to tell a simple story without flashy special effects and practice until the results are acceptable. The features and functions of more elaborate editing systems will become familiar with experience. Technology promises to provide continual changes and improvements to video editing systems. To be successful, you must learn to make the best use of new technology and adapt to the changes.

Review Questions

Please answer the following questions on a separate sheet of paper. Do not write in this book.

1. What is the difference between linear editing systems and non-linear editing systems?
2. What information is included on an EDL? When is the EDL created?
3. Summarize the guidelines for editing and action.
4. What is a matched dissolve?
5. What is an edit transition?
6. Explain how cutaways are used in the editing process.
7. Identify the steps involved in the NLE process.
8. What is the purpose of trimming?
9. What is the benefit of using several tracks of audio while editing?

Activities

1. Watch a few prime time television programs. Make a list of all the matched cuts you notice. Be prepared to describe the matched cuts in class.
2. Compare the cut rates of two different types of programs. Is the cut rate of one faster or slower than the other? Does the cut rate serve a particular purpose in either program?

STEM and Academic Activities

Technology

1. What advancements have been made in video editing software products available on personal computers in the last 10 years? What video editing functions are you familiar with on your home computer?

Mathematics

2. The pace of most prime time television programs averages one cut every 7 seconds. How many cuts are in a 30-minute program, with 10 minutes subtracted for spots?

3. Watch a program that makes use of cutaway shots. List each and identify the purpose of each cutaway shot. Examples of purposes include picking up the pace of the program, providing a reaction shot, supporting the message of the program, covering an audio edit.

4. Find an example of an edit between shots or scenes that was intentionally jarring or surprising. Why did you find the edit jarring or surprising? How did this edit contribute to the program?

www.emmysfoundation.org The Academy of Television Arts and Sciences Foundation offers programs to educate the next generation of television and emerging technology professionals by honoring academic excellence, creating learning opportunities, mentoring students and faculty, and building partnerships with creative and business leaders in the industry.

2010 COLLEGE TELEVISION AWARDS GALA

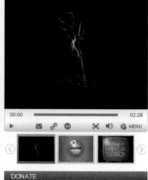

VIDEOS

00:00   02:28

TELEVISION ACADEMY FOUNDATION

WITH TELEVISION, THE SKY IS THE LIMIT . . .

Explore this site to find out more about the Foundation's programs and how you can get involved!

ACHIEVEMENT

INTERNSHIP

SCHOLARSHIP

ENRICHMENT

Welcome to the online home of the Academy of Television Arts & Sciences Foundation. Established in 1959 as the charitable arm of the Television Academy, the Academy of Television Arts & Sciences Foundation is dedicated to using the artistry of television to preserve and celebrate the history of television, and educate those who will shape its future.

We invite you to explore some of the Foundation's renowned programs including the Archive of American Television, Student Internship Program, College Television Awards, Fred Rogers Memorial Scholarship, TV Faculty Seminar and more.

DONATE

Your assistance and support will provide resources to our community and help those who will be the future of the television industry.

DONATE

JOIN THE FOUNDATION'S MAILING LIST

name@mail.com   GO

PRIVACY POLICY | TERMS AND CONDITIONS | ABOUT US | CONTACT US
ARCHIVE OF AMERICAN TELEVISION | DONATE
EMMYS.COM | EMMYS TV | JOURNEYS BELOW THE LINE | THEATRE RENTAL

FOLLOW SITE

SHARE

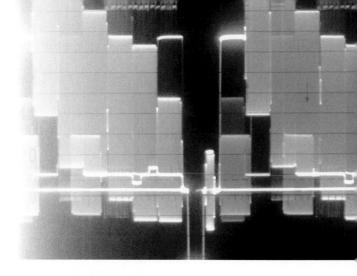

Chapter 25

Getting Technical

Objectives

After completing this chapter, you will be able to:

- Summarize how the television picture is produced.
- Explain the function and importance of sync to video equipment during production.
- Identify the differences between standard definition television and high-definition television.
- Understand the difference between interlace and progressive scan technology.
- Recall how each of the digital television technologies discussed create an image on screen.

Introduction

Even though current television technology is being tremendously impacted by the digital revolution, the most common consumer television receivers found throughout the country are still analog. Digital television signals were broadcast in conjunction with analog television signals from 2000 until 2009. The Federal Communications Commission mandated that analog signals cease to be broadcast as of June 12, 2009.

This chapter provides an overview of the technical aspects of the analog video signal as a general perspective for typical production personnel. The basics of digital video systems are also presented, including some of the most common digital video receiver options. Having a basic understanding of the video signal is helpful when troubleshooting problems during a program shoot.

This chapter is not intended to be an introduction to video engineering. A video engineer is extremely knowledgeable in the intricacies of television signals from an electronics perspective. Students with an aptitude for electronics may find a rewarding career as a video engineer.

Important Terms

cathode ray tube (CRT)
DLP (digital light processing)
field
frame
genlock
high-definition (HD)
HiDef
HDV
interlace
LCD (liquid crystal display)
letterbox
OLED (organic light emitting diode)
plasma
progressive scan technology
standard definition (SD)
sync generator
vectorscope
waveform monitor

Analog Television

A *CRT (cathode ray tube)* is a type of video display device that creates the on-screen image using a cathode that fires electrons at a phosphor-coated surface. When a phosphor is struck by an electron, it glows for a fraction of a second. Consumers commonly refer to a CRT television as a "picture tube" television. If we were to saw the picture tube of an analog CRT television set in half lengthwise, we could see the phosphor coating inside of the front piece of glass. The cathode is positioned at the back of the picture tube and "fires" a ray or beam of electrons through the tube, **Figure 25-1**. When the electrons hit the back of the television screen, they produce a glowing dot. Wire coils wrapped around the back of the tube serve as electromagnets that vary their magnetic strength to "steer" the beam of electrons. The trajectory of the beam can be bent after firing to sweep the entire phosphor-coated surface of the screen, instead of just the very center of the screen. The electromagnet, in essence, changes the course of the electron beam from a straight pitch to the screen, to a "curve ball."

ASSISTANT ACTIVITY

Stand closely to a television screen and notice that each picture is actually made up of glowing vertical bars or dots. All of the glowing dots create a picture just like all the pieces of a mosaic create a larger picture. Now try to imagine just how many dots there are in one picture!

Closer inspection of the television screen reveals that the glowing dots are arranged in horizontal lines. There are 525 lines of dots on an analog television screen, but only 486 lines are visible. The remaining 39 lines are at the very top and bottom of the screen. If you have ever seen a television picture "roll," (the image visually scrolls up or down, on a television screen), the black bar between each rolling picture contains the 39 lines that are not normally visible on the screen. The 39 lines are black, having no luminance, but they contain information about how the television set creates the image. Those lines also include the information to place closed

Figure 25-1. The cathode "fires" a ray of electrons at the phosphor-coated television screen, which produces glowing dots.

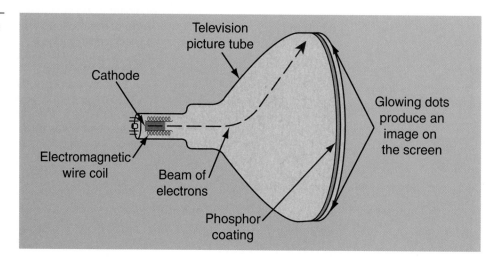

captioning on the screen when the option is activated by the viewer. This information is relevant to the responsibilities of a video engineer and will be addressed in advanced television production courses.

Developing the Technology

During the early development of television systems, electrons were sprayed on the back of a phosphor-coated screen to create a picture from the resulting glowing dots. There were understandable limitations to this technology. The pioneers of television determined that the standard used in this country would be to horizontally spray each individual line until all 525 lines were sprayed. The beam would then return to the top of the screen and begin again. This system is very much like typing a paper on a typewriter. The typewriter "sprays" the letters across the page until it reaches the end of the right edge of the page, returns to the left edge on the next line and sprays letters to the end of the right edge, returns to the left edge on the next line, and so on.

By the time the cathode ray sprayed the bottom line of the television screen, the top of the screen had already dimmed out to black. Remember that phosphor glows only for a fraction of a second. No matter what they tried, the television pioneers could not move the beam fast enough to go all the way to the bottom of the screen and return to the top to respray before the top line dimmed out. There were just too many dots on the screen to be lit up.

It was then decided to try spraying every other line with the cathode ray. This allows the beam of electrons to reach the bottom of the screen in half the time. The cathode ray then returns to the top of the screen and sprays the lines skipped on the first pass. So, the first pass sprays all the odd lines (1, 3, 5, 7) and the second pass sprays the even lines (2, 4, 6, 8). The concept of alternately firing odd and even lines is called *interlace*, **Figure 25-2**. After some experimentation, it was determined that the second set of lines was sprayed just before the first set dimmed out. This gives the illusion of a continuous picture. In fact, an interlaced image is two pictures that "jump" up and down a fraction of an inch at such a fast pace that they appear to be one image. The picture clarity does suffer, unfortunately, with this process. The television pioneers decided that firing half of the lines in each pass

interlace: The process in which the cathode ray fires all the odd lines on a television screen in one pass and returns to the top of the screen to fire all the even lines in the second pass.

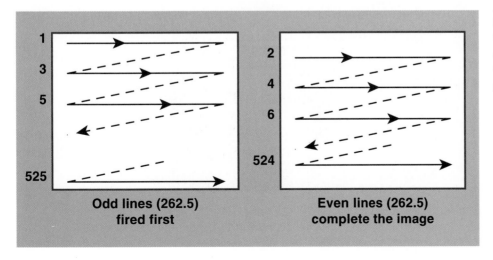

Figure 25-2. Interlace is the process of firing all the odd lines and returning to the top of the screen to fire all the even lines.

still produced an acceptable picture quality and they made it the national standard. This standard came to be known by the initials of the committee that created it: NTSC (National Television Standards Committee).

Frames and Fields

field: One entire pass of the electron beam, spraying every other line from the top to the bottom of the screen.

frame: Two consecutive passes of the cathode ray; every line on the television screen sprayed with electrons. Two complete fields create one frame.

The amazing part of the interlace process is the speed at which it must occur. Remember that only every other line is fired in a single pass of the cathode ray. One entire pass of the electron beam, spraying every other line from the top to the bottom of the screen, is called a *field*. After one field is fired, the beam returns to the top to fire the second field, **Figure 25-3**. Two complete fields create one *frame*. There are 60 fields of 262.5 lines per second or 30 frames of 525 lines per second. This amounts to a total of 15,750 lines fired every second! Each field is slightly different than the preceding field. The slight change in odd and even fields, along with the rapid flashing of the fields, creates the illusion of motion pictures. This amazing technology is more than 70 years old and currently remains the standard for most television sets in this country.

Sync

If the picture tube is like an amazingly fast ray gun that fires electrons, then the television camera can be compared to an amazingly fast vacuum cleaner wand that sucks up the electrons to be fired later at the phosphor on the screen. This preposterous analogy works in explaining the sync signal.

Figure 25-3. Each pass of the cathode ray "sprays" half of the lines on the television screen with electrons, which creates a field. Two consecutive fields sprays every line on the screen and creates a frame.

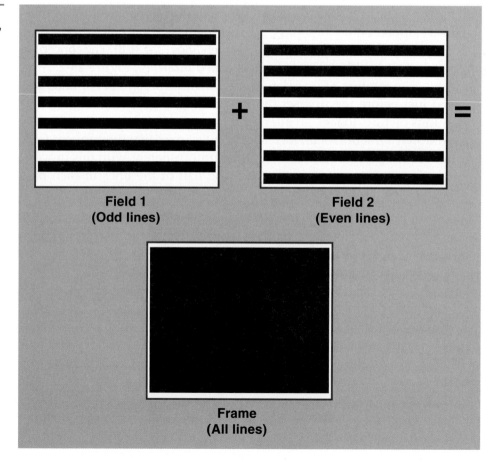

Field 1
(Odd lines)

Field 2
(Even lines)

Frame
(All lines)

If the "vacuum wand" in the camera is pointed at the upper-right corner of the image, how does the "ray gun" in the television set know to aim itself at the upper-right corner of the screen? If the cathode ray were aimed at a different place on the screen, the television screen would display a collection of multicolored dots resembling confetti. The sync pulse is the means by which the cathode ray in the television picture tube is tied to video camera's "vacuum wand."

Even though the scanning rate on each piece of equipment is identical, individual pieces of equipment are powered up at different times. Therefore, the "vacuum wand" in camera 1 may be pointed at the top right corner and the "vacuum wand" in camera 2 may be pointed at the middle of the picture. If the SEG is used to switch between camera 1 and camera 2, the picture will roll or otherwise glitch because the cathode ray is forced to skip large chunks of the picture. The sync pulse overrides the internal timing of each piece of equipment and sets everything scanning in the same place at the same time.

The importance of sync becomes greater as additional cameras are added to a shoot. All the cameras on a shoot must be tied to one sync signal, otherwise the image will jump or be briefly distorted every time the TD switches from one camera to another. The *sync generator* is a device that provides a synchronization signal to all the video equipment in the television production facility. In small studio facilities, the sync generator is usually built into the SEG, which is often used to provide the signal to all the necessary gear. This output signal is usually labeled "sync out," "black burst," or "genlock." The sync generators built into some cameras must be turned off to allow the camera to "listen" to the external sync source. *Genlock* is a module that overrides the internal sync of an individual piece of equipment to synchronize the equipment with the signal provided by the sync generator. In larger facilities, every piece of gear in the facility is commonly connected to a separate, large sync generator. A large sync generator puts out a sync signal referred to as "house sync," because it covers the entire production house.

sync generator: A device that provides a synchronization signal to all the video equipment in the television production facility.

genlock: A module built into video equipment that overrides the internal sync of an individual piece of equipment to synchronize the equipment with the signal provided by the sync generator.

VISUALIZE THIS

Imagine a marching band without a drum section to keep the beat while marching in a parade. After the band stops playing a piece of music, they would be marching in total silence before the next song begins. Without a drum beat, it would not take long for band members to get out of step with each other. The drum beat, like the sync pulse, keeps everyone in step.

waveform monitor: A piece of testing equipment that measures every aspect of an image, including the brightness and darkness of the video signal.

vectorscope: A piece of testing equipment that graphically measures the way the colors are structured in the video signal.

Monitoring Video Signal Quality

The waveform monitor and vectorscope are two pieces of testing equipment that allow an engineer to examine the quality and strength of the video signal. The *waveform monitor*, **Figure 25-4**, measures every aspect of an image, as well as the brightness and darkness of the video signal. The *vectorscope*, **Figure 25-5**, graphically measures the way the colors are structured in the signal.

Color bars are a series of colors on a chart, including black and white, which can be shot by a camera or may be internally generated by the camera, SEG, CCU, or several other pieces of gear. See **Figure 25-6**. The purpose of color bars is to verify that all the cameras on a shoot are "seeing" the same shade of a color when pointed at the same object. The output from two or more cameras can be simultaneously brought into the waveform monitor and vectorscope. This allows the operator to adjust the CCU for each camera to match the output to an industry standard. The colors on the color bar chart should be memorized. The colors, displayed from left to right, are:

- White
- Yellow
- Cyan (a light blue)
- Green
- Magenta
- Red
- Blue (navy blue)
- Black

Figure 25-4. The waveform monitor measures every aspect of the television signal except color strength.

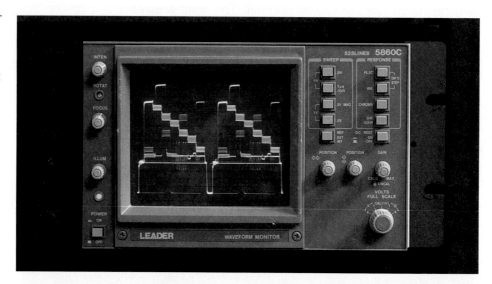

Figure 25-5. The vectorscope graphically measures a signal's color strength and adherence to industry standard settings.

Figure 25-6. The color bars are used to provide an industry-wide standard for color reproduction.

When the color bar chart is activated on a camera, the colors should appear on the monitor. If the colors do not appear in order, it is likely that the monitor's color or tint settings need adjustment. It is important to make the necessary adjustments before beginning a shoot, otherwise colors in the program will be skewed on the monitor. If you view the monitor and see an incorrect color scheme, you might assume the camera is not functioning properly and choose to delay or cancel the shoot unnecessarily. In fact, the tint or color knob on the monitor may have been accidentally bumped and simply needs to be adjusted.

Digital Television Technology

The Federal Communications Commission decreed that all television stations in the country must cease broadcasting analog signals and broadcast only digital signals, based on the guidelines and timeframe the FCC provided. This sounds like a very good change, and it is, for the most part. However, every piece of analog-based television equipment in production houses, studios, cable systems, and private homes will fade into disuse as digital systems provide improved quality and abilities. This means that every television set in every home will be replaced by a digital television, as analog sets are no longer manufactured. Those who do not want to purchase a digital television set may either purchase an ASTC tuner or subscribe to a cable television company that offers analog service. An ASTC (Advanced Standards Television Committee) tuner is a conversion box connected to the television that converts digital broadcast signals to analog signals. Cable television companies convert the digital signals to analog before they are sent through the cable system to customers. However, cable companies will likely offer analog service only as long as it is profitable for them to do so. As soon as demand for analog service dips low enough to be unprofitable, the service will no longer be offered.

Video Format

Many broadcast programs are now broadcast in 16:9 format. If a 16:9 format program is viewed on a 4:3 screen, it appears in letterbox format. An image displayed in *letterbox* means that the original program was shot in a wide screen video or film format (16:9 aspect ratio or higher) instead of the

letterbox: A video display format that allows a program originally shot using a wide screen film or video format to be displayed in 16:9 aspect ratio.

analog television 4:3 aspect ratio, **Figure 25-7**. To squeeze a very wide picture onto an almost square screen without distorting the image, the entire picture is shrunk until the left and right edges of the image are touching the left and right edges of the screen. The top and bottom of the image also move toward the center of the screen. This creates an image that is short and wide.

The 16:9 television format is now standard. 4:3 aspect ratio television screens are no longer manufactured for the consumer market. For a short

Figure 25-7. A letterboxed image would appear considerably smaller on a 4:3 format television. Notice that the 16:9 format television is in the shape of a letterboxed image.

4:3 format television

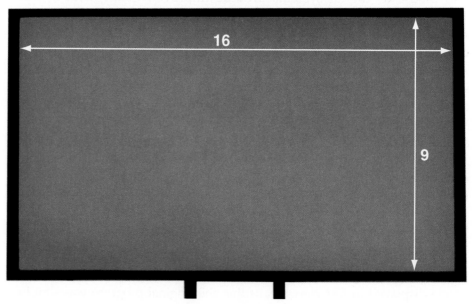

16:9 format television

time, 4:3 aspect ratio digital televisions were manufactured. On these televisions, the picture is still letterboxed. On 4:3 aspect ratio analog television sets with an ASTC tuner, the converter box offers the options of either 100% letterbox images or 4:3 aspect ratio (the left and right sides of the picture are cut off). If watching a program in 4:3 aspect ratio (typically produced prior to 2000) on a 16:9 television screen, the image will not be letterboxed, but display black vertical bars on the left and right sides of the screen to fill up the space on the 16:9 screen. If the 4:3 image were stretched to fill the entire 16:9 screen, the image would likely be comically distorted.

The shape of a 16:9 television more closely matches the shape of the broadcast picture. Therefore, the television picture does not appear to be letterboxed at all. Normally, there is not a black bar at the top and bottom of the screen because the entire television screen is in the letterbox shape. However, some films are shot in aspect ratios considerably larger than 16:9. Even when displayed on a 16:9 screen, these films are likely to be letterboxed so the entire image fits on the screen from left to right.

Talk the Talk

When referring to aspect ratio in writing, the first number is always the width and the second number is always the height. Additionally, the two numbers are always written with a colon between them (4:3 or 16:9). Do not use an *x* between the numbers when writing an aspect ratio.

When speaking an aspect ratio aloud, say "four by three" or "sixteen by nine."

Progressive Scan

Televisions with *progressive scan technology* are capable of painting every line on the screen, instead of every other line, before the top lines on the screen fade out, **Figure 25-8**. This technology doubles the sharpness and clarity of the picture. Moreover, 16:9 television screens are shaped very similarly to 35mm motion picture film, which allows progressive scan televisions to rival the film theater in picture quality.

An HD camera with progressive scan technology may offer an option of recording at "native" frame rate. With progressive scan technology, there

progressive scan technology: The process in which every line on a television screen is fired in one pass, instead of every other line. The system moves fast enough to fire every line on the screen before the lines at the top of the screen begin to fade.

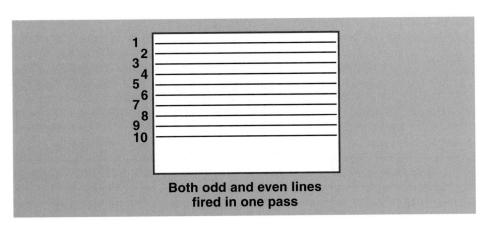

Both odd and even lines fired in one pass

Figure 25-8. Progressive scan technology allows the beam of electrons to move rapidly enough to scan every line on the screen and return to the top line before it has dimmed out.

are still two fields that create one frame. However, unlike analog television, both of the fields in a frame are identical. Each second still contains 60 fields, but there are only 30 different pictures. Using progressive scan, there are 30 *pairs* of pictures. Native frame technology records only one field for each pair of fields. This doubles the amount of video that can be recorded onto the same amount of hard drive space. In playback, each single field is automatically doubled by the native frame technology.

Image Definition

standard definition (SD): A digital television in 4:3 aspect ratio format that displays images using 720 x 480 lines of resolution.

high-definition (HD): A digital television in 16:9 aspect ratio format. 1080 x 720 and 1920 x 1080 are common resolution formats. Also noted as *HiDef*.

Standard definition (SD) is the term applied to a digital television in 4:3 aspect ratio format. Digital television uses scanning lines, just like analog television. However, digital television has clearly organized rows and columns of pixels. In SD, there are 720 columns of pixels and 480 rows of pixels; noted as 720 x 480.

High-definition (HiDef or HD) is a digital television in 16:9 aspect ratio format. There are two widely accepted formats of HD television:

- 1080 x 720 (1080 columns of pixels wide and 720 rows of pixels tall)
- 1920 x 1080 (1920 columns of pixels wide and 1080 rows of pixels tall)

Talk the Talk

When referring to the number of pixels on a television screen in writing, the first number is the width and the second number is the height. Additionally, the two numbers are always written with an *x* and standard spacing between them (1080 x 720).

When speaking the number of pixels aloud, say "seven twenty by four eighty," "ten eighty by seven twenty," or "nineteen twenty by ten eighty."

These expressions, written and spoken, are standard among industry professionals.

The sharpness of a video picture is determined by the number of pixels that make up the image. To obtain the total number of pixels in an image, multiply the number of pixel columns by the number of pixel rows. The higher the number of pixels, the sharper the image. Image quality is strikingly different between SD and HD video. Note the difference in total pixels between the SD format and the HD formats presented in the chart that follows:

Image formats and Pixels		
Image Format	**Pixel Dimensions**	**Total Pixels**
SD	720 X 480	345,600
HD	1080 X 720	777,600
HD	1920 X 1080	2,073,600

HDV recording technology uses MPEG compression to record a 1440 x 1080 image using only 19 megabits of hard drive space per second. SD recording typically requires 25 megabits of hard drive space per second, compressed HD recording requires 100+ megabits per second, and full uncompressed HD requires 1.5 gigabits per second. This means that an HDV camera can record a "compressed HD format" quality picture using less hard drive space than an SD camera. "Compressed HD format" means that when the HDV signal is compressed and then uncompressed in an editor, a noticeable picture quality loss occurs when converting between various video resolutions or formats. In true HD, converting from one video resolution to another does not have any ill effects on the image. Many moderately priced consumer and professional cameras are HDV cameras.

HDV: Video recording technology that uses MPEG compression to record a 1440 x 1080 image using only 19 megabits of recording memory space per second.

Video Recording Options

Many camcorders have options to record in formats other than 4:3. There may be a switch or a menu setting option called "letterbox." Choosing this setting, however, does not guarantee the footage will be recorded in widescreen format. The "letterbox" option may merely be a 4:3 format image with a black bar at the top and bottom of the screen. This produces a lower quality, smaller picture because fewer pixels are used to create the image. Always check the owner's manual for details on specific features and operations. Video cameras and available features are constantly changing.

Recording options labeled "squeeze," "anamorphic," or "stretch" indicate that the images will be recorded in 16:9 format. If an image recorded in this format is viewed on a 4:3 screen, the people and objects in the shot will be very tall and thin because the left and right sides of the screen are "squeezed" into the frame. If viewed on a 16:9 screen, however, the images appear normal. A non-linear editor with the proper software enabled can normalize squeezed footage during the editing process. Normalizing the footage results in higher quality images because more of the pixels are used to create the image.

PRODUCTION NOTE

With television stations broadcasting digital signals, analog VCRs cannot record the broadcast digital signals. Digital VCRs have this capability, but DVRs and DVD recorders are completely replacing VHS recording technology. It is not necessary to replace existing DVD players, as long as the model has a digital output—older DVD players only have an analog output.

Digital Televisions

All digital televisions use pixels to create the picture. Pixels are arranged in lines, which may be scanned in an interlace (every other line) or a progressive (every line in order) pattern. To differentiate between progressive and interlace systems, television sets are given a designation to indicate the number of horizontal scan lines followed by the lowercase letter *p* or *i*, such as 720p or 1080i. Each pixel is colorized to create a tiny dot on the screen. Combining hundreds of thousands of dots creates a picture. Televisions with the highest number of pixels displayed create the best possible image. The number of pixels is typically expressed as "total resolution" on the specifications sheet in the television's instruction manual.

LCD Televisions

LCD (liquid crystal display) televisions contain a thin layer of liquid crystal between the pixel layer and the surface of the television screen. The liquid crystal layer controls each color contained in each pixel in the television image. Each pixel contains three tiny vertical bars—one each of red, green, and blue (RGB). A bright white light source (either fluorescent or LED light) inside the back of an LCD television illuminates the color bars in each pixel. The color bars act like color filters, turning the white light from the source into colored light. Until an electrical charge is applied, the liquid crystal layer in front of each color bar is clear and displays the full intensity of the color bar on screen. An electrical charge applied to the liquid crystal layer affects the opacity of the liquid crystals, which changes or blocks the intensity of colored light displayed and creates the image on the television screen, **Figure 25-9**. The amount of electric charge can vary, which varies the opacity of the liquid crystal layer. By combining the three pixel colors, RGB, and varied opacity levels of liquid crystals, any color and shade can be created.

LCD (liquid crystal display): A television that contains a thin layer of liquid crystal between the pixel layer and the surface of the television screen. The liquid crystal layer controls each color contained in each pixel in the television image.

Figure 25-9. In simple terms, the picture on an LCD screen is created when light passes through the color bars in the pixel layer and electrical charges in the liquid crystal layer change the colored light, which create image pixels displayed on the television screen.

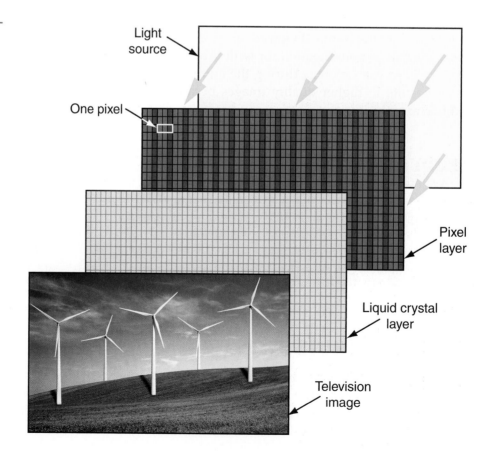

VISUALIZE THIS

If light through the red bar in a pixel is bright red (with a clear liquid crystal), light through the green bar is completely opaqued (blocked by a liquid crystal), and light through the blue bar is bright blue (with a clear liquid crystal), the pixel will appear bright purple (a combination of full intensity red and blue). If the electrical charge applied to the liquid crystals in front of the red and blue bars changes and darkens the liquid crystals to 50%, the pixel will appear a deeper royal purple.

LCD televisions are very thin and can be hung on the wall like a picture with special mounting hardware. However, LCD televisions have a relatively narrow viewing angle. The image on an LCD screen is clearest when viewed directly from the center to no more than 40° off the center of the screen. When viewed from greater angles to the side of the television, the image rapidly deteriorates. LCD televisions have difficulty reproducing deep, dark blacks on screen; dark gray tones are usually the closest LCD televisions can display. However, most consumers do not even notice the dark gray instead of deep black in the images displayed.

Plasma Televisions

Plasma televisions use phosphor and rare gases to light the pixels that create an image. The pixels inside a plasma television are a combination of three tiny fluorescent light cells—one each of red, green, and blue (RGB). Each cell is coated with phosphor and contains a mixture of neon, xenon, and other gases, called *plasma*. See **Figure 25-10**. The fluorescent light cells within each pixel receive an electrical charge that excites the plasma

plasma: A television that uses phosphor and rare gases to light the pixels that create an image.

One Pixel

Phosphor coating inside each cell

Each cell is filled with plasma gas

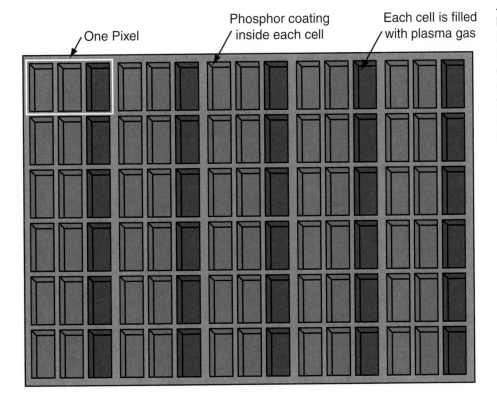

Figure 25-10. The pixel layer in plasma televisions is comprised of phosphor-coated cells filled with plasma gas. The cells function as tiny fluorescent lights when excited by an electrical charge.

gases. The gases generate light photons when excited, which interact with the phosphor coating and emit colored light. Varying the strength of the electrical charge changes the intensity of the color and light output. The RGB combination determines the output color of the pixel.

Plasma televisions are very thin and can be hung on the wall like a picture with special mounting hardware. A plasma screen is very bright and the image is more detailed compared to an LCD screen. However, the overall output of a plasma screen compared to an LCD screen may not be noticeable to most consumers. The image on a plasma television screen is clearest when viewed directly from the center to no more than 80–85° off the center of the screen. Viewing the screen from greater angles to the side of the television causes the image to rapidly deteriorate. Seating arranged in a semicircle provides the best quality image on a plasma screen for multiple viewers.

DLP Televisions

DLP (digital light processing) televisions use a light projection system to create the image on screen. A light source inside the DLP television projects very bright light through color filters and onto a digital micromirror device (DMD) chip, **Figure 25-11**. The DMD chip contains millions of tiny mirrors on the surface that reflect colored light through the projection lens and on to the back of the television screen. DLP televisions are available with either a single DMD chip or with three DMD chips (one for each color—RGB). A three-chip DLP television has considerably better color and clarity than the single chip screen, but also has a higher price tag.

The image on a DLP is best when viewed directly from the center to no more than 60° off the center of the screen. There are almost no size limitations to DLP televisions. As long as the internal projection bulb or bulbs are bright enough, the on-screen image can be huge. The projection lamps do, however, have a finite lifespan and are rather expensive to replace.

DLP (digital light processing): A television that uses a light projection system to create the image on screen.

Figure 25-11. The DLP Cinema® chip is a DMD chip used in DLP televisions. The mirrored surface on the chip measures approximately 11mm x 8mm and contains hundreds of thousands of tiny, individual mirrors. (*Texas Instruments Incorporated*)

OLED Televisions

OLED televisions are incredibly thin and energy efficient. *OLEDs (organic light emitting diodes)* are a type of LED in the form of a layer of film made from organic compounds. The compounds emit light when stimulated by an electrical current. OLED televisions have several advantages over other television technologies, which result in superior picture quality:

- OLEDs generate light, so the televisions do not require a backlight. This allows the televisions to reproduce true black tones more realistically and reduces power consumption.
- The display on OLED televisions can refresh at a much faster rate than LCD screens.
- OLED technology can produce contrast ratios of 1,000,000:1—much higher than traditional LCD displays.
- Quality viewing on an OLED screen extends as far out as 90° from the center of the screen.

The organic compounds used in OLED technology do have a limited lifespan compared to other television technologies. Blue OLEDs, in particular, have the shortest lifespan in an RGB display. Additionally, the organic compounds can be damaged by water, such as condensation and humidity.

OLED (organic light emitting diode): A television that uses a type of LED in the form of a layer of film made from organic compounds to create the image on screen.

VISUALIZE THIS

Because an OLED screen is so thin, it can be flexible. A flexible screen opens the door to brand new uses of video. In the *Harry Potter* movies, the newspaper read by witches and wizards contained moving pictures with the printed articles. Imagine if your daily newspaper had video embedded right on the page!

Each of the digital televisions presented create the television picture in different ways, which affects the best field for viewing, **Figure 25-12**. Consider the type of digital television you will be watching when choosing "the best seat in the house"!

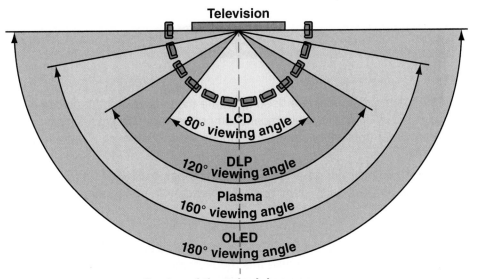

Figure 25-12. This illustration presents the best viewing area for an undistorted image on each digital television discussed in this chapter.

Television

80° LCD viewing angle

120° DLP viewing angle

Plasma 160° viewing angle

OLED 180° viewing angle

Center of the television screen

Wrapping Up

Upon completion of the FCC's mandated digital conversion, digital 16:9 television formats became a unified standard of television technology. This milestone marked a new video age. Comprehensive knowledge of upcoming technology is vital to success in the television industry. Articles published in various trade magazines are good resources in keeping current with industry technology. The reach of technology is constantly expanding—all the technology we know today may be obsolete tomorrow!

Review Questions

Please answer the following questions on a separate sheet of paper. Do not write in this book.

1. How many fields create one frame?
2. What is the function of the sync signal?
3. List the colors on a color bar chart, as displayed from left to right.
4. Summarize the difference in total number of pixels between SD and HD video formats.
5. Explain how progressive scan technology differs from interlace.
6. How do the liquid crystals in an LCD television change the pixel colors?
7. How do the pixels on plasma televisions emit light?

Activities

1. Research the development of the cathode used in television picture tubes. Report on other methods and devices that were created for the same purpose.
2. Search the Federal Communications Commission Web site for facts about digital broadcasting. Make note of any items related to the conversion that you did not know or that you found surprising. Be prepared to share this information with the class.

STEM and Academic Activities

Science

1. Phosphor substances are used to create the picture on many types of televisions. What phosphorescent compounds are used in televisions? What other applications are common for phosphorescent compounds?

Technology

2. How do 3-D televisions create a three-dimensional image on the screen? Explain how 3-D televisions operate differently from traditional television displays.

Engineering

3. A vectorscope measures how colors are structured in a video signal. How does a vectorscope take this measurement? How is the data received from the vectorscope used by a technician or engineer?

4. Watch a movie on DVD in both standard and letterbox formats. List scenes that were different from one format to the other. Did the differences in scenes from one format to the other affect the plot or action in the movie?

5. After June 12, 2009, people could no longer watch analog TV. How did the conversion from analog broadcast signals to digital broadcast signals affect television sales? What accommodations were made for people that could not afford to purchase a new television?

Congratulations!

You have successfully completed this text! Hopefully, you have created a dictionary of terms that can be used as a reference throughout your career. You have learned an entirely new language and should be practicing its usage every day.

This is not the end of your course of study to enter and succeed in the broadcast journalism and television production industries. You are now armed with enough basic information about the entire system to launch into many different directions of study. Never stop learning about new gear and technologies. The technology will not stop changing because you finally mastered it. Technology always marches onward. Remember that knowledge, enthusiasm, hard work, networking, "talking the talk," flexibility, and responsibility are the keys to entry and longevity in this industry. Good luck to you!

Acknowledgements

The greatest acknowledgements in completing this book must be reserved for the most important people in my life. My son, Michael, had to share me with this edition of the book for many months and did not complain. My wife, Agnes, is the reason you are reading this book. She inspired me to write it, she stepped into the "Dad" duties around the house while I was writing it, and she let me "try out" some of the particularly difficult sections on her. This edition was so much more difficult to write and I was often discouraged. Because the technology changes so rapidly, capturing it for a textbook was a bit like trying to catch smoke—the more I tried to catch it, the more it seemed to fan out over a continually expanding area. Agnes reminded me of all the positive feedback I've received from teachers and students about how much the first edition helped them. This is how she motivated me to continue with the project when I was tired and wanted to quit. Agnes was also the first of many who encouraged me expand the text even further to include broadcast journalism. Finally, I'd like to thank the hundreds of students and teachers I have had the privilege of speaking to in workshops and at conventions all over the country. You have all shared so many of your thoughts with me, and I hope you recognized some of them as you read the text.

Thank you to everyone involved in the creation of this text.

The publisher would like to acknowledge the following companies and individuals for their contributions to the illustrations and photos used within this text:

Anderson Districts I & II/Career and Technology Center (Williamston, SC)
Brian Franco
Chuck Pharis Video
Countryside High School, Clearwater, FL
EVTV/Eastview High School (Apple Valley, MN)
EZFX Inc.
Fayetteville High School (Fayetteville, AR)
Glidecam Industries, Inc.
Iconix Video, Inc.
K-AHS/Austin High School (Austin, TX)
Kaleidoscope TV News/King Career Center (Anchorage, AK)
Keystone High School (LaGrange, OH)
Lake Charles-Boston Academy of Learning (Lake Charles, LA)
Lowel-Light Mfg., Inc.
Matthews Studio Equipment
Mole-Richardson Co., Hollywood, CA
Oakleaf Junior High (Orange Park, FL)
OC-TV/Ocean City High School (Ocean City, NJ)
ONW...Now/Olathe Northwest High School (Olathe, KS)
PowerProduction Software
Primera Technology, Inc.
Pro Cyc. Inc.
RCA
South County Secondary School, Lorton, VA
Texas Instruments Incorporated
The Neilsen Company
The Tiffen Company
Titan TV/Centennial High School (Frisco, TX)
VHSTV/Voorhees High School (Glen Gardner, NJ)
Vinten Broadcast Ltd.

That's a Wrap!

The goal of every production should be to produce the best program possible. However, so many things go into a really good production and sometimes the simple things get overlooked. The simple things can often ruin an otherwise great production. The following information addresses the most common technical errors (not journalistic issues) made in student productions and offers simple pre-production, production, and post-production steps that can help make a production an award winner.

Graphics

- Do not misspell anything.
- Use a television-friendly font and color. Do not use fancy letters.
- Avoid the color red. Much of your audience will be watching on an analog television set, which has a difficult time reproducing the color red. Additionally, when viewed on the Internet, the color red bleeds horribly.
- The font should be large enough to be easily read on a small display screen. Move at least 10 feet away from your computer screen and view the graphic. If it is difficult to read at that distance, recreate the graphic until you can see it clearly from 10 feet away.
- Any graphic should remain on screen long enough for to be read aloud <u>twice</u>.
- Use no more than 5 words per line and no more than 5 lines of type per page.
- Put an edge or border on letters to help them stand out from the background.
- Use a font color that contrasts sufficiently with background so the letters do not get lost in the background.
- Writing anything with a pen on a piece of paper is unsatisfactory and unreadable. If the audience cannot read what is written, it is a waste of time and frustrates the audience.
- If you must handwrite a graphic, use a dark marker and write large enough to be readable by the audience. Remember that viewers will not be seated as close to their televisions as you are to your computer monitor.
- If you must use words on a piece of paper, create them with a word processor using a large, bold font that looks professional.

Audio

- Be careful not to cut off the beginning and ending of words when editing. Re-edit until it is right.

- Music should fade in and fade out. It should not end abruptly because a video edit needs to occur before a time limit is exceeded. Many PSAs have been ruined by music that abruptly stops.
- When background music is used with spoken words, the music must be significantly lower in volume than the spoken words. A good rule of thumb: the music should be ¼ the volume of the spoken words.
- Avoid music with lyrics. The lyrics in the music will compete with the spoken audio.
- Closely monitor audio when recording to be certain that the audio level does not change from scene to scene.
- When recording spoken words, speak slowly and clearly enough so that anyone is able understand your words. Let a few people view your piece to verify that they can clearly hear every word spoken. An audience who has never seen the piece before can offer helpful constructive criticism.

Video

- Do not use a fade in a PSA. It slows the pacing and wastes valuable seconds that could be used delivering your message.
- Never shoot with a bright object in the background of a shot, such as a window, the sky, or a white wall. This will pop the contrast ratio and make the talent in the shot appear too dark. Talent with dark skin may become a totally black silhouette.
- When possible, use neutral to medium-dark colored backgrounds (not bright backgrounds).
- If there is not enough light at a location, either shoot someplace else or set up lights. Shooting in the dark is a waste of time.
- Remember that dark images cannot be lightened in post-production without serious side effects.
- If an image or video is viewed over the Internet, transmission through the Internet will make images darker than they were originally shot. You must compensate for this when shooting.
- A black and white "artsy" image does not often work on modern day television. Stay away from black and white images unless there is an overwhelming reason to use them.
- A majority of the editing transitions found on an NLE editor should never be used because they will be seen as a gimmick. Use an unusual editing transition *only* when that type of transition is absolutely warranted by the content of the video.

Technical Evaluation

In both an educational environment and the television production industry, the technical attributes of every production must meet a minimum threshold for acceptance. In the classroom, a program that does not meet the minimum technical requirements is returned to students to be fixed. In the television production industry, programs that do not meet the established technical requirements are not broadcast.

The checklist that follows presents criteria for evaluating the technical aspects of productions.

Checklist for Program Acceptance

Administrative

- ☑ The program proposal, outline, and script are approved.
- ☑ No vulgar language used in the program.
- ☑ No inappropriate music used in the program.
- ☑ Any copyright issues are resolved and written documentation is provided to the executive producer.

Lighting

- ☑ Good lighting used in the program.
- ☑ The iris is open or closed appropriately.

Video

- ☑ Video quality meets or exceeds the minimum standards.
- ☑ The video in the program supports the audio.
- ☑ The audio in the program supports the video.

Audio

- ☑ The audio is at an appropriate level (−3 to +3 Db for analog, −20 to −25 Db for digital).
- ☑ The audio is synced with the video.

Camerawork

- ☑ A steady picture is presented, whether a tripod was used or the camera was hand-held.
- ☑ The video is white balanced (no blue or orange tint).
- ☑ The contrast ratio is appropriate (no deep blacks or glowing whites).
- ☑ The camera angles used meet any classroom-defined shot standards.
- ☑ The shots are in focus.
- ☑ Glitches are not evident in the program.
- ☑ Multiple camera angles are used with many cutaways to make the program visually interesting.
- ☑ Jump cuts and errors in continuity are not evident.

Post-Production

- ☑ Color bars are displayed at the beginning of the program.
- ☑ Five seconds of black follow the color bars at the beginning of the program.
- ☑ Five seconds of black is recorded after the program ends.
- ☑ All CG and credits are spelled correctly and are displayed in an appropriate font and color.
- ☑ The final credit says "Production Company Name and Date."
- ☑ All edit transitions used are appropriate for the program.

Interviewing and Research

- ☑ Appropriate identification of stakeholders and/or experts is provided.
- ☑ The in-depth information presented came from valid research and interviews.

News Writing/Reporting (as applicable)

- ☑ The program follows the appropriate elements of broadcast style.
- ☑ Brief SOTs provide perspective through opinion, prediction, experience, or reaction.
- ☑ The program information is accurate.
- ☑ The program information is balanced.
- ☑ The delivery communicates the story effectively.

Overall

- ☑ There is no significant deviation from the approved script.
- ☑ The program has a clear and understandable flow.
- ☑ The program has a clear message or purpose with interest/relevance to the intended audience.
- ☑ The program meets the minimum and maximum length requirements.
- ☑ The production observes legal guidelines and the code of ethics.
- ☑ The director is not in the program.

Talent Information Sheet

Talent Information Sheet

In preparing for your television appearance, you have probably thought of several questions that you would like answered before arriving at the television studio. This information sheet answers many common questions and provides useful tips on performing and looking your best on the television screen.

Common Questions

What can I expect when I get to the studio?

When you enter the studio, it may seem like all the activity around you is organized chaos—"hurry up and wait" may seem to be the theme of the day. While you should arrive completely prepared for your part in the program, the television crew will probably have to make on-the-spot adjustments for your unique performance. Expect some last minute lighting changes and adjustment of set pieces. Bear with the activity of the studio staff, as their goal is to make you look your best!

What do I wear?

When deciding what to wear for your television appearance, there are several things to consider:
- The colors black, white, and red are not flattering in large quantities on television. If you must wear black or white, do not wear a great amount of either color and do not wear these colors side-by-side (such as a white shirt and black suit).
- If your complexion is dark, do not wear light-colored clothing. Light colors will make you look even darker and may silhouette you. If you have a very light complexion, avoid dark clothing. Dark colors will cause your skin to shine with a science-fiction type glow.
- Wear colors of medium shade and near the middle of the color spectrum. Choose light gray and pastel tones rather than white and dark gray, and dark blue, brown, or dark green rather than black. If you have an olive complexion, however, avoid wearing green.
- Avoid extreme contrasts between items of clothing and between your clothing and the set backgrounds. Avoid small, busy patterns—such as herringbone—and vertical or horizontal thin stripes. These patterns will appear to vibrate in

rainbows of color on the television screen. Horizontal stripes also make you appear considerably heavier than you actually are.

- Both men and women should wear a button-up item of clothing, like a shirt, jacket, or vest. These articles of clothing allow a clip-on mic to be clipped to an edge of the clothing. Mics should not be placed at the neck of a pullover article of clothing.
- Do not wear skin-tight clothing. The TV camera adds the appearance of about ten pounds onto a person. This additional "weight" is unattractive in skin-tight clothing.
- Do not wear a shirt with emblems, logos, symbols, or writing printed on it. The writing is probably not in a font style or size that is readable to the audience. Additionally, wearing a particular logo or company insignia on the program may create complicated copyright issues.
- Avoid wearing short sleeves. Elbows are not pretty on television.
- Clothing should be pressed. Wrinkles in clothing are magnified by the television camera.

Is makeup used on both men and women?

Yes, makeup is applied to both men and women who appear in a television program. In addition to your face, makeup must be applied to your neck, ears, and perhaps even your hands. People with extremely pale-complexions need to have a darker shade of base makeup applied. Most men do not need more than a base layer applied. Teenagers usually appear better on camera if they have translucent powder applied to take away an oily shine.

Should I apply more facial makeup than I usually wear?

No. Heavy makeup is only needed in theater. If you wear eyeliner, do not line all the way around your eyes. Using this much eyeliner gives you a beady-eyed, villainous look and diminishes the effect of the most expressive feature on your face. Also, avoid both glossy and very red shades of lipstick. Remember, that red is not flattering on television. Lipstick with too much gloss creates visible light reflections that are very distracting to viewers.

Should I wear larger pieces of jewelry so they can be seen on television?

Not necessarily. Avoid large, shiny metallic pieces of jewelry. These pieces reflect light back into the camera, which creates a distraction for the viewer. Large dangling earrings wiggle with the natural movement of your head and body while you speak, which is very distracting on television. Large jewelry also gives nervous hands something to fiddle with, which is an undesirable activity while on camera.

Where do I look when I'm in front of the camera?

In an interview or talk show type of program, for example, you should look at the people you are talking to—just as you would in real life. If you are acting in a scene, look at your fellow actors and completely ignore the cameras. If you are giving a campaign speech or delivering news, look directly at the lens of the camera—that is where your audience is.

General Guidelines

The following items are general guidelines to keep in mind while at the television studio and while giving your performance. Following these simple rules will help the recording process run more smoothly.

Please be quiet in the studio!

This is a common rule in any television studio. The technical crew will be very busy with last minute details, such as setting mic levels, and they need to be able to hear each other. Any questions you have should be directed to the floor manager.

Watch your step!

Wires and cables will likely be running along the floor. Be aware of the wires in your walking path so you do not trip and fall. Please do not stand on or step on the wires and cables, as they break rather easily. Repairing broken wires or cables will cause a delay in the program.

If you make a mistake during your performance, do not yell "Cut!"

Try not to let your facial expression or actions show that you have made a mistake. Do not roll your eyes, make a disgusted face, or giggle. The studio staff may be able to fix errors when editing the program. If necessary, *one of the staff members* will call "Cut" and set everyone up to begin again at a point before the mistake occurred. However, the staff cannot fix a mistake in the editing room if you suddenly change your behavior, persona, actions, or if you make sudden sounds.

Pay attention to the floor manager.

The floor manager is in charge of starting and stopping action in the studio. Be attentive to what he says and follow his direction.

If the floor manager says "Cut," it means that recording has been stopped by the director. This is usually due to an error somewhere. If it was something you did, you will be told. However, it was likely due to a technical error on the production side of the cameras. Please do not ask what happened. The production staff will not have time to explain at that moment. If you ask later, they will be glad to tell you. Until that time, you should remain quiet and in position. Listen for the floor manager to let you know at which point you should begin again. Return to that spot physically and verbally. Quietly wait for the floor manager to give you the cue to start again.

Use the microphone properly.

If possible, the mic cord will run under your clothing and be clipped to your lapel. Never put the mic right next to your mouth and speak into it. Do not blow into the mic or tap it. Be aware of where the mic cord is and do not step on the cord, roll over the cord with the wheels of a chair or stool, or get the cord tangled on the legs or wheels of a chair or stool. When asked to do a sound check, do not be shy—just speak normally. Speaking in a normal voice gives the technician an accurate audio level reading. For the sound check, it may be helpful to recite what you are going to say when recording begins.

Glossary

1/8″ connector: See *mini connector.* (7)

1/4″ connector: See *phone connector.* (7)

1/2″ tape: A reel format videotape found only in low-end, industrial equipment. (5)

1″ tape: A reel format videotape available in three formats: Type A, Type B, and Type C. Type C was the most common format. (5)

2″ tape: A reel format videotape used on older machines called quadruplex recorders. (5)

3200° Kelvin: The temperature of white light in degrees Kelvin. Also noted as *3200K* or "32K" when spoken. (15)

A

actors: Individuals who participate in a drama or comedy program, performing as someone or something other than themselves. (8)

ad: A television advertisement for a product or service. Also commonly called a *spot.* (1)

adapter: A connector that changes the type, or connector end, of an existing connector. (7)

ad-lib: A performance technique; when talent speaks lines or performs actions that are not in the script or have not been rehearsed. (19)

affiliate: A broadcast station that has aligned itself with a particular network. The network provides a certain number of hours of daily programming. The affiliate is responsible for providing the remainder of programming to fill the daily schedule. (1)

anchor: The person who delivers the news from the news desk set in a studio. (2)

angle: The approach or point of view used to tell a story. (10)

aperture: The opening, adjusted by the iris, through which light passes into the lens. (3)

arc: Moving the camera in a curved truck around the main subject in the shot—the main subject never leaves the frame of the picture. (4)

arc left (AL): Rolling the camera, tripod, and dolly in a circle to the camera operator's left (clockwise) around the subject of a shot. (4)

arc right (AR): Rolling the camera, tripod, and dolly in a circle to the camera operator's right (counterclockwise) around the subject of a shot. (4)

artifacts: Tiny, rectangular distortions that appear on the screen in digital video formats when a portion of the digital signal is corrupted. (5)

aspect ratio: The relationship of the width of the television screen to the height of the television screen, as in 4:3 (four by three) or 16:9 (sixteen by nine). (14)

assignment editor: The person who schedules necessary equipment and personnel to cover the stories for the day's newscast. (2)

assistant director (AD): See *production assistant.* (2)

attribution: Crediting the source of information used in a story. (10)

audio: The portion of a program that you can hear. Audio includes narration, spoken lines, sound effects, and background music. (2)

audio booth: A room in the studio that contains all of the equipment capable of adding sound to the program. (16)

audio console: A unit that includes many different pieces of audio gear, including the microphone mixer, audio cassette players, CD players, and turntables. (16)

audio delay edit: An edit that cuts to the video portion of the next scene before the corresponding audio of the new scene is heard by the audience. (24)

audio engineer: The person responsible for the audio/sound quality on the production and related equipment. (2)

audio mixer: A piece of equipment that takes the sounds from a variety of sources, such as mics, a CD player, or tape player, and combines them into a single sound signal that is sent to the recorder. (6)

audition: The process in which a director makes casting decisions for a program by watching and listening to prospective performers. (20)

auto-focus: A common feature on consumer cameras that keeps only the center of the picture in focus. (3)

auto-iris circuit: A feature on many consumer and professional cameras that automatically examines the light levels coming into the camera and adjusts the iris according to generic standards of a "good" picture. (3)

automatic gain control (AGC): A circuit found on most consumer video cameras that controls the audio level during the recording process. (6)

B

B-roll: Footage that includes shots of anything visual mentioned during the interview or that is related to the topic, and any natural sound associated with the story. (11)

back light: A lighting instrument that is placed above and behind the talent or object in a shot, at the twelve o'clock position, to separate the talent or object from the background. (15)

background: 1—All the information gathered through research prior to conducting an interview. (11) 2—The material or object(s) that are placed behind the talent in a shot. (19)

background light: A lighting instrument that is pointed at the background of a set. (15)

background music: Music in a program that helps to relay or emphasize the program's message by increasing its emotional impact. (13)

background sound: Type of environmental sound that is not the focus of or most important sound in a shot. (6)

bank: Two buses on the control panel of a SEG that are electronically connected to each other. (22)

barndoors: Fully moveable black metal flaps attached to the front of a lighting instrument; used to block or reshape the light. (15)

barrel adapter: A type of adapter that has the same type of connector and connector end on both sides. (7)

base: The first layer of makeup applied—usually covers the entire face, neck, ears, back of the hands, and bald spot (if applicable). Also called *foundation.* (21)

basic hang: The initial process of hanging instruments over the set according to the light plot and plugging them into the raceway. Also called a *rough hang.* (15)

beat: A specific area (topics or geographic location) regularly covered by a reporter. (9)

Beta SX: A ½″ videotape that uses digital MPEG compression. (5)

Betacam: A ½″ format, broadcast-quality videotape. (5)

Betacam SP: A ½″ videotape format that used to be the best format for professional television use, but digital video formats are challenging this format in professional markets. (5)

big talking face: See *lecture.* (8)

bin: A folder on a non-linear editing computer that contains all of the captured footage for a program. A thumbnail icon of the first frame of each video clip contained in the bin may be viewed in a window on the computer monitor. (24)

black video: A bona-fide video signal that has no lightness. (22)

blending: Incorporating a layer of makeup into the areas surrounding it by brushing the makeup with the fingers or a brush. (21)

BNC connector: A type of connector commonly used in television production. The female and male versions lock together securely with a simple 1/4-turn twist. (7)

boom: A pole that is held over the set with a microphone attached to the end of the pole. (6)

bounce lighting: A lighting technique where a lighting instrument is not pointed directly at the subject of the shot, but the light is bounced off of another object, such as a ceiling, wall, or the ground. (15)

boundary mic: A microphone used to pick up a sound on a stage or in a large room and is most commonly a condenser type. Boundary mics are usually placed on a table, floor, or wall to "hear" the sound that is reflected off hard surfaces. (6)

broadcast: The television signal travels through the air from one antenna to another antenna. (1)

broadcasting rights: Permission to broadcast copyrighted material to the public. (13)

bus: A row of buttons on the control panel of a SEG that access different inputs and functions. (22)

bust shot: See *medium close-up (MCU).* (4)

C

C-clamp: A clamp in the shape of a "C" that is used to attach lighting instruments to the grid. (15)

cable end connector: A connector found on the end of a length of cable. (7)

cablecasting rights: Permission to cablecast copyrighted material to the public. (13)

camcorder: A portable camera/recorder combination. (3)

camera control unit (CCU): A piece of equipment that controls various attributes of the video signal sent from the camera to the video recorder, and is usually placed in the control room or the master control room. Also commonly called a *remote control unit (RCU).* (3)

camera head: The portion of the video camera that contains all the electronics needed to convert the reflection of light from the subject into an electronic signal. (3)

camera line: See *vector line.* (19)

camera monitor: A monitor that displays the image shot by the corresponding camera. (16)

camera operator: The person who runs the piece of equipment that captures the video images of the program. (2)

camera rehearsal: See *dry run.* (20)

cannon connector: See *XLR connector.* (7)

capture: Non-linear editing process of copying all the program footage to a computer hard drive or to a server. Also called *digitize.* (24)

cardioid mic: A mic with a pick-up pattern that captures sound from primarily one direction. Also called a *uni-directional mic* or *directional mic.* (6)

cast: The collective name given to all the talent participating in a production. (2)

cast breakdown by scene: A list of the program's cast members with the corresponding scene numbers in which they appear. (20)

cathode ray tube (CRT): A video display device that creates an image using a cathode that fires electrons at a phosphor-coated surface, which results in glowing dots on the screen. (25)

CG operator: The person who creates the titles for the program using a character generator. (2)

character generator (CG): A device that creates (generates) letters (characters), primarily for titles. (14)

character makeup: Makeup application technique used to make a performer look like someone or something other than the performer's own persona. (21)

charge coupled device (CCD): A component of the camera head into which light enters and is converted into an electronic, or video, signal. The video signal exits on the opposite side of the CCD and enters the rest of the camera. (3)

chassis mount connector: A connector that is built into a piece of equipment. (7)

cheating out: Positioning the talent's body to slightly face that camera to give the audience a better view. (19)

chromakey: A type of key effect where a specific color is blocked from the key camera's input. (23)

chrominance: The color portion of the video signal, which includes the hue and color saturation. (23)

circle wipe: A video effect where a circle grows or shrinks, replacing one picture with another. (23)

clip: A captured scene or piece of video that can be used when compiling the completed program. (24)

clip control: A knob on the control panel of a SEG that adjusts the amount of luminance in a video signal that is sent to the SEG. Also called *key level control.* (23)

close: The conclusion of a story. (10)

close-up (CU): A shot that captures a subject from the top of the head to just below the shoulders. Also called a *narrow angle shot.* (4)

closed circuit television (CCTV): Television where the signal is sent through wires and serves only an extremely small, private predetermined area. (1)

commercial broadcast television: This type of television production facility is "for-profit." The television signal is sent via a transmitter tower through the air and is free for anyone with an antenna to receive it. (1)

concert style music video: A type of music video in which the audience sees the band perform the music that is heard. (8)

condenser mic: A type of mic that requires an external power supply (usually a battery) to operate. The generating element is a thin piece of metal foil or coated film. Also called an *electret condenser mic.* (6)

confidence monitor: A monitor connected to the output of the video recorder. Seeing the image on this monitor ensures that the video recorder received the signal. (16)

connectors: Metal devices that attach cables to equipment or to other cables. (7)

content specialist: A person who works with the scriptwriter and is considered to be an expert in the program's subject matter. (2)

contrast ratio: The relationship between the brightest object and the darkest object in the television picture. (14)

control room: A room in the studio containing several monitors and the special effects generator. In smaller facilities, the control room also houses the audio mixer, all of the sound equipment, video recorders, the CG, CCUs, and even the light board. (16)

control track: A series of inaudible pulses recorded onto a tape that regulates the speed of the tape in playback. (5)

convertible camera: A camera with a variety of accessory packages available to make it operational in a studio, as a portable field camera, or both. (3)

Copyright Law: Set of laws that protect the creators of original materials from having their materials and creative work used without proper permission and compensation. (12)

corner insert: A video effect where a small image is positioned in any corner, usually the upper right or left corner, of the full screen image. All four sides of the smaller image rest inside the frame of the larger picture. Also called an *over-the-shoulder graphic* in news programs. (23)

corner wipe: A video effect where a small image is positioned in any corner, usually the upper right or left corner, of the full screen image. Two edges of the smaller image touch the frame of the larger picture. (23)

corporate television: See *industrial television*. (1)

countdown: The procedure used by a production's floor manager to initiate action on the set and cue the performers by counting down from "10." (20)

cover music: A band's rendition of another band's copyrighted song. (13)

crab dolly: A four-wheeled cart that travels on a lightweight track and enables the camera to smoothly capture movement shots while being pushed or pulled along the track. Also called a *track dolly*. (17)

crawl: Words that appear either at the top or the bottom of the screen and move from the right edge of the screen to the left, without interrupting the program in progress. (14)

credits: The written material presented before and after programs, listing the names and job titles of the people involved in the program's production. (14)

crème makeup: An oil-based makeup product that easily blends with other colors. (21)

crew: Production personnel that are normally not seen by the camera, which generally includes equipment operators. (2)

cross-camera shooting: A two-camera shooting technique in which the camera on the left shoots the person on the right of the set and the camera on the right shoots the person on the left of the set. (19)

cross-key lighting: A lighting technique that covers more than one person or object in the lighting spread using only two key lights and one back light. (15)

cue: A signal that directs something specific to happen. (2)

cut: 1—A command given by the director that indicates all production activity, talent performances and crew activities, should stop immediately. (20) *2*—An instantaneous picture change that occurs on screen. Used synonymously with *take* in this context. (22)

cut bar: A feature on a SEG that provides a quick way of cutting between the input selected on one bus and the input selected on another bus in the bank. Also called a *cut button*. (22)

cut button: See *cut bar*. (22)

cut rate: See *pace*. (24)

cutaway: A shot that is not a key element in the action. It is commonly used to bridge what would otherwise be a jump cut. (19)

cyclorama (cyc): An indistinct, solid-color background that is typically used for limbo shooting and chromakey shooting. (18)

D

D-9: See *Digital S*. (5)

deck: The common term used for a video recorder/player. (5)

delegation control: A switch or button on the control panel of a SEG that changes a bank from one function to another. For example, the delegation control can switch the selected bank from operating as a mix bank to operate as an effects bank. (22)

depth of field (DOF): The distance between the closest point to the camera that is in focus and the furthest point from the camera that is also in focus. (4)

diaphragm: See *generating element*. (6)

diffusion: A translucent material that is placed in front of a lighting instrument to soften and reduce the intensity of light, without altering the color temperature. (15)

Digital Betacam (Digi-Beta): A ½″ videotape with higher quality than Betacam SP and the capability of recording of digital signals instead of analog signals. (5)

digital intermediate: A high quality digital version of a motion picture created by digitally scanning motion picture film. (14)

Digital S: A ½″ digital videotape format that is broadcast quality. Also known as *D-9*. (5)

digital video effect (DVE): Video effects that are created using digital technology; based on the ability to alter an image by manipulating each individual pixel. (23)

digital video recorder (DVR): A device that records a digital signal directly onto either a hard drive or a solid-state memory module inside or connected to the DVR unit. (5)

digitize: See *capture*. (24)

dimmer: A device attached to the power control of a lighting instrument that regulates the amount of electricity that flows to the lamp. (15)

DIN connector: A term that refers to any type of connector with four or more holes/pins. (7)

diopter adjustment: A knob or lever that adjusts the magnifier on the viewfinder to compensate for differences in vision. (3)

directional mic: See *cardioid mic*. (6)

director: The person who is in charge of the creative aspects of the program and interacts with the entire staff. (2)

dissolve: A video effect where one picture slowly disintegrates while another image slowly appears. At no time is the screen solid black (or any other solid color). Also called a *lap dissolve*. (22)

distribution: The final phase of production, which includes DVD authoring, DVD/videotape duplication, and distribution to the end user. (2)

distribution amplifier (DA): A machine used to amplify a signal before it is split and sent to multiple outputs. (24)

DLP (digital light processing): A television that uses a light projection system to create the image on screen. (25)

documentary: A program format that is essentially a research paper for television. The audio in the program may include both on-camera and off-camera narration. The video footage used in the program is determined by the topic research and should support the audio of the program. (8)

dolly: 1—A three-wheeled cart onto which the feet of a tripod are mounted. A dolly allows smooth camera movements to be performed. (3) 2—Physically moving the camera, its tripod, and dolly perpendicularly toward or away from the set. (4)

dolly grip: Member of the crew who pushes or pulls a crab dolly along the track during production. (17)

dolly in (DI): Smoothly pushing the camera directly forward toward the set. (4)

dolly out (DO): Pulling the camera backward while facing the set. (4)

drag: Resistance to movement created by tripod head mount. (3)

drama: A program format that includes both dramas and comedies and requires actors to portray someone or something other than themselves. (8)

dramatic aside: A performance technique; when a performer steps out of character and directly addresses the audience. (19)

dropout: A tiny white dot seen on the television screen when the medium has fallen off an analog videotape and the video head passes over an "empty spot" on the tape. (5)

dry run: A program rehearsal session that includes the talent, technical director, audio engineer, camera operators, and director. Also called a *camera rehearsal.* (20)

dub: A copy of the master program recording. Also called *dup.* (24)

dubbing: The process of copying a video recording. (5)

dup: See *dub.* (24)

DVCam: A 6mm digital format that is proprietary to Sony Corporation. (5)

DVCPRO: A 6mm, metal particle tape used as a professional digital video format. (5)

DVCPRO100: The high definition format of DVCPRO tape. (5)

DVCPRO50: A 6mm digital format with even higher quality than DVCPRO. (5)

DVD (Digital Video Disc): An optical disc that can store a very large amount of digital video data, as well as text and/or music. (5)

dynamic mic: A very rugged type of mic that has good sound reproduction ability. The generating element is a diaphragm that vibrates a small coil that is housed in a magnetic field. (6)

E

edit decision list (EDL): A list noting which take of each scene should be used in the final program and the location of each take on the raw footage tape. (24)

edit point: The exact location in the footage where an edit should occur. The edit "out point" is the edit point at the end of a scene. The edit "in point" is the edit point at the beginning of a scene. (24)

edit through black: An edit in which a cut is made during the period of black on screen between a fade out and a fade in. Also called *kiss black.* (24)

edit transition: The way in which one scene is edited to end and the next scene is edited to begin. (24)

editing: 1—The process of placing individual recorded scenes in logical order. (2) 2—The process of selecting the best portions of raw video footage and combining them into a coherent, sequential, and complete program. Editing also includes post-production additions of music and sound effects, as well as effects used as scene transitions. (24)

editing suite: A cubicle or small room where the program is put through postproduction processing, such as video and audio editing, voice-over, music and sound effects recording, and graphics recording. (16)

editor: 1—The person responsible for putting the various pieces of the entire program together. The editor removes all the mistakes and bad takes, leaving only the best version of each scene, and arranges the individual scenes into the proper order. (2) 2—A collective term that refers to the systems and equipment used to edit program footage. (24)

educational television: Television that aims to inform the public about various topics. This includes television programming that supports classroom studies and replays classroom sessions. (1)

effects bank: A bank on a SEG that allows each signal to be processed individually. (22)

EFP (electronic field production): A shoot in which the video crew and production staff are in total control of the events and action. (16)

electret condenser mic: See *condenser mic*. (6)

ENG (electronic news gathering): The process of shooting information, events, or activity that would have happened whether a reporting/production team was there with a camera or not. (16)

equipment breakdown: A list of each scene in a program with all the equipment needed to shoot each scene. (20)

error in continuity: An error that occurs during editing where a sequence of shots in the finished product contains physically impossible actions or items. (19)

essential area: The area of an image or shot that must be seen on any television set, regardless of aspect ratio or age, and must include all the words in a graphic. (14)

establishing shot: A specific type of extreme long shot used to tell the audience where and when the program takes place. (4)

evergreen: A story that is appropriate to be broadcast at any time, regardless of season or time of day. (9)

executive producer (EP): The person, or people, who provides the funding necessary to produce the program. (2)

export: The process of copying the completed program from the hard drive of the NLE computer for duplication and distribution. (24)

extended package: A 2–4 minute story that is shot and edited before a newscast and typically provides more in-depth coverage of a specific story. (9)

extreme close-up (ECU/XCU): A shot of an object that is so magnified that only a specific part of the object fills the screen. (4)

extreme long shot (ELS/XLS): The biggest shot a camera can capture of the subject matter. Also called a *wide angle (WA) shot*. (4)

F

fade: A visual effect where the video image either slowly appears from a solid-colored screen (usually black) or slowly disintegrates from an image to a solid-colored screen (usually black). (22)

fade in: A video effect where a totally dark picture gradually transitions into a fully visible picture. Also called *up from black*. (22)

fade out: A video effect where the image slowly goes to a black screen as a scene of a program ends. Also called *fade to black*. (22)

fade to black: See *fade out*. (22)

fader bar: See *fader lever*. (22)

fader handle: See *fader lever*. (22)

fader lever: A control on a SEG, usually a T-shaped handle, that controls the strength of that signal coming from each bus. Also called a *fader bar* or *fader handle*. (22)

Fair Use: A section of the Copyright Law that provides guidelines for the limited use of copyrighted materials without obtaining permission from the copyright holder(s). (12)

fast lens: A camera lens that can produce a large aperture and let a great deal of light into the camera. (3)

F-connector: A type of connector that carries an RF signal and is commonly found on the back of consumer VCRs and televisions. (7)

feature: See *news feature package*. (9)

feature package: See *news feature package*. (9)

feedback: A high-pitched squeal that occurs when a microphone picks up the sound coming from a speaker that is carrying that microphone's signal. (6)

female connector: See *jack*. (7)

field: One entire pass of the electron beam, spraying every other line from the top to the bottom of the screen. (25)

fill camera: A video camera shooting the live image used as the fill source in a key effect. (23)

fill light: A lighting instrument that is placed opposite the key light and above the talent to provide illumination on the other side of the talent's face or object in the shot. (15)

fill source: The video source that provides the background image in a key effect; the image may be live, pre-recorded, or computer generated. (23)

film chain: See *telecine*. (14)

film island: See *telecine*. (14)

film scanner: A digital device designed to copy/scan motion picture film. (14)

film-style shooting: A type of single-camera shooting in which a scene is shot many times with the camera moving to a different position each time to capture the scene from various angles. The finished scene is edited together to look like it was shot with several cameras. (17)

FireWire connector: A type of connector designed to carry digital signals and available with 4-pin (audio and video only) and 6-pin (audio, video, and power) connections. This connector is also known as *IEEE 1394*. (7)

fishpole boom: Type of boom that must be physically held over the heads of talent. (6)

flag: A flexible metal rod with a flat piece of metal attached to the end; used to block light from hitting certain objects on the set. (15)

flat: A scenery unit that is usually a simple wood frame with a painted plywood shell. (16)

flood light: A soft light instrument that provides general lighting in a large area. (15)

floor director: See *floor manager.* (2)

floor manager: The person who is the director's "eyes and ears" in the studio. The floor manager relays the director's commands to the studio personnel. Also commonly called *floor director.* (2)

floor stand: A lighting support with three or four legs and a long vertical pole to which a lighting instrument is attached. (15)

fluid head: A mounting assembly on some tripods that stabilizes the camera using the pressure between two pieces of metal and a thick fluid that provides additional resistance to movement. (3)

fluorescent lamp: Type of lamp that functions when electricity excites a gas in the lamp, which causes the material coating the inside of the lamp to glow (fluoresce) with a soft, even light. (15)

focal length: The distance (measured in millimeters) from the optical center, or focal point, of the lens assembly to the back of the lens assembly. (3)

focal point: See *optical center.* (3)

focus: The act of rotating the focus ring on a camera lens until the lines of contrast in the image are a sharp as possible. (3)

foreground: The area between the talent and the camera. (19)

foreground music: Music in a program that is the subject of the production. (13)

format script: A program script that is very brief and used for programs in which the order of events is predetermined and the sequence of each episode is consistent. (8)

foundation: See *base.* (21)

four shot: A shot that captures four items. (4)

four-point lighting: A lighting technique that uses four lighting instruments for each person or object photographed: two key lights and two fill lights. The two key lights are positioned diagonally opposite each other, and the two fill lights are placed in the remaining two corners. (15)

fps: The rate at which individual pictures are displayed in a motion picture and on television, expressed as frames per second. (14)

frame: *1*—The actual edge of the video picture; the edge of the picture on all four sides. (2) *2*—Two consecutive passes of the cathode ray; every line on the television screen sprayed with electrons. Two complete fields create complete one frame. (25)

framing: Involves placing items in the camera's frame by operating the camera and tripod. (2)

Fresnel: A hard light instrument that is lightweight and easily focused. (15)

friction head: A mounting assembly on some tripods that stabilizes the camera using the pressure created when two pieces of metal are squeezed together by a screw. (3)

f-stop: A camera setting that determines the amount of light passing through the lens by controlling the size of the iris. (3)

G

gaffer: The lighting director's assistant who often does the actual hauling of heavy instruments up and down ladders. (2)

gain: The strength of a video or audio signal. (3)

gel: A heat resistant, thick sheet of plastic placed in front of a lighting instrument to turn white light from a lamp into a colored light. (15)

generating element: A thin surface inside the mic that vibrates when hit by sound waves in the air and creates an electrical signal. Also called a *diaphragm.* (6)

genlock: A module built into video equipment that overrides the internal sync of an individual piece of equipment to synchronize the equipment with the signal provided by the sync generator. (25)

glitch: A momentary "trashing" of the video signal (such as a roll, tear, or briefly appearing video noise). (22)

graphic artist: The person responsible for all the artwork required for the production. This includes computer graphics, traditional works of art, charts, and graphs. (2)

graphics: All of the "artwork" seen in a program, including the paintings that hang on the walls of a set, the opening and closing program titles, computer graphics, charts, graphs, and any other electronic representation that may be part of a visual presentation. (14)

great depth of field: When a camera's depth of field is as large as possible. (4)

grid: A pipe system that hangs from the studio ceiling and supports the lighting instruments. (15)

grip: A person who moves the equipment, scenery, and props on a studio set. (2)

group shot: A shot that incorporates any number of items above four. (4)

H

hand-held mic: A mic that is designed to be held in the hand, rather than placed on a boom or clipped to clothing. Also called a *stick mic.* (6)

hand shots: A type of B-roll shot that features a close-up of someone's hands. (19)

hard lead: The first line of a story that begins the story abruptly and immediately presents the most important information. (10)

hard light: Type of illumination used in a studio that creates sharp, distinct, and very dark shadows. (15)

hard news: Type of news story that contains information that viewers need to have immediately; characterized by seriousness and timeliness. (9)

HDMI connector: This connector is designed to carry high definition video and audio, as well as power. HDMI stands for High Definition Multimedia Interface. (7)

HDV: Video recording technology that uses MPEG compression to record a 1440 x 1080 image using only 19 megabits of recording memory space per second. (25)

head: A 15 second "lead-in" recorded at the beginning of every take. (5)

head room: The space from the top of a person's head to the top of the television screen. (4)

helical scan: The pattern in which a video signal is placed onto a videotape. The videotape is wrapped around the video head and, because the head is slanted, the video signal is recorded diagonally on the tape. Also called *slant track.* (5)

high angle shot: Shooting talent with the camera positioned higher in the air and pointing down at an angle. (4)

high impedance (HiZ): A type of mic that is typically inexpensive, low-quality, and cannot tolerate cable lengths longer than 8′. (6)

high-definition (HD): A digital television in 16:9 aspect ratio format. 1080 x 720 and 1920 x 1080 are common resolution formats. Also called *HiDef.* (25)

highlight: Makeup that is three or four shades lighter than the area to which it is applied. (21)

home video: Videotaped records of family events and activities taken by someone using a consumer camcorder. (1)

honeycomb: A device that attaches to fluorescent instruments to reduce the shape and size of the light beam, making the light more directional and easier to control. (15)

horizontal wipe: A video effect where a vertical line moves across the screen horizontally, replacing one picture with another. (23)

hot: *1*—The state of a video camera when the image captured by the camera is being recorded. (3) *2*—A term used to describe an image or shot that is very bright. (14)

hypercardioid mic: A directional mic with a narrower and longer pick-up pattern than a cardioid mic. (6)

I

IEEE 1394: See *FireWire connector.* (7)

IFB: Interrupted feedback; a line of communication between the anchors and the producer in the control room. An earpiece worn by the anchor is connected to the producer's headset, allowing the producer to speak directly to an anchor while the anchor is on the air live. (9)

incandescent lamp: Type of lamp that function when electricity is applied and makes a filament inside the lamp glow brightly. (15)

industrial television: Television that communicates relevant information to a specific audience, such as job training videos. Also commonly called *corporate television.* (1)

input: A port or connection on a video device through which a signal enters the device, such as the "audio in" port. (5)

instrument: The device into which a lamp is installed to provide illumination on a set. (15)

interlace: The process in which the cathode ray fires all the odd lines on a television screen in one pass and returns to the top of the screen to fire all the even lines in the second pass. (25)

interview: A program format that involves a conversation between an interviewer and an interviewee. (8)

iris: A component of a lens that is comprised of blades that physically expand and contract, adjusting the aperture size. (3)

J

jack: *1*—A connector with one or more holes designed to receive the pins of a male (plug) connector. Also called a *female connector.* (7) *2*—A triangle-shaped brace that is fastened to the back of a flat to provide stable, upright support. (16)

jib: A type of camera mount that allows the camera to be raised high over the set and swung in any direction. (3)

jump cut: A sequence of shots that constitutes an error in editing. This error can occur during production when cutting between similar sized camera shots of the same object or during post-production when shots are edited together. The result is an on-screen object or character that appears to jump from one position to another. (19)

K

Kelvin Color Temperature Scale: A scale developed by the scientist Lord Kelvin for measuring color temperatures of light in degrees Kelvin. (15)

key: *1*—A video effect where a portion of the picture is electronically removed and replaced with another image. (22) *2*—A video effect where two images, or the output from two different sources, are displayed on screen at the same time. Each image is displayed with 100% of its original intensity. (23)

key camera: The video source (camera) that provides the shape to be cut into the background image in a key effect. The portions removed from the key camera image (usually areas of blue or green) will be replaced by the fill source image. (23)

key level control: See *clip control*. (23)

key light: The lighting instrument that provides the main source of illumination on the person or object in a shot. (15)

kiss black: See *edit through black*. (24)

knee shot: See *medium long shot (MLS)*. (4)

L

lamp: Part of a lighting instrument that glows when electricity is supplied. (15)

lap dissolve: See *dissolve*. (22)

lapel mic: The smallest type of mic that can be worn by talent and is attached to clothing at or near the breastbone with a small clip or pin. Sometimes referred to as a *lav*. (6)

large-scale video production companies: Facilities with sufficient staff and equipment to produce multi-camera, large-budget programming shot on location or in studios for broadcast networks or cable networks. (1)

lav: See *lapel mic*. (6)

LCD (liquid crystal display): A television that contains a thin layer of liquid crystal between the pixel layer and the surface of the television screen. The liquid crystal layer controls each color contained in each pixel in the television image. (25)

lead: *1*—The very first sentence of a story. (10) *2*—Basic information provided by the interviewee that is recorded at the beginning of every interview. The lead typically includes the interviewee's name and proper spelling, title (if pertinent and applicable), and contact information. (11)

lead room: See *nose room*. (4)

lecture: A program format in which the talent speaks and the camera shoots almost entirely in a medium close-up. Also known as *big talking face (BTF)* and *talking head*. (8)

lecture/demonstration: A program format that provides action and makes use of props in addition to lecture. Examples of this format include cooking shows, how-to shows, and infomercials. (8)

lens: An assembly of several glass discs placed in a tube attached to the front of a camera. (3)

letterbox: A video display format that allows a program originally shot using a wide screen film or video format to be displayed in 16:9 aspect ratio. (25)

light hit: A white spot or star shaped reflection of a lighting instrument or sunlight off of a highly reflective surface on the set. (15)

light plot: A diagram developed by the lighting designer that indicates the placement of lighting instruments on the set of a program. (15)

lighting director: The person who decides the placement of lighting instruments, the appropriate color of light to use, and which lamps should be used in the instruments. (2)

limbo lighting: A lighting technique in which the background of the set is lit to create the illusion of a solid-color, indistinct background. (15)

limited public forum: Public property or media that is made available for a specified use; the topic or content of speech is restricted to the business at hand or objectives of the particular group. (12)

line level: The level of audio between pieces of audio equipment. For example, the level of audio going from the output of a CD player to the input on an amplifier. (6)

linear editing system: Videotape-based editing equipment in which the best takes from the raw footage are copied in the order the audience will see them to a tape in the record VCR. (24)

live shot: A news story that is introduced by an anchor and delivered through a live feed by a reporter on location. (9)

local origination: Programming made in a specific geographic area, to be shown to the public in that same geographic area. (1)

location: Any place, other than the studio, where production shooting is planned. (16)

location breakdown: A list of each location included in a program with the corresponding scene numbers that take place at that location. (20)

location survey: An assessment of a proposed shoot location that includes placement of cameras and lights, available power supply, equipment necessary, and accommodations needed for the talent and crew. (16)

long shot (LS): A shot that captures a subject from the top of the head to the bottom of the feet and does not include many of the surrounding details. (4)

low angle shot: A shot created by placing the camera anywhere from slightly to greatly below the eye level of the talent and pointing it up toward the talent. (4)

low impedance (LoZ): A type of mic that is costly, high-quality, and can tolerate long cable lengths. (6)

lower third key: A video effect created when an image or text is displayed in the lower third of the screen using a key effect. Also called a *lower third*. (23)

lower third super: A video effect created when an image or text is displayed in the lower third of the screen using a superimposition. (23)

lower third: See *lower third key*. (23)

luminance: A measure of the brightness or lightness of a video image. (14)

M

macro: A lens setting that allows the operator to focus on an object that is very close to the camera, almost touching the lens. (4)

magazine: A program format comprised of feature packages, each addressing a different story for seven to eleven minutes. (8)

mainstream media: Television news programming that is expected to provide a fair and unbiased presentation of facts, without any particular viewpoint. (9)

maintenance engineer: The person who keeps all the production equipment functioning at its optimum performance level. (2)

makeup: Any of the cosmetics applied to a performer's skin to change or enhance their appearance. (2)

makeup artist: The person responsible for applying cosmetics to the talent's face and body, giving them the intended appearance in front of the camera. (2)

male connector: See *plug*. (7)

master control room: A room in a production facility where all the hardware is located, including video recorders and other equipment needed to improve and process the video and audio signals. (16)

matched cut: A type of edit in which a similar action, concept, item, or a combination of these is placed on either side of a cut. (24)

matched dissolve: A type of edit in which a similar action, concept, item, or a combination of these is placed on either side of a dissolve. (24)

medium close-up (MCU): A shot that frames a subject from the top of the head to a line just below the chest. Also called a *bust shot*. (4)

medium long shot (MLS): A shot that includes the top of a subject's head to a line just above or just below the knee. Also called a *knee shot*. (4)

medium shot (MS): A shot that captures a subject from the top of the head to a line just above or below the belt or waistline. Also called a *mid shot*. (4)

mic level: The level of audio that comes from a microphone. It is designed to be sent to the "mic in" on a recorder or mixer. (6)

mic mixer: A piece of equipment that combines only the microphone signals into a single sound signal. (6)

microphone (mic): The piece of equipment that picks up sounds in the air and sends them to the mixer or recorder. (6)

mid shot: See *medium shot (MS)*. (4)

middle ground: The area in which the action of a program typically takes place and where the most important items in a picture are usually positioned. (19)

mini connector: A connector that is ⅛″ in diameter and single-pronged. It is most commonly found on headsets used with mp3 players and other personal audio devices. Also called a *⅛″ connector*. (7)

mini-DV: A metal evaporated tape, 6mm digital video format used by many industrial video producers. (5)

minimum object distance (MOD): The closest an object can be to the camera and still be in focus. (4)

mix bank: A single bank on an SEG that contains the cut, fade, and dissolve effects. (22)

moiré: An effect caused by shooting certain patterns, usually fabrics, in which the television system reproduces the pattern with a rainbow of colors or moving lines displayed in the patterned area. (18)

monitor: A television set that can receive only pure video and audio signals. (5)

monitor/receiver: A hybrid television that can receive pure video and audio signals, as well as RF signals. (5)

montage: A production device that allows a gradual change in a relationship or a lengthy time passage to occur in a very short amount of screen time by showing a series of silent shots accompanied by music. (8)

multi-camera shooting: A technique of remote shooting where multiple cameras are used. (17)

music video: A program format in which all or most of the audio is a song. (8)

N

narrow angle shot: See *close-up (CU)*. (4)

natural sound (nat sound): Environmental sound that enhances a story and is important to the shot. (6)

network: A corporation that bundles a collection of programs (sports, news, and entertainment) and makes the program bundles available exclusively to its affiliates. Generally, networks produce some of their own programming, but do not produce all of their own programs. (1)

news: Information people want to know, information they should know, or information they need to know. (9)

news director: The person responsible for the structure of the newsroom, for personnel matters (performance evaluations and hiring and firing employees), managing the budget, and the overall effectiveness of the newsroom. The news director is also the final authority on which stories will air during a news broadcast. (2)

news feature package: A package covering soft news stories that are connected to current events. Also called a *feature package* or *feature*. (9)

news package: A package that covers hard news/current events. (9)

newscast: A program format that is a collection of individual news stories. (8)

nickel cadmium (NiCad): A type of rechargeable battery commonly used to power cameras. (17)

nod shots: A cutaway shot often used in interview programs and usually recorded after the interviewee has left the set. In a nod shot, the interviewer does not say anything, but simply "nods" naturally as if listening to the answer to a question. (8)

non-linear editing system (NLE): Video editing equipment that is based on digital technology and uses high-capacity computer hard drives to store and process video and audio. The raw footage is converted to a digital format, copied to a computer's hard drive, and may then be arranged and otherwise manipulated. (24)

non-mainstream media: Television news programming that is expected to express a particular point of view. (9)

non-public forum: Either public or private property or media that is not typically used or made available for public expression. Regulation on speech is allowable in a non-public forum, but must be reasonable and not intentionally exclude any particular or opposing viewpoint. (12)

nose room: The space from the tip of a person's nose to the side edge of the frame. Also called a *lead room*. (4)

O

off-camera narration: Program narration provided by talent that is heard, but not seen by the viewer. Also called *voiceover (VO)*. (6)

OLED (organic light emitting diode): A television that uses a type of LED in the form of a layer of film made from organic compounds to create the image on screen. (25)

omni-directional mic: A mic with a pick-up pattern that captures sound from nearly every direction equally well. (6)

on-camera narration: Program narration provided by on-screen talent (seen by the camera). (6)

optical center: The physical location within the lens assembly where an image is inverted. Also called the *focal point*. (3)

outline script: A program script that usually has a word-for-word introduction and conclusion, but an outline for the body of the script. (8)

output: A port or connection on a video device through which the signal leaves the deck and travels to another piece of equipment, such as the "video out" port. (5)

outro: The salutation at the end of a story; opposite of an intro. (9)

over-the-shoulder graphic: See *corner insert*. (23)

over-the-shoulder shot (OSS): A shot in which the back of one person's head and shoulder are in the foreground of the shot, while a face shot of the other person in the conversation is in the background. (4)

P

P2: A static memory card that is proprietary to Panasonic and used in certain high end cameras. (5)

pace: The frequency of cuts or edits per minute during a program. Also called *cut rate*. (24)

package: A story that is about 1½–2 minutes in length, contains its own intro and outro, is edited, and can be inserted into a live program at any time the producer chooses. (9)

pan: Moving only the camera to scan the set horizontally, while the dolly and tripod remain stationary. (4)

pan handle: A device attached to the back of the tripod head that allows the camera operator to move the tripod head with the camera attached while standing behind the tripod. (3)

pan left (PL): Moving the camera to the camera operator's left to scan the set, while the dolly and tripod remain stationary. (4)

pan right (PR): Moving the camera to the camera operator's right to scan the set, while the dolly and tripod remain stationary. (4)

pancake makeup: A powder makeup foundation that is water-soluble. (21)

panel discussion: A program format that presents a group of people gathered to discuss topics of interest. Daytime talk shows are an example of this format. (8)

parabolic reflector mic: A very sensitive mic that looks like a satellite dish with handles and is designed to pick up sounds at a distance. (6)

passive talent release: A document that serves as a general notice indicating that, from time to time, organizations outside the school system may request permission to video record inside the school building. Parents acknowledge the release by *not* responding to the notice. (12)

patter: The spontaneous on-air conversation or small talk between anchors or anchors and reporters. (9)

pedestal: Raising or lowering the camera on the pedestal of a tripod, while facing the set. The tripod and dolly remain stationary. (4)

pedestal column: A column in the center of a tripod used to raise or lower the camera. (3)

pedestal control: A crank on the side of the pedestal column that twists a gear to raise and lower the pedestal column. (3)

pedestal down (PedD): Lowering the camera on the pedestal of a tripod, while facing the set. The tripod and dolly remain stationary. (4)

pedestal up (PedU): Raising the camera on the pedestal of a tripod, while facing the set. The tripod and dolly remain stationary. (4)

personality feature: Type of human interest story that focuses on one person and why that person is newsworthy. (9)

phone connector: A connector that is ¼″ in diameter and single-pronged, with a little indentation near the end of the prong. This type of connector is commonly found on the cord used with large stereo headphones. Also called a *¼″ connector*. (7)

phono connector: A connector commonly found on the back of quality home entertainment system components. The female phono connector is usually a chassis mounted connector. The male phono connector is usually a cable end connector with a single, center prong surrounded by a shorter crown. Also called an *RCA connector*. (7)

photog: See *photographer*. (2)

photographer: The cameraperson in the field, on location with a reporter in a news operation. Also commonly called *photog* or *shooter*. (2)

photojournalist: A photographer who regularly performs duties of both the photographer, as well as the reporter. (2)

pick-up pattern: A term that describes how well a mic hears sounds from various directions. (6)

pixel: One of the millions of little dots that make up the picture on a television screen, in a photograph, and any other type of digital image display medium. (23)

PL259 connector: A connector that is similar to the F-connector, but much larger. The male end has a single prong with a large nut to tighten. (7)

plasma: A television that uses phosphor and rare gases to light the pixels that create an image. (25)

plug: A connector with one or more pins that are designed to fit into the holes of a jack (female connector). Also called a *male connector*. (7)

pop filter: A barrier made of shaped wire covered with a piece of nylon that is placed between a sensitive mic and the talent to avoid damage to the diaphragm of the mic. (6)

pop the contrast ratio: When the brightness or darkness of objects in a shot exceeds the contrast ratio limitations of video. (14)

post-production: Any of the activities performed after a program has been shot. This includes music beds, editing, audio overdubs, titles, and duplication. (2)

potentiometer (pot): A knob or a slider control that regulates the strength a signal. (6)

power level: The audio level from the output on an amplifier to the speaker. (6)

pre-focus: A three-step process to focus a zoom lens. 1) Zoom in on the furthest object on the set that must be in focus in the shot. 2) Focus the camera on that object. 3) Zoom the lens back out. (4)

pre-production: Any activity on a program that occurs prior to the time that the cameras begin rolling. This includes production meetings, set construction, costume design, music composition, scriptwriting, and location surveys. (2)

preview monitor: A monitor that allows the director to set up an effect on the SEG before the audience sees it. (16)

private property: Property that is owned by an individual or private organization. Permission is required to be on the premises. (12)

processing amplifier (proc amp): A machine commonly used when recording or duplicating videotapes that corrects some color and brightness problems in the video signal passing through it. (24)

producer: In a non-news environment, the producer purchases materials and services in the creation of a finished program. In a broadcast news facility, the producer coordinates the content and flow of a newscast. (2)

production: The actual shooting of the program. (2)

production assistant (PA): The person who provides general assistance around the studio or production facility. The PA is commonly hired to fill a variety of positions when key personnel are sick, out of town, working on another project, or otherwise unavailable. In many facilities, the production assistant position is synonymous with the *assistant director (AD)* position. (2)

production manager: The person who handles the business portion of the production by negotiating the fees for goods, services, and other contracts and by determining the staffing requirements based on the needs of each production. (2)

production meeting: A meeting with the entire crew in which the director lays out the program's main message and either the director or producer assigns each task involved in the production to members of the crew. (16)

production switcher: A video switcher that is used to cut between live camera shots while a program is being recorded. The term indicates a particular use of a video switcher. (22)

production switching: The process of cutting between cameras. (2)

production team: Everyone involved in the production, both staff and talent. (2)

production values: The general aesthetics of the show. (2)

profile shot: A shot in which the talent's face is displayed in profile. (4)

program monitor: A monitor that displays the image going to the recorder. (16)

program proposal: A document created by the scriptwriter that contains general information about the program, including the basic idea, applicable format, message to be imparted to the audience, intended audience, budget considerations, shooting location considerations, and rough shooting schedule. Used to present the program to the executive producer to obtain permission and funding for the production. (8)

progressive scan technology: The process in which every line on a television screen is fired in one pass, instead of every other line. The cathode ray moves fast enough to fire every line on the screen before the lines at the top of the screen begin to fade. (25)

prop: Any item handled by the performers during a production, other than furniture. (18)

prop list: A list of each prop needed for a production. (20)

prop plot: A list of all the props used in a program sorted by scene. (20)

property release: A signed document that grants a video team permission to shoot on private property. (12)

prosthetic: A cosmetic appliance, usually made of foam, latex, or putty, that can be glued to the skin with special adhesives. (21)

public domain: A status designation applied to material that is no longer copyrighted due to the passage of time (relative to the date of creation) or when rights are relinquished by the copyright holder. (12)

public forum: An environment or location, typically public property or media, where an individual can stand and publicly speak their mind. The content discussed in a public forum is not restricted, but the speech cannot incite a riot, violence, or similar activity. (12)

public property: Property that is owned by local, state, or national government organizations. It is *usually* legal to be on the premises of public property. (12)

public service announcement (PSA): A program that is 30 or 60 seconds in length and aims to inform the public or to convince the public to do (or not to do) something in the interest of common good. (8)

pull focus: See *rack focus.* (4)

Q

Quadruplex (Quad): A very large, older videotape recorder that uses 2″ tape. (5)

R

raceway: The system of electrical cables and outlets used to power lighting instruments on the grid. The raceway either hangs beside the grid pipes or is mounted to the ceiling above the grid. (15)

rack focus: The process of changing focus on a camera while that camera is hot. Also called *pull focus.* (4)

RCA connector: See *phono connector.* (7)

reaction shot: A shot that captures one person's face reacting to what another person is saying or doing. (4)

reader: A story, written by a reporter or anchor, that does not have video to accompany the story. The anchor simply reads the text on the teleprompter aloud for the viewing audience to hear. (9)

receiver: A television set that can receive only RF signals. (5)

recording rights: Permission to record music from a live performance. (13)

release: A grant of permission that is commonly provided in written form with signatures of all the people involved. (12)

remote control unit (RCU): See *camera control unit (CCU).* (3)

remote shoot: Any production shooting that takes place outside of the studio. (16)

reporter: The individual responsible for gathering information from many sources, including research and interviews, for writing news stories, and often editing their own stories. (2)

reporter track: Everything spoken by the reporter in a package. (10)

re-recording rights: Permission to copy copyrighted material from its current format to a video medium. (13)

RF: Radio frequency signal that is a combination of both audio and video. (5)

RF converter: A small module inside the VCR that combines pure video and audio into one radio frequency. (5)

ribbon mic: The most sensitive type of mic used in television. A thin ribbon of metal surrounded by a magnetic field serves as the generating element. (6)

robo operator: The person who remotely operates all of the cameras and the robotic camera mounts from a single location in the studio or control room. (2)

roll: Titles in a program that move up the screen. (14)

room tone: The sound present in a room or at a location before human occupation. (6)

rough hang: See *basic hang*. (15)

rule of thirds: A composition rule that divides the screen into thirds horizontally and vertically, like a tic-tac-toe grid placed over the picture on a television set. Almost all of the important information included in every shot is located at one of the four intersections of the horizontal and vertical lines. (4)

rundown: The organization of stories and sequence of a newscast in written form. (9)

S

scene breakdown by cast: A list of each scene number in a program with all the cast members needed for each scene. (20)

scenery: Anything placed on a set that stops the distant view of the camera. Outside the studio, scenery may be a building or the horizon. (2)

scoop: A common type of flood light with a half-spheroid shape that produces a great deal of light. (15)

scrim: A wire mesh or woven material placed in front of an instrument to reduce the intensity of light. (15)

script: An entire program committed to paper, including dialog, music, camera angles, stage direction, camera direction, and computer graphics (CG) notations. (8)

script breakdown: The process of analyzing a program's script from many different perspectives. (20)

scriptwriter: The person responsible for placing the entire production on paper. (2)

selective depth of field: A technique of *choosing* to have a shallow depth of field in a shot or scene. (4)

set decorator: See *set dresser*. (18)

set design: A scale drawing of the set, as viewed from above, that illustrates the location of furniture, walls, doors, and windows. (18)

set dresser: The person responsible for selecting the furniture, wall and window coverings, accent accessories, and all the other design elements that complete a program's set. Also called a *set decorator*. (18)

set dressing: All the visual and design elements on a set, such as rugs, lamps, wall coverings, curtains, and room accent accessories. (18)

shadow: Makeup that is three or four shades darker than the area to which it is applied. (21)

shallow depth of field: A depth of field technique that moves the audience's attention to the one portion of the picture that is in focus. (4)

shooter: See *photographer*. (2)

shooting for the edit: The process in which a director plans exactly how each scene of a program will transition from the scenes that immediately precede it and to the scenes that follow it. Production shooting then follows the director's plan. (20)

shot: An individual picture taken by a camera during the process of shooting program footage. (4)

shot log: See *take log*. (20)

shot sheet: A numerical listing of each shot to be captured by each camera in a multi-camera shoot. Shot sheets are developed specifically for each camera. (4)

shotgun mic: A directional mic with an extremely narrow pick-up pattern. (6)

shutter: A circuit on a video camera that regulates how long the CCD is exposed to light coming through the lens. (3)

single-camera shooting: A technique of remote shooting that involves only one camera and is most often used for event recording. (17)

slant track: See *helical scan*. (5)

slate: A board or page that is held in front of the camera to note the scene number, the take number, and several other pieces of information about the scene being shot. (20)

slow lens: A lens that is capable of small aperture settings and lets little light into the camera. (3)

small-scale video production companies: Businesses with limited staff and equipment resources. They thrive on producing videos of weddings, commercials for local businesses, home inventories for insurance purposes, seminars, legal depositions, and real estate videos. (1)

soft cut: A fast dissolve that appears to be a cut, but is not as sharp as a cut. (22)

soft lead: The first line of a story that communicates the general idea of a story, but does not offer any facts. (10)

soft light: Type of illumination used in a studio that creates indistinct shadows. (15)

soft news: Type of news story that contains information viewers may find interesting, but not necessarily information they *need* to know. (9)

soft wipe: Any wipe effect in which the wipe line is out of focus. (23)

SOT: Sound on tape; footage of a principal player connected to a story, which includes voice/audio that supports the story. Also called *sound bite*. (9)

sound bite: See *SOT*. (9)

special effects: Anything the audience sees in a video picture that did not really happen in the way it appears on the screen. (2)

special effects generator (SEG): A piece of production equipment with the basic functions of a video switcher, as well as the capability of producing various video effects. (22)

spirit gum: A type of adhesive commonly used to apply prosthetic items. (21)

split: Non-linear editing operation in which program footage is separated into individual clips. (24)

split screen: A wipe that is stopped part of the way through its move, dividing the screen into two or more parts. (23)

spot: See *ad*. (1)

spotlight: Type of hard light instrument that creates a circle of light in varying diameters. (15)

staff: Production personnel that work behind the scenes and generally includes management and designers. (2)

staging: The arrangement of items, such as furniture, props, and talent, in a shot. (19)

standard definition (SD): A digital television in 4:3 aspect ratio format that displays images using 720 x 480 lines of resolution. (25)

stand-up: Footage in a package that depicts a reporter standing in front of the camera, speaking directly to the viewers from the location of a story. (9)

stick mic: See *hand-held mic*. (6)

story style music video: A type of music video in which the audience hears the music, but does not see the band perform. Instead, actors act out a story line that is supported by the lyrics of the song. (8)

storyboards: Sketches that portray the way the image on television should look in the finished program. (8)

straight makeup: Makeup application technique used to correct or hide blemishes, make the complexion more even, and generally help people look attractive and like themselves under bright television lights. (21)

streaming rights: Permission to stream material on the Internet with settings that do not allow the material to be downloaded or recorded. (13)

strike: To dismantle or tear down a set that is no longer needed. (18)

studio camera: A television camera placed on a tripod or studio pedestal for exclusive use within the studio. (3)

studio pedestal: A large, single column on wheels that supports the camera and is pneumatically or hydraulically controlled. (3)

subjective camera: A hand-held camera technique, in which the camera itself becomes the eye of one cast member. The viewers see the world through the eyes of that character. (3)

subscriber television: Fee-for-service programming where customers pay scheduled fees based on the selected programming package. The television signals are transported by satellite transmission, by underground cables, or a combination of both. (1)

super: See *superimposition*. (23)

Super VHS (S-VHS): A low-end, industrial ½″ videotape format that is superior to VHS. (5)

supercardioid mic: A directional mic with a narrower pick-up pattern than a hypercardioid mic. (6)

superimposition: A video effect where two images, or the output from two different sources, are placed on screen at the same time. Each image is less than 100% of its original intensity, but when combined, the result is 100% of the intensity possible on the television screen. Also called a *super*. (23)

surveillance television: A form of CCTV that is usually, but not always, used for security purposes. The cameras used in the system are always interconnected to a closed circuit television system. (1)

S-VHS connector: The consumer term for a Y/C connector. (7)

swish pan: See *whip pan*. (19)

switcher: See *video switcher*. (22)

sync generator: A device that provides a synchronization signal to all the video equipment in the television production facility. (25)

synchronization rights: Permission to synchronize video with the music. (13)

syndication: The process of making a specified number of program episodes available for "lease" to other networks or individual broadcast stations, after the current network's contract for the program expires. (1)

T

T-connector: A connector that is shaped like the capital letter *T* and is made entirely of metal. The three ends of a T-connector are used to split one signal into two signals, or to combine two signals into one. (7)

tabloid media: Television news programming that stretches and exaggerates facts by dealing with sensational stories; generally considered more entertainment than news. (9)

tail: A 10 second "lead-out" recorded at the end of each scene. (5)

take: 1—Each recording of an individual scene. The take number increases by *1* each time an individual scene is shot. (20) *2*—See *cut.* (22)

take log: A written list of each scene and take number that have been shot and recorded on a particular tape, disc, or other recording medium. Also called a *shot log.* (20)

talent: Anyone seen by the camera, whether or not they have a speaking or any other significant role in the program, as well as individuals who provide only their vocal skills to the production. (2)

talent release: A document that gives video producers permission to photograph the talent and/or to use audio of the talent's voice. (12)

talking head: See *lecture.* (8)

target: Photosensitive surface of a charge coupled device (CCD). (3)

technical director (TD): A member of the production team whose primary job function is to follow the camera script or verbal commands from the director and operate the video switcher accordingly. (22)

telecine: A device that facilitates the transfer of film images onto videotape. Telecines are used, for example, to transfer theatrical motion pictures to DVDs for purchase or rental. Also called a *film chain* or *film island.* (14)

teleprompter: A computer screen positioned in front of the camera lens that displays dialogue text in large type. This allows the talent to look directly at the lens of the camera and read the text. (19)

test record: The process of using the video recorder to record audio and video signals before the session recording begins to ensure the equipment is functioning properly and to indicate any necessary adjustments. (5)

three shot: A shot that frames three items. (4)

three-point lighting: A common lighting technique that uses three lighting instruments for each person or object photographed: a key light, a fill light, and a back light. Also called *triangle lighting.* (15)

tighten: See *zoom in (ZI).* (3)

tilt: Pointing only the front of the camera (lens) vertically up or down while the dolly and tripod remain stationary. (4)

tilt down (TD): Pointing the camera lens down toward the ground, while the dolly and tripod remain stationary. (4)

tilt up (TU): Pointing the camera lens up toward the ceiling, while the dolly and tripod remain stationary. (4)

time base corrector (TBC): A machine that corrects any quality-related imperfections in the video and audio signals caused by mechanical errors associated with the VCR's functionality. (24)

time coding: A system of assigning each frame of a video a specific number, like an address. (24)

titles: The letters and characters generated by a CG that are displayed on-screen. (14)

track dolly: See *crab dolly.* (17)

tracking control: A knob on a professional VCR that is used to manually adjust the tape tracking speed. (5)

Trademark Law: A set of laws that protects a company's brand identification in an effort to avoid confusion in the marketplace. These laws ensure that when a consumer sees a logo or label on a product, the consumer knows who makes that product. (12)

transformative use: Using a work (image or other material) for an entirely different purpose than it was originally created and intended to be used. (12)

transitional device: An effect that is used as a means of getting from one scene to another. (23)

transitory digital transmission rights: Permission to place material on the Internet in a format that permits downloading and recording from the Internet. (13)

treatment: A narrative written from a program outline that tells the program's story in paragraph form. (8)

triangle lighting: See *three-point lighting.* (15)

trimming: Non-linear editing process of determining the exact place an edit should occur and cutting the clip to remove unnecessary footage. (24)

tripod: A three-legged stand that supports a camera. (3)

tripod head: The assembly at the top of the pedestal column to which the camera attaches. (3)

TRT: Total running time; industry abbreviation. (9)

truck: Moving the camera, its tripod, and dolly to the left or right in a motion that is parallel to the set. (4)

truck left (TL): To move the camera, its tripod, and dolly sideways and to the camera operator's left while facing the set. (4)

truck right (TR): To move the camera, its tripod, and dolly sideways and to the camera operator's right while facing the set. (4)

two shot: A shot that includes two items of primary importance. (4)

U

uni-directional mic: See *cardioid mic.* (6)

up from black: See *fade in.* (22)

USB connectors: A durable digital connector for video, audio, and power. The electrical contacts are enclosed in a metal housing and buffered by a plastic plate. (7)

V

variable focal length lens: A camera lens in which the optical center can vary its position within the lens assembly, varying the focal length measurement as well. Also called a *zoom lens*. (3)

vector line: An imaginary line, parallel to the camera, that bisects a set into a foreground and a background. Also called a *camera line*. (19)

vectorscope: A piece of testing equipment that graphically measures the way the colors are structured in the video signal. (25)

vertical wipe: A video effect where a horizontal line moves up or down the screen vertically, replacing one picture with another. (23)

VHS (Video Home System): A ½″ videotape format that emerged as the preferred standard for consumer VCRs. (5)

VHS-C: A ½″ videotape format that is shorter than regular VHS and is, therefore, packaged in a compact cassette. (5)

video: The portion of the program that you can see. (2)

video delay edit: An edit that cuts to the audio portion of the next scene before the corresponding video of the new scene is seen by the audience. (24)

video engineer: The person who manages the video equipment and is ultimately responsible for the technical quality of the video signal. (2)

video heads: The components inside a VCR that lay down the video signal onto a tape when in record mode. When a VCR is in playback mode, the video heads pick up the video signal from a tape. (5)

video noise: The white dots seen on a screen if the videotape is blank or if the heads are dirty. (5)

video operator: The individual responsible for recording the master video file in a tapeless television production environment. (2)

video switcher: A piece of equipment to which several video sources are connected. One signal from various video sources is selected and sent to the video recorder for recording. Also called a *switcher*. (22)

viewfinder: A small video monitor attached to the camera that allows the camera operator to view the images in the shot. (3)

visualization: The ability to mentally picture the finished program. (8)

VO: Voiceover; a type of story that incorporates B-roll video rolled-in from the control room, in addition to the script read by the anchor. (9)

voice track: The audio portion of a program created through dialogue or narration. (6)

voiceover (VO): See *off-camera narration*. (6)

volume unit meter (VU meter): A meter on either an audio or mic mixer that indicates signal strength. (6)

VO-SOT: Voiceover-sound on tape; a type of story in which the audience sees B-roll video and hears both the anchor reading from the teleprompter and footage of a comment from a principal player in the story. (9)

VTR interchange: The ability of a tape that was recorded on one machine to be played back on another machine. (5)

VTR operator: The person in charge of recording the program onto videotape by correctly operating the VTR equipment. (2)

W

waveform monitor: A piece of testing equipment that measures every aspect of an image, including the brightness and darkness of the video signal. (25)

whip pan: An extremely fast camera pan. Also called a *swish pan*. (19)

white balance: A function on cameras that forces the camera to see an object as white, without regard to the type of light hitting it or the actual color of the object. (4)

wide angle (WA) shot: See *extreme long shot (ELS/XLS)*. (4)

widen: See *zoom out (ZO)*. (3)

windscreen: A covering, usually foam, placed over a mic to reduce the rumble or flapping sound created when wind blows across the mic. (17)

wipe: A video effect where a line, or multiple lines, moves across the screen and replaces one picture with another. (22)

wireless mic: A mic that uses a short cable to connect the mic to a radio transmitter with an antenna, or the transmitter may be built into the mic itself. The transmitter wirelessly sends the signal to the receiver, which sends the mic signal through a short cable to the recorder. (6)

word-for-word script: A program script in which every word spoken by the talent is written out. (8)

X

XLR connector: A connector for microphones that usually has 3-pins, but can have 4-pins, 5-pins, and other pin configurations. The male and female ends lock together with a hook. Also called a *cannon connector*. (7)

Y

Y/C connector: A video input and output connector that is characterized by four tiny round pins and a rectangular plastic stabilizing pin. (7)

Y/C: The professional name for the signal placed onto S-VHS videotape. (5)

Y-connector: A connector that has three wires with a connector on the end of each. All the wires are tied together in the middle. The three ends of a Y-connector are used to split one signal into two signals, or to combine two signals into one. (7)

Z

zebra stripes: A special function of some viewfinders that displays black and white diagonal stripes on any object in a shot that is too brightly lit. (3)

zoom in (ZI): The act of rotating a ring on the zoom lens so that the center of the picture appears to be moving toward the camera. Also called *tighten.* (3)

zoom lens: 1—The particular piece of glass within the lens assembly that moves forward and back, magnifying or shrinking the image accordingly. This individual lens is the focal point, or optical center, of the zoom lens assembly. (3) 2—See *variable focal length lens.* (3)

zoom lenses: Camera lens assembly that is capable of magnifying an image merely by twisting one of the rings on the outside of the lens housing.

zoom out (ZO): The act of rotating a ring on the zoom lens so that the center of the picture appears to be moving away from the camera. Also called *widen.* (3)

Index